AQA Science Physics

New GCSE

Jim Breithaupt

Pauline Anning

Bev Cox

Gavin Reeder

Series Editor
Lawrie Ryan

Nelson Thornes

Published in 2011 by:
Nelson Thornes Ltd
Delta Place
27 Bath Road
CHELTENHAM
GL53 7TH
United Kingdom

12 13 14 15 / 10 9 8 7 6 5 4

A catalogue record for this book is available from the British Library.

AQA examination questions are reproduced by permission of the Assessment and Qualifications Alliance.

ISBN 978 1 4085 0832 9

Cover photograph: iStockphoto (background); David Deas/Getty Images (boy)

Page make-up by Wearset Ltd, Boldon, Tyne and Wear

Printed in China

Photo acknowledgements

How Science Works
H1.1 AFP/Getty Images; H1.2 iStockphoto; H2.1 Roger Bamber/Alamy; H2.2 Martyn F. Chillmaid; H3.1 iStockphoto; H3.2 Getty Images; H3.3 Martyn F. Chillmaid; H4.1 Bongarts/Getty Images; H4.2 Cordelia Molloy/Science Photo Library; H9.1 NASA Hubble Space Telescope Collection: NASA, ESA and The Hubble Heritage Team (STScI/AURA); H10.1 iStockphoto; HSQ1 iStockphoto; HSQ2 Martyn F. Chillmaid; HSQ3 Fotolia.

Unit 1
P1.1.1 Ted Kinsman/Getty Images; P1.1.3 Tony Craddock/Science Photo Library; P1.1.4 Photolibrary/Tsuneo Nakamura; P1.2.1 AP/PA Photos; P1.3.1 Charles D. Winters/Science Photo Library; P1.4.1 Fotolia; P1.4.3 Gary Ombler/Getty Images; P1.5.1 iStockphoto; P1.6.2 Spohn Matthieu/Getty Images; P1.6.5 iStockphoto; P1.7.1 Cordelia Molloy/Science Photo Library; P1.7.2 iStockphoto; P1.7.3 iStockphoto; P1.7.5a iStockphoto; P1.7.5b iStockphoto; P1.8.2 G&D Images/Alamy; P1.SQ1 Fotolia; P2.1.1 SNCF; P2.1.4 Photolibrary/Superstock; P2.2.1 Fotolia; P2.2.3 iStockphoto; P2.3.1 iStockphoto; P2.3.2a iStockphoto; P3.1.1 Martyn F. Chillmaid/Science Photo Library; P3.1.2 Emmeline Watkins/Science Photo Library; P3.1.3 Getty Images; P3.2.1 www.powerstudies.com; P3.2.2 Getty Images; P3.2.3 iStockphoto; P3.3.1 Jim Breithaupt; P3.3.3 iStockphoto; P3.4.2 Ted Kinsman/Science Photo Library; P3.4.3a Fotolia, P3.4.3b iStockphoto, P3.4.3c Dimplex; P3.SQ1 iStockphoto; P4.1.2 Adam Gault/Science Photo Library; P4.1.3 PA Archive/Press Association Images; P4.2.1 Skyscan/Science Photo Library; P4.2.3 iStockphoto; P4.2.4 Canada Press/PA Photos; P4.3.2 G. Brad Lewis/Science Photo Library; P4.3.4 Peter Menzel/Science Photo Library; P4.4.3 Fotobank/Rex Features; P4.4.4 Photolibrary; P4.5.2 Fotolia; P4.5.4 Tony Gwynne/Alamy; P4.6.1 iStockphoto; P4.6.3a iStockphoto; P4.6.3b iStockphoto; P5.1.1 iStockphoto; P5.3.1 Photolibrary; P5.3.5 Shout/Rex Features; P5.4.1a Sciencephotos/Alamy; P5.4.3 Photolibrary/Peter Arnold Images; P5.4.5 iStockphoto; P5.5.2 NASA/ESA/STSCI/Hubble Heritage Team/Science Photo Library; P5.5.3 iStockphoto; P5.6.1 Fotolia; P5.7.1 Photolibrary/Imagebroker; P6.3.1 iStockphoto; P6.3.2 iStockphoto; P6.4.1 NASA/ESA/Getty Images; P6.5.1 Mark Garlick/Science Photo Library; P6.5.2 NASA/Science Photo Library.

Unit 2
P1.1.1 Getty Images; P1.1.4 Martyn Chillmaid; P1.2.1 iStockphoto; P1.2.3 Fotolia; P1.3.2a Data Harvest; P1.3.2b Martyn Chillmaid; P2.3.3 Rob Melnychuk/Getty Images; P2.4.3 iStockphoto; P2.6.1 Martyn F. Chillmaid/Science Photo Library; P2.7.1 Martin Hospach/Getty Images; P2.7.2 AFP/Getty Images; P2.7.3 iStockphoto; P2.7.4 Cordelia Molloy/Science Photo Library; P3.1.1 iStockphoto; P3.1.2 AFP/Getty Images; P3.3.2 iStockphoto; P3.4.1 Getty Images; P3.5.3 AFP/Getty Images; P3.6.1 Photolibrary; P3.7.1 Fstop/Getty Images; P3.7.2 iStockphoto; P4.4.1a SSPL/Science Museum; P5.1.1 Charles D. Winters/Science Photo Library; P5.2.3 iStockphoto; P5.3.1a Fotolia; P5.3.1b iStockphoto; P5.3.3 Fotolia; P5.4.1 SSPL/Science Museum; P5.4.2 Cordelia Molloy/Science Photo Library; P6.1.2 Popperfoto/Getty Images; P6.2.2 SSPL/Science Museum; P6.6.3 Fotolia; P7.2.3 Photolibrary; P7.3.2 Getty Images; P7.3.3 Getty Images; P7.4.2 NOAO/AURA/NSF/T. Rector and B.A. Wolpa; P7.4.3 Physics Today Collection/American Institute of Physics/Science Photo Library; P7.4.4 NASA Image of the Day Collection; P7.5.3 X-ray: NASA/CXC/CfA/W. Forman et al.; Optical: DSS; P7.6.1 NASA/ESA/JPL/Arizona State University; P7.6.3 NASA/JPL/Cornell University.

Unit 3
P1.1.1a AJ Photo/Hop Americain/Science Photo Library; P1.1.1b iStockphoto; P1.1.2 iStockphoto; P1.1.4 Martyn F. Chillmaid/Science Photo Library; P1.2.2b iStockphoto; P1.3.1 Giphotostock/Science Photo Library; P1.4.5 David M. Martin, MD/Science Photo Library; P1.5.2 iStockphoto; P1.5.7 Fotolia; P1.8.2 Ian Hooton/Science Photo Library; P1.8.4 Adam Gault/Science Photo Library; P2.2.1a American Stock Archive/Getty Images; P2.2.1b Fred Dufour/AFP/Getty Images; P2.4.1 iStockphoto; P2.4.4 Optare plc; P2.4.5 iStockphoto; P2.5.1 iStockphoto; P2.5.3a iStockphoto; P2.6.1 Hrvoje Polan/AFP/Getty Images; P2.6.3a George Rose/Getty Images; P2.6.3b Cordelia Molloy/Science Photo Library; P2.7.1 Ria Novosti/Science Photo Library; P3.1.1 Cordelia Molloy/Science Photo Library; P3.1.2 Alex Bartel/Science Photo Library; P3.3.1 B. Boissonnet/Science Photo Library; P3.3.5 Science Photo Library; P3.5.1 ImageBroker.net/Photolibrary; P3.6.2 iStockphoto; P3.6.3 Geoff Tompkinson/Science Photo Library; EQ4 iStockphoto.

Physics   Contents

Welcome to AQA GCSE Physics!

This book has been written for you by the people who will be marking your exams, very experienced teachers and subject experts. It covers everything you need to know for your exams and is packed full of features to help you achieve the very best that you can.

Questions in yellow boxes check that you understand what you are learning as you go along. The answers are all within the text so if you don't know the answer, you can go back and reread the relevant section.

Figure 1 Many diagrams are as important for you to learn as the text, so make sure you revise them carefully.

Key words are highlighted in the text. You can look them up in the glossary at the back of the book if you are not sure what they mean.

Where you see this icon, you will know that this part of the topic involves How Science Works – a really important part of your GCSE and an interesting way to understand 'how science works' in real life.

Where you see this icon, there are supporting electronic resources in our Kerboodle online service.

Learning objectives

Each topic begins with key questions that you should be able to answer by the end of the lesson.

AQA Examiner's tip

Hints from the examiners who will mark your exams, giving you important advice on things to remember and what to watch out for.

Did you know ... ?

There are lots of interesting, and often strange, facts about science. This feature tells you about many of them.

⬭ links

Links will tell you where you can find more information about what you are learning.

Activity

An activity is linked to a main lesson and could be a discussion or task in pairs, groups or by yourself.

Maths skills

This feature highlights the maths skills that you will need for your GCSE Physics exams with short, visual explanations.

Practical

This feature helps you become familiar with key practicals. It may be a simple introduction, a reminder or the basis for a practical in the classroom.

Anything in the Higher Tier boxes must be learned by those sitting the Higher Tier exam. If you'll be sitting the Foundation Tier exam, these boxes can be missed out.

The same is true for any other places which are marked Higher or [H].

Higher

Summary questions

These questions give you the chance to test whether you have learned and understood everything in the topic. If you get any wrong, go back and have another look.

And at the end of each chapter you will find …

Summary questions

These will test you on what you have learned throughout the whole chapter, helping you to work out what you have understood and where you need to go back and revise.

AQA Examination-style questions

These questions are examples of the types of questions you will answer in your actual GCSE exam, so you can get lots of practice during your course.

Key points

At the end of the topic are the important points that you must remember. They can be used to help with revision and summarising your knowledge.

H1

How does science work? ⓚ

Learning objectives

- What is meant by 'How Science Works'?
- What is a hypothesis?
- What is a prediction and why should you make one?
- How can you investigate a problem scientifically?

⚭ **links**

You can find out more about your ISA by looking at H10 The ISA at the end of this chapter.

This first chapter looks at 'How Science Works'. It is an important part of your GCSE because the ideas introduced here will crop up throughout your course. You will be expected to collect scientific **evidence** and to understand how we use evidence. These concepts will be assessed as the major part of your internal school assessment.

You will take one or more 45-minute tests. These tests are based on **data** you have collected previously plus data supplied for you in the test. They are called **Investigative Skills Assignments (ISA)**. The ideas in 'How Science Works' will also be assessed in your examinations.

How science works for us

Science works for us all day, every day. You do not need to know how a mobile phone works to enjoy sending text messages. But, think about how you started to use your mobile phone or your television remote control. Did you work through pages of instructions? Probably not!

You knew that pressing the buttons would change something on the screen (**knowledge**). You played around with the buttons, to see what would happen (**observation**). You had a guess based on your knowledge and observations at what you thought might be happening (**prediction**) and then tested your idea (**experiment**).

Perhaps 'How Science Works' should really be called 'How Scientists Work'.

Science moves forward by slow, steady steps. When a genius such as Einstein comes along, it takes a giant leap. Those small steps build on knowledge and experience that we already have.

The steps don't always lead in a straight line, starting with an observation and ending with a conclusion. More often than not you find yourself going round in circles, but each time you go around the loop you gain more knowledge and so can make better predictions.

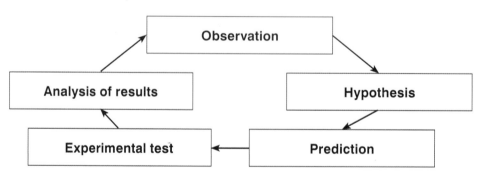

Each small step is important in its own way. It builds on the body of knowledge that we have, but observation is usually the starting point. In 1581 Galileo, during a long service at Pisa Cathedral, observed a lamp swinging from a long chain attached to the ceiling. This started him thinking about what factors affect the time period of a pendulum. After many tests on pendulums, he was able to come up with a **hypothesis** that would enable him to predict what the time period would be for any given length of pendulum.

Figure 1 Albert Einstein was a genius, but he worked through scientific problems in the same way as you will in your GCSE

Activity

Bouncing balls

No matter how good a basketball player you are, if the ball is not properly inflated, you cannot play well. As the balls get used during the game it is possible that some of them will get soft. They should all bounce the same.

How high the ball bounces will depend on lots of variables. It will depend on:

- what the ball is made of
- how much air has been pumped in
- what the temperature of the ball is
- what the floor surface is made of, and
- how hard you throw it.

It is impossible to test all of these during a match. The simple way is to drop a ball from a certain height and see how high it bounces. Can you work out a way to see how changing the height from which a ball is dropped can affect how high it bounces? This could then be used as a simple test during the match to see whether the balls are good enough.

You can use the following headings to discuss your ideas. One person should be writing your thoughts down, so that you can discuss them with the rest of your class.

- What prediction can you make about the height the ball is dropped from and the height it will bounce to?
- What would you vary in each test? This is called the independent variable.
- What would you measure to judge the effect of varying the independent variable? This is called the dependent variable.
- What would you need to keep unchanged to make this a fair test?
- Write a plan for your investigation.
- These are called control variables.

Figure 2 Playing basketball

Summary questions

1 Copy and complete the paragraph using the words below:

experiment   knowledge   conclusion   prediction   observation

You have learned before that a cup of tea loses energy if it is left standing. This is a piece of .... You make an .... that dark-coloured cups will cool faster. So you make a .... that if you have a black cup, this will cool fastest of all. You carry out an .... to get some results, and from these you make a ....

??? Did you know ... ?

The Greeks were arguably the first true scientists. They challenged traditional myths about life. They put forward ideas that they knew would be challenged. They were keen to argue the point and come to a reasoned conclusion.

Other cultures relied on long-established myths and argument was seen as heresy.

Key points

- **Observations** are often the starting point for an investigation.

- A **hypothesis** is a proposal intended to explain certain facts or observations.

- A **prediction** is an intelligent guess, based on some **knowledge**.

- An **experiment** is a way of testing your prediction.

Fundamental ideas about how science works

Learning objectives

- How do you spot when an opinion is not based on good science?

- What is the importance of continuous and categoric variables?

- What does it mean to say that evidence is valid?

- What is the difference between a result being repeatable and a result being reproducible?

- How can two sets of data be linked?

AQA Examiner's tip

Read a newspaper article or watch the news on TV. Ask yourself whether any research presented is valid. Ask yourself whether you can trust that person's opinion and why.

Science is too important for us to get it wrong

Sometimes it is easy to spot when people try to use science poorly. Sometimes it can be funny. You might have seen adverts claiming to give your hair 'body' or sprays that give your feet 'lift'!

On the other hand, poor scientific practice can cost lives.

Some years ago a company sold the drug thalidomide to people as a sleeping pill. Research was carried out on animals to see whether it was safe. The research did not include work on pregnant animals. The **opinion** of the people in charge was that the animal research showed the drug could be used safely with humans.

Then the drug was also found to help ease morning sickness in pregnant women. Unfortunately, doctors prescribed it to many women, resulting in thousands of babies being born with deformed limbs. It was far from safe.

These are very difficult decisions to make. You need to be absolutely certain of what the science is telling you.

a Why was the opinion of the people in charge of developing thalidomide based on poor science?

Deciding on what to measure: variables

Variables are physical, chemical or biological quantities or characteristics.

In an investigation, you normally choose one thing to change or vary. This is called the **independent variable**.

When you change the independent variable, it may cause something else to change. This is called the **dependent variable**.

A **control variable** is one that is kept the same and is not changed during the investigation.

You need to know about two different types of these variables:

- A **categoric variable** is one that is best described by a label (usually a word). The 'colour of eyes' is a categoric variable, e.g. blue or brown eyes.

- A **continuous variable** is one that we measure, so its value could be any number. Temperature (as measured by a thermometer or temperature sensor) is a continuous variable, e.g. 37.6 °C, 45.2 °C. Continuous variables can have values (called quantities) that can be found by making measurements (e.g. light intensity, flow rate, etc.).

b Imagine you were testing a solar cell, what would be better: putting a light bulb into the circuit to see how bright it was, or using a voltmeter to measure the potential difference?

Figure 1 Road sign that uses solar cells

Making your evidence repeatable, reproducible and valid

When you are designing an investigation you must make sure that other people can get the same results as you. This makes the evidence you collect **reproducible**.

A measurement is **repeatable** if the original experimenter repeats the investigation using the same method and equipment and obtains the same results.

A measurement is reproducible if the investigation is repeated by another person, or by using different equipment or techniques, and the same results are obtained.

You must also make sure you are measuring the actual thing you want to measure. If you don't, your data can't be used to answer your original question. This seems very obvious but it is not always quite so easy. You need to make sure that you have controlled as many other variables as you can, so that no one can say

Figure 2 Student recording a range of temperatures – an example of a continuous variable

that your investigation is not **valid**. A measurement is valid if it measures what it is supposed to be measuring with an appropriate level of performance.

  c State one way in which you can show that your results are valid.

How might an independent variable be linked to a dependent variable?

Looking for a link between your independent and dependent variables is very important. The pattern of your graph or bar chart can often help you to see whether there is a link.

But beware! There may not be a link! If your results seem to show that there is no link, don't be afraid to say so. Look at Figure 3.

The points on the top graph show a clear pattern, but the bottom graph shows random scatter.

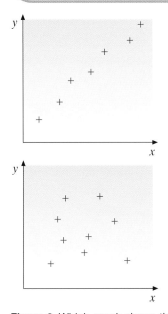

Figure 3 Which graph shows that there might be a link between *x* and *y*?

?？? Did you know … ?

Aristotle, a brilliant Greek scientist, once proclaimed that men had more teeth than women! Do you think that his data collection was reproducible?

Summary questions

1 Copy and complete the paragraph using the words below:

continuous independent categoric dependent

Stefan wanted to find out which was the strongest supermarket plastic carrier bag. He tested five different bags by adding weight to them until they broke. The type of bag he used was the .... variable and the weight that it took to break it was the .... variable. The 'type of bag' is called a .... variable and the 'weight needed to break' it was a .... variable.

2 A researcher claimed that the metal tungsten 'alters the growth of leukaemia cells' in laboratory tests. A newspaper wrote that they would 'wait until other scientists had reviewed the research before giving their opinion.' Why is this a good idea?

Key points

● Be on the lookout for non-scientific opinions.

● Continuous data give more information than other types of data.

● Check that evidence is reproducible and valid.

H3

Starting an investigation

Learning objectives

- How can you use your scientific knowledge to observe the world around you?

- How can you use your observations to make a hypothesis?

- How can you make predictions and start to design an investigation?

Figure 1 A wind turbine

Observation

As humans we are sensitive to the world around us. We can use our senses to detect what is happening. As scientists we use observations to ask questions. We can only ask useful questions if we know something about the observed event. We will not have all of the answers, but we know enough to start asking relevant questions.

If we observe that the weather has been hot today, we would not ask whether it was due to global warming. If the weather was hotter than normal for several years, we could ask that question. We know that global warming takes many years to show its effect.

When you are designing an investigation you have to observe carefully which variables are likely to have an effect.

a Would it be reasonable to ask whether the wind turbine in Figure 1 generates less electricity in the rain? Explain your answer.

Amjid was waiting to cross at a zebra crossing. A car stopped to let him cross when a second car drove into the first car, without braking. Being a scientist, Amjid tried to work out why this had happened . . . while the two drivers argued! He came up with the following ideas:

- The second driver was tired.
- The second car had faulty brakes.
- The first car stopped too quickly.
- The second car was driving too fast.
- The second car was travelling too close.
- The second car had worn tyres.
- The first car had no brake lights.

b Discuss each of these ideas and use your knowledge of science to decide which three are the most likely to have caused the crash.

Observations, backed up by really creative thinking and good scientific knowledge can lead to a hypothesis.

Testing scientific ideas

Scientists always try to think of ways to explain how things work or why they behave in the way that they do.

After their observations, they use their understanding of science to come up with an idea that could explain what is going on. This idea is sometimes called a **hypothesis**. They use this idea to make a prediction. A prediction is like a guess, but it is not just a wild guess – it is based on previous understanding.

A scientist will say, 'If it works the way I think it does, I should be able to change **this** (the independent variable) and **that** will happen (the dependent variable).'

Predictions are what make science so powerful. They mean that we can work out rules that tell us what will happen in the future. For example, electricians can predict how much current will flow through a wire when an electric cooker is connected. Knowing this, they can choose the right thickness of cable to use.

When a steady wind blows against a structure like a bridge or tall chimney, it can cause 'vortex shedding.' This exerts an oscillating force on the structure, a force that can be predicted.

Figure 2 The Tacoma Narrows Bridge in the USA twisting just before it collapsed

  c Look at the photograph in Figure 2. Note down anything that you find interesting. Use your knowledge and some creative thought to suggest a hypothesis based on your observations.

Not all predictions are correct. If scientists find that the prediction doesn't work, it's back to the drawing board! They either amend their original idea or think of a completely new one.

Starting to design a valid investigation

observation ✚ knowledge ➡ hypothesis ➡ prediction ➡ investigation

We can test a prediction by carrying out an **investigation**. You, as the scientist, predict that there is a relationship between two variables.

The independent variable is one that is selected and changed by you, the investigator. The dependent variable is measured for each change in your independent variable. Then all other variables become control variables, kept constant so that your investigation is a fair test.

If your measurements are going to be accepted by other people, they must be **valid**. Part of this is making sure that you are really measuring the effect of changing your chosen variable. For example, if other variables aren't controlled properly, they might be affecting the data collected.

  d Look at Figure 3. Darren was investigating the light given out by a 12 V bulb. He used a light meter in the laboratory that was set at 10 cm from the bulb. What might be wrong here?

Figure 3 Testing a light bulb

Summary questions

1 Copy and complete the paragraph using the words below:
  controlled dependent independent knowledge prediction hypothesis
  An observation linked with scientific .... can be used to make a .... A .... links an .... variable to a .... variable. All other variables need to be ....

2 What is the difference between a prediction and a guess?

3 Imagine you were testing whether the length of a wire affected its resistance. The current through the wire might cause it to get hot.
  a How could you monitor the temperature?
  b What other control variables can you think of that might affect the results?

Key points

- Observation is often the starting point for an investigation.
- Testing predictions can lead to new scientific understanding.
- You must design investigations that produce valid results if you are to be believed.

H4

Planning an investigation

Fair testing

A **fair test** is one in which only the independent variable affects the dependent variable. All other variables (called control variables) should be kept the same. If the test is not fair, the results of your investigation will not be valid.

Sometimes it is very difficult to keep control variables the same. However, at least you can **monitor** them, so that you know whether they have changed or not.

a How would you set up an investigation to see how the wing setting on the rear of a car affected its top speed down the straight?

Figure 1 How do wing settings on the rear wing of a sports car affect its top speed?

Surveys

Not all scientific investigations involve deliberately changing the independent variable.

If you were investigating the effect that using a mobile phone may have on health you wouldn't put a group of people in a room and make them use their mobile phones to see whether they developed brain cancer!

Instead, you might conduct a **survey**. You might study the health of a large number of people who regularly use a mobile phone and compare their health with those who never use a mobile phone.

You would have to choose people of the same age and same family history to test. The larger the sample size you test, the better your results will be.

Control group

Control groups are used in investigations to try to make sure that you are measuring the variable that you intend to measure. When investigating the effects of a new drug, the control group will be given a **placebo**. This is a 'pretend' drug that actually has no effect on the patient at all. The control group think they are taking a drug but the placebo does not contain the drug. This way you can control the variable of 'thinking that the drug is working' and separate out the effect of the actual drug.

Usually neither the patient nor the doctor knows until after the trials have been completed which of the patients were given the placebo. This is known as a **double-blind trial**.

Risks and hazards

One of the first things you must do is to think about any potential **hazards** and then assess the **risk**.

Everything you do in life presents a hazard. What you have to do is to identify the hazard and then decide the degree of risk that it gives. If the risk is very high, you must do something to reduce it.

For example, if you decide to go out in the pouring rain, lightning is a possible hazard. However, you decide that the risk is so small that you will ignore it and go out anyway.

If you decide to cross a busy road, the cars travelling along it at high speed represent a hazard. You decide to reduce the risk by crossing at a pedestrian crossing.

Figure 2 The hazard is the busy road; we reduce the risk by using a pedestrian crossing

 How Science Works

Energy testing

Imagine you were using an electric immersion heater to see how much energy it supplied to a beaker of water when switched on.

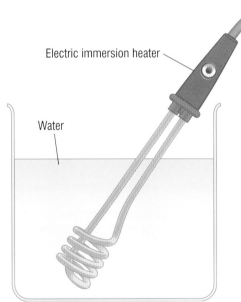

Electric immersion heater

Water

- What are the **hazards** that are present?
- What control measures could you take to reduce the **risk** from these hazards?

AQA **Examiner's tip**

Before you start your practical work you must make sure that it is safe. What are the likely hazards? How could you reduce the risk caused by these hazards? This is known as a **risk assessment**. You may well be asked questions like this on your ISA paper.

Key points

- Care must be taken to ensure fair testing – as far as is possible.
- Control variables must be kept the same during an investigation.
- Surveys are often used when it is impossible to carry out an experiment in which the independent variable is changed.
- Control groups allow you to make a comparison.
- A risk assessment must be made when planning a practical investigation.

Summary questions

1 Copy and complete the paragraph using the words below:

investigation hazards assessment risks

Before you carry out any practical ...., you need to carry out a risk .... You can do this by looking for any potential .... and making sure that the .... are as small as possible.

2 Explain the difference between a control group and a control variable.

3 Briefly describe how you would go about setting up a fair test in a laboratory investigation. Give your answer as general advice.

H5

Designing an investigation

Learning objectives

- How do you make sure that you choose the best values for your variables?
- How do you decide on a suitable range?
- How do you decide on a suitable interval?
- How do you ensure accuracy and precision?

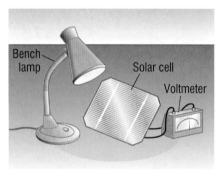

Figure 1 Measuring the output voltage from a solar cell

Choosing values of a variable

Trial runs will tell you a lot about how your early thoughts are going to work out.

Do you have the correct conditions?

An experiment to measure the output voltage from a solar cell might not give an output at all. Perhaps the light isn't bright enough, perhaps the surface area of the cell is too small, or maybe the voltmeter that you are using cannot measure such a small voltage.

Have you chosen a sensible range?

Range means the maximum and minimum values of the independent or dependent variables. It is important to choose a suitable range for the independent variable, otherwise you may not be able to see any change in the dependent variable.

For example, if the results are all very similar, you might not have chosen a wide enough range of light intensities.

Have you got enough readings that are close together?

The gap between the readings is known as the **interval**.

For example, you might alter the light intensity by moving a lamp to different distances from the solar cell. A set of 11 readings equally spaced over a distance of 1 metre would give an interval of 10 centimetres.

If the results are very different from each other, you might not see a pattern if you have large gaps between readings over the important part of the range.

Accuracy

Accurate measurements are very close to the **true value**.

Your investigation should provide data that is accurate enough to answer your original question.

However, it is not always possible to know what that true value is.

How do you get accurate data?

- You can repeat your measurements and your mean is more likely to be accurate.
- Try repeating your measurements with a different instrument and see whether you get the same readings.
- Use high-quality instruments that measure accurately.
- The more carefully you use the measuring instruments, the more accuracy you will get.

Precision, resolution, repeatability and reproducibility

A **precise** measurement is one in which there is very little spread about the mean value.

If your repeated measurements are closely grouped together, you have precision. Your measurements must be made with an instrument that has a suitable **resolution**. Resolution of a measuring instrument is the smallest change in the quantity being measured (input) that gives a perceptible change in the reading.

It's no use measuring the time for a fast reaction to finish using the seconds hand on a clock! If there are big differences within sets of repeat readings, you will not be able to make a valid conclusion. You won't be able to trust your data!

How do you get precise data?

- You have to use measuring instruments with sufficiently small scale divisions.
- You have to repeat your tests as often as necessary.
- You have to repeat your tests in exactly the same way each time.

If you repeat your investigation using the same method and equipment and obtain the same results, your results are said to be **repeatable.**

If someone else repeats your investigation in the same way, or if you repeat it by using different equipment or techniques, and the same results are obtained, it is said to be **reproducible.**

You may be asked to compare your results with those of others in your group, or with data from other scientists. Research like this is a good way of seeing whether your results are reproducible.

A word of caution!

Precision depends only on the extent of random errors – it gives no indication of how close results are to the true value. Just because your results show precision does not mean they are accurate.

a Draw a thermometer scale reading 49.5 °C, showing four results that are both accurate and precise.

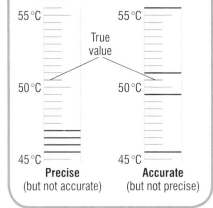
Summary questions

1 Copy and complete the paragraph using the words below:

range   repeat   conditions   readings

Trial runs give you a good idea of whether you have the correct .... to collect any data; whether you have chosen the correct .... for the independent variable; whether you have enough ....; and if you need to do .... readings.

2 Use an example to explain how a set of repeat measurements could be accurate, but not precise.

3 Explain the difference between a set of results that are reproducible and a set of results that are repeatable.

Key points

- You can use a trial run to make sure that you choose the best values for your variables.
- The range states the maximum and minimum values of a variable.
- The interval is the gap between the values of a variable.
- Careful use of the correct equipment can improve accuracy and precision.
- You should try to reproduce your results carefully.

H6

Making measurements

Learning objectives

- Why do results always vary?

- How do you choose instruments that will give you accurate results?

- What do we mean by the resolution of an instrument?

- What is the difference between a systematic error and a random error?

- How does human error affect results and what do you do with anomalies?

Figure 1 Matt testing how fast a trolley goes down a ramp

AQA Examiner's tip

If you are asked what may have caused an error, never answer simply 'human error' – you won't get any marks for this.

You need to say what the experimenter may have done to cause the error, or give more detail, e.g. 'Human reaction time might have caused an error in the timing when using a stopwatch.'

Using instruments

Try measuring the temperature of a beaker of water using a digital thermometer. Do you always get the same result? Probably not! So can we say that any measurement is absolutely correct?

In any experiment there will be doubts about actual measurements.

a Look at Figure 1. Suppose, like this student, you tested the time it takes for one type of trolley to run down a ramp. It is unlikely that you would get two readings exactly the same. Discuss all the possible reasons why.

When you choose an instrument you need to know that it will give you the accuracy that you want. You need to be confident that it is giving a true reading.

Perhaps you have used a simple force meter in school for measuring force. How confident were you that you had measured the true force? You could use a very expensive force meter to calibrate yours. The expensive force meter is more likely to show the true reading that is accurate – but are you really sure?

You also need to be able to use an instrument properly.

Instruments that measure the same thing can have different sensitivities. The **resolution** of an instrument refers to the smallest change in a value that can be detected. This is one factor that determines the precision of your measurements.

Choosing the wrong scale can cause you to miss important data or make silly conclusions. We would not measure the distance to Jupiter in centimetres – we would use kilometres.

b Match the following timers to their best use:

Used to measure	Resolution of timer
Time taken to sail around the world	milliseconds
Timing a car rolling down a slope	seconds
Time taken for a bullet to travel to its target	minutes
Timing a pizza to cook	hours

Errors

Even when an instrument is used correctly, the results can still show differences.

Results may differ because of **random error**. This is most likely to be due to a poor measurement being made. It could be due to not carrying out the method consistently.

If you repeat your measurements several times and then calculate a mean, you will reduce the effect of random errors.

The **error** might be a **systematic error**. This means that the method was carried out consistently but an error was being repeated. A systematic error will make your readings be spread about some value other than the true value. This is because your results will differ from the true value by a consistent amount each time a measurement is made.

No number of repeats can do anything about systematic errors. If you think that you have a systematic error, you need to repeat using a different set of equipment or a different technique. Then compare your results and spot the difference!

A **zero error** is one kind of systematic error. Suppose that you were trying to measure the length of your desk with a metre rule, but you hadn't noticed that someone had sawn off half a centimetre from the end of the ruler. It wouldn't matter how many times you repeated the measurement, you would never get any nearer to the true value.

Look at the table. It shows the two sets of data that were taken from the investigation that Matt did. He tested five different trolleys. The bottom row is the time that was expected from calculations:

Type of trolley used	A	B	C	D	E
Time taken for trolley to run down ramp (seconds)	12.6	23.1	24.8	31.3	38.2
	12.1	15.2	24.3	32.1	37.6
Calculated time (seconds)	10.1	13.1	22.1	30.1	35.3

c Discuss whether there is any evidence of random error in these results.

d Discuss whether there is any evidence of systematic error in these results.

Anomalies

Anomalous results are clearly out of line. They are not those that are due to the natural variation you get from any measurement. These should be looked at carefully. There might be a very interesting reason why they are so different. You should always look for anomalous results and discard them before you calculate a mean, if necessary.

● If anomalies can be identified while you are doing an investigation, it is best to repeat that part of the investigation.

● If you find anomalies after you have finished collecting data for an investigation, they must be discarded.

Did you know ... ?

Two radio astronomers, Robert Wilson and Arno Penzias used radio telescopes to search outer space. They wanted to cut out all background noise. They were troubled by low levels of 'noise' and could not find its source, so they investigated this anomaly further. They thought the noise might be from a big city, but that was ruled out. It wasn't any nearby electrical equipment. They found and eliminated some of the noise, but still some remained. They found pigeons on the antenna – but having sent them away by post they returned! Regrettably they had them shot – still the noise! The noise came from all directions, constantly. It would have been easy to ignore it – but they persisted and discovered that it was noise left over from the Big Bang. They had discovered evidence for the early stages of the universe!

Key points

● Results will nearly always vary.

● Better quality instruments give more accurate results.

● The resolution of an instrument refers to the smallest change that it can detect.

● Human error can produce random and/or systematic errors.

● We examine anomalies as they might give us some interesting ideas. If they are due to a random error, we repeat the measurements. If there is no time to repeat them, we discard them.

Summary questions

1 Copy and complete the paragraph using the words below:

accurate discarded random resolution systematic use variation

There will always be some .... in results. You should always choose the best instruments that you can in order to get the most .... results. You must know how to .... the instrument properly. The .... of an instrument refers to the smallest change that can be detected. There are two types of error: .... and .... Anomalies due to random error should be ....

2 What kind of error will most likely occur in the following situations?
   a Asking everyone in the class to measure the length of the bench.
   b Using a ruler that has a piece missing from the zero end.

Presenting data

Learning objectives

- How do you calculate the mean from a set of data?
- How do you use tables of results?
- What is the range of the data?
- How do you display your data?

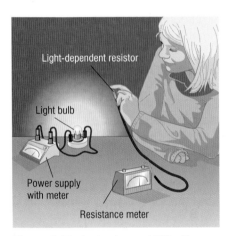

Figure 1 Student using an LDR with a light bulb

For this section you will be working with data from this investigation:

Mel shone a bulb on to a light-dependent resistor (LDR). She recorded how quickly energy was transferred to the bulb and the resistance of the LDR. Then she changed the rate of energy transferred to the bulb by altering the setting on the power supply and repeated the experiment.

The room was kept as dark as possible while she made the readings.

Tables

Tables are really good for getting your results down quickly and clearly. You should design your table **before** you start your investigation.

Your table should be constructed to fit in all the data to be collected. It should be fully labelled, including units.

You may want to have extra columns for repeats, calculations of means or calculated values.

Checking for anomalies

While filling in your table of results you should be constantly looking for anomalies.

- Check to see whether any reading in a set of repeat readings is significantly different from the others.
- Check to see whether the pattern you are getting as you change the independent variable is what you expected.

Remember, a result that looks anomalous should be checked out to see whether it really is a poor reading.

Planning your table

Mel had decided on the values for her independent variable. We always put these in the first column of a table. The dependent variable goes in the second column. Mel will find its values as she carries out the investigation.

So she could plan a table like this:

Rate of energy transferred to the bulb (W)	Resistance of LDR (Ω)
0.5	
1.4	
2.6	
4.8	
8.4	

Or like this:

Rate of energy transferred to the bulb (W)	0.5	1.4	2.6	4.8	8.4
Resistance of LDR (Ω)					

All she had to do in the investigation was to write the correct numbers in the second column to complete the top table.

Mel's results are shown in the alternative format in the table below.

Rate of energy transferred to the bulb (W)	0.5	1.4	2.6	4.8	8.4
Resistance of LDR (Ω)	4000	3000	1000	350	150

The range of the data

Pick out the maximum and the minimum values and you have the range of a variable. You should always quote these two numbers when asked for a range. For example, the range of the dependent variable is between 150 Ω (the lowest value) and 4000 Ω (the highest value) – and don't forget to include the units!

a What is the range for the independent variable and for the dependent variable in Mel's set of data?

Maths skills

The mean of the data

Often you have to find the **mean** of each repeated set of measurements.

The first thing you should do is to look for any anomalous results. If you find any, miss these out of the calculation. Then add together the remaining measurements and divide by how many there are.

For example:
- Mel takes four readings, 350 Ω, 355 Ω, 420 Ω and 345 Ω.
- 420 Ω is an anomalous result and so is missed out. So 350 + 355 + 345 = 1050.
- 1050 ÷ 3 (the number of valid results) = 350 Ω.

The repeat values and mean can be recorded as shown below:

Rate of energy transferred to the bulb (W)	Resistance of LDR (Ω)			
	1st test	2nd test	3rd test	Mean

Displaying your results

If one of your variables is categoric, you should use a **bar chart**. If you have a continuous independent and a continuous dependent variable, a **line graph** should be used. Plot the points as small 'plus' signs (+).

Summary questions

1 Copy and complete the paragraph using the words below:

categoric   continuous   mean   range

The maximum and minimum values show the .... of the data. The sum of all the values in a set of repeat readings divided by the total number of these repeat values gives the .... Bar charts are used when you have a .... independent variable and a continuous dependent variable. Line graphs are used when you have .... independent and dependent variables.

2 Draw a graph of Mel's results from the top of this page.

H8 Using data to draw conclusions

Learning objectives

- How do we best use charts and graphs to identify patterns?
- What are the possible relationships we can identify from charts and graphs?
- How do we draw conclusions from relationships?
- How can we decide if our results are good and our conclusions are valid?

Identifying patterns and relationships

Now that you have a bar chart or a line graph of your results you can begin to look for patterns. You must have an open mind at this point.

First, there could still be some anomalous results. You might not have picked these out earlier. How do you spot an anomaly? It must be a significant distance away from the pattern, not just within normal variation. If you do have any anomalous results plotted on your graph, circle these and ignore them when drawing the **line of best fit**.

Now look at your graph. Is there a pattern that you can see? When you have decided, draw a line of best fit that shows this pattern.

A line of best fit is a kind of visual averaging process. You should draw the line so that it leaves as many points slightly above the line as there are points below. In other words it is a line that steers a middle course through the field of points.

The vast majority of results that you get from continuous data require a line of best fit.

Remember, a line of best fit can be a straight line or it can be a curve – you have to decide from your results.

You need to consider whether your graph shows a linear **relationship**. This simply means, can you be confident about drawing a straight line of best fit on your graph? If the answer is yes – is this line positive or negative?

a Say whether graphs **i** and **ii** in Figure 1 show a positive or a negative linear relationship.

Look at the graph in Figure 2. It shows a positive linear relationship. It also goes through the origin (0, 0). We call this a **directly proportional** relationship.

Your results might also show a curved line of best fit. These can be predictable, complex or very complex! Look at Figure 3 below.

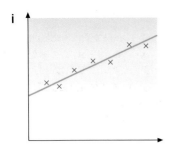

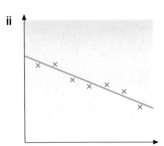

Figure 1 Graphs showing linear relationships

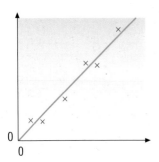

Figure 2 Graph showing a directly proportional relationship

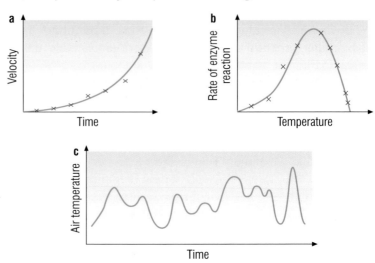

Figure 3 a Graph showing predictable results **b** Graph showing complex results **c** Graph showing very complex results

Drawing conclusions

If there is a pattern to be seen (for example as one variable gets bigger the other also gets bigger), it may be that:

- changing one has caused the other to change
- the two are related, but one is not necessarily the cause of the other.

Activity

Looking at relationships

Some people think that watching too much television can cause an increase in violence.

The table shows the number of television sets in the UK for four different years, and the number of murders committed in those years.

Year	Number of televisions (millions)	Number of murders
1970	15	310
1980	25	500
1990	42	550
2000	60	750

Plot a graph to show the relationship.

- Do you think this proves that watching television causes violence? Explain your answer.

Your conclusion must go no further than the evidence that you have.

Poor science can often happen if a wrong decision is made here. Newspapers have said that living near electricity substations can cause cancer. All that scientists would say is that there is possibly an association.

Did you know …?

Pythagoras of Samos declared that 'Everything is number'. He believed that everything in the universe can be explained by simple mathematical relationships. He went on to discover the relationship between the length of a string and the sound it produces when it vibrates.

He developed this idea into a theory that the Sun, the Moon and the planets produced a sort of music that kept them in their orbits!

Evaluation

You will often be asked to evaluate either the method of the investigation or the conclusion that has been reached. Ask yourself: Could the method have been improved? Is the conclusion that has been made a valid one?

Summary questions

1. Copy and complete the paragraph using the words below:

   anomalous   complex   directly   negative   positive

   Lines of best fit can be used to identify .... results. Linear relationships can be .... or .... If a straight line goes through the origin of a graph, the relationship is .... proportional. Often a line of best fit is a curve which can be predictable or ....

2. Nasma knew about the possible link between cancer and living near to electricity substations. She found a quote from a National Grid Company survey of substations:

   'Measurements of the magnetic field were taken at 0.5 metre above ground level within 1 metre of fences and revealed 1.9 microteslas. After 5 metres this dropped to the normal levels measured in any house.'

   Discuss the type of experiment and the data you would expect to see to support a conclusion that it is safe to build houses over 5 metres from an electricity substation.

Key points

- Drawing lines of best fit helps us to study the relationship between variables.
- The possible relationships are linear, positive and negative; directly proportional; predictable and complex curves.
- Conclusions must go no further than the data available.
- The reproducibility of data can be checked by looking at other similar work done by others, perhaps on the internet. It can also be checked by using a different method or by others checking your method.

H9 Scientific evidence and society

STAR IN SCANDAL SHOCK

We Find Out What They Don't Want You To Know... And WE TELL YOU!

MOBILE PHONE TUMOUR RISK?

Swedish researchers found that the risk of developing an ear tumour increased if you used a mobile phone. The study was of 750 people. This type of tumour affects one in 100 000 people and the risk increased four times if you used the phone for more than 10 years.

A spoke...
that

Did you know ...?

A scientist who rejected the idea of a causal link between smoking and lung cancer was later found to be being paid by a tobacco company.

AQA Examiner's tip

If you are asked about bias in scientific evidence, there are two types:

- The measuring instruments may have introduced a bias because they were not calibrated correctly.
- The scientists themselves may have a biased opinion (e.g. if they are paid by a company to promote their product).

Now you have reached a conclusion about a piece of scientific research. So what is next? If it is pure research, your fellow scientists will want to look at it very carefully. If it affects the lives of ordinary people, society will also want to examine it closely.

You can help your cause by giving a balanced account of what you have found out. It is much the same as any argument you might have. If you make ridiculous claims, nobody will believe anything you have to say.

Be open and honest. If you only tell part of the story, someone will want to know why! Equally, if somebody is only telling you part of the truth, you cannot be confident about anything they say.

a 'X-rays are safe, but should be limited' is the headline in an American newspaper. What information is missing? Is it important?

You must be on the lookout for people who might be biased when representing scientific evidence. Some scientists are paid by companies to do research. When you are told that a certain product is harmless, just check out who is telling you.

b Suppose you wanted to know about safe levels of noise at work. Would you ask the scientist who helped to develop the machinery or a scientist working in the local university? What questions would you ask, so that you could make a valid judgement?

EAR DEFENDERS MUST BE WORN AT ALL TIMES

We also have to be very careful in reaching judgements according to who is presenting scientific evidence to us. For example, if the evidence might provoke public or political problems, it might be played down.

Equally, others might want to exaggerate the findings. They might make more of the results than the evidence suggests. Take as an example the siting of mobile phone masts. Local people may well present the same data in a totally different way from those with a wider view of the need for mobile phones.

c Check out some websites on mobile phone masts. Get the opinions of people who think they are dangerous and those who believe they are safe. Try to identify any political bias there might be in their opinions.

Science can often lead to the development of new materials or techniques. Sometimes these cause a problem for society where hard choices have to be made.

Scientists can give us the answers to many questions, but not to every question. Scientists have a contribution to make to a debate, but so do others such as environmentalists, economists and politicians.

The limitations of science

Science can help us in many ways but it cannot supply all the answers. We are still finding out about things and developing our scientific knowledge. For example, the Hubble telescope has helped us to revise our ideas about the beginnings of the universe.

There are some questions that we cannot answer, maybe because we do not have enough reproducible, repeatable and valid evidence. For example, research into the causes of cancer still needs much work to be done to provide data.

Figure 1 The Hubble space telescope can look deep into space and tell us things about the universe's beginning from the formations of early galaxies

There are some questions that science cannot answer at all. These tend to be questions where beliefs, opinions and ethics are important. For example, science can suggest what the universe was like when it was first formed, but cannot answer the question of why it was formed.

VILLAGERS PROTEST AGAINST WIND FARM

There was considerable local opposition from local villagers to building a wind farm near the A14 road in Cambridgeshire. Planners turned down the application after seven months of protests by local residents. Some described it as being like 16 football pitches rotating in the sky. Others were concerned at the effect on the value of their houses. Friends of the Earth were, in principle, in favour. The wind farm company said that it would provide energy for 20,000 homes.

Summary questions

1 Copy and complete the paragraph using the words below:

status   balanced   bias   political

Evidence from scientific investigations should be given in a .... way. It must be checked for any .... from the experimenter. Evidence can be given too little or too much weight if it is of .... significance. The .... of the experimenter is likely to influence people in their judgement of the evidence.

2 Collect some newspaper articles to show how scientific evidence is used. Discuss in groups whether these articles are honest and fair representations of the science. Consider whether they carry any bias.

3 Extract from a newspaper report about Sizewell nuclear power station:

A radioactive leak can have devastating results but one small pill could protect you. Our reporter reveals how for the first time these life-saving pills will be available to families living close to the Sizewell nuclear power station.

Suppose you were living near Sizewell power station. Who would you trust to tell you whether these pills would protect you from radiation?

Key points

- Scientific evidence must be presented in a balanced way that points out clearly how valid the evidence is.

- The evidence must not contain any bias from the experimenter.

- The evidence must be checked to appreciate whether there has been any political influence.

- The status of the experimenter can influence the weight placed on the evidence.

H10

The ISA

There are several different stages to the ISA.

Stage 1

Your teacher will tell you the problem that you are going to investigate, and you will have to develop your own hypothesis. They will also set the problem in a context – in other words, where in real life your investigation could be useful. You should have a discussion about it, and talk about different ways in which you might solve the problem. Your teacher should show you the equipment that you can use, and you should research one or two possible methods for carrying out an experiment to test the hypothesis. You should also research the context and do a risk assessment for your practical work. You will be allowed to make one side of notes on this research, which you can take into the written part of the ISA.

Figure 1 Doing practical work allows you to develop the skills needed to do well in the ISA

You should be allowed to handle the equipment and you may be allowed to carry out a preliminary experiment.

Make sure that you understand what you have to do – now is the time to ask questions if you are not sure.

How Science Works

Section 1 of the ISA

At the end of this stage, you will answer Section 1 of the ISA. You will need to:

- develop a hypothesis
- identify one or more variables that you need to control
- describe how you would carry out the main experiment
- identify possible hazards and say what you would do to reduce any risk
- make a blank table ready for your results.

a What features should you include in your written plan?

b What should you include in your blank table?

Stage 2

This is where you carry out the experiment and get some results. Don't worry too much about spending a long time getting fantastically accurate results – it is more important to get some results that you can analyse.

After you have got results, you will have to compare your results with those of others. You will also have to draw a graph or a bar chart.

c How do you decide whether you should draw a bar chart or a line graph?

Stage 3

This is where you answer Section 2 of the ISA. Section 2 of the ISA is all about your own results, so make sure that you look at your table and graph when you are answering this section. To get the best marks you will need to quote some data from your results.

How Science Works

Section 2 of the ISA

In this section you will need to:

- say what you were trying to find out
- compare your results with those of others, saying whether you think they are similar or different
- analyse data that is given in the paper. This data will be in the same topic area as your investigation
- use ideas from your own investigation to answer questions about this data
- write a conclusion
- compare your conclusion with the hypothesis you have tested.

You may need to change or even reject your hypothesis in response to your findings.

AQA *Examiner's tip*

When you are comparing your conclusion with the hypothesis, make sure that you also talk about the **extent** to which your results support the hypothesis. Which of these answers do you think would score the most marks?

- My results support the hypothesis.
- In my results, as *x* got bigger, *y* got bigger, as stated in the hypothesis.
- In my results, as *x* got bigger, *y* got bigger, as stated in the hypothesis, but unlike the hypothesis, *y* stopped increasing after a while.

Key points

- **When you are writing the plan make sure that you include details about:**
  - **the range and interval of the independent variable**
  - **the control variables**
  - **the number of repeats.**
- **Try to put down at least two possible hazards, and say how you are going to minimise the risk from them.**
- **Look carefully at the hypothesis that you are given – this should give you a good clue about how to do the experiment.**
- **Always refer back to the hypothesis when you are writing your conclusion.**

Summary questions

1 Copy and complete the paragraph using the words below:

control   independent   dependent

When writing a plan, you need to state the variable that you are deliberately going to change, called the .... variable. You also need to say what you expect will change because of this; this is called the .... variable. You must also say what .... variables you will keep constant in order to make it a fair test.

Summary questions

1 Put these words into order. They should be in the order that you might use them in an investigation.

design; prediction; conclusion; method; repeat; controls; graph; results; table; improve; safety; hypothesis

2 a How would you tell the difference between an opinion that was scientific and a prejudiced opinion?

b Suppose you were investigating the loss of energy from a beaker of hot water. Would you choose to investigate a categoric or a continuous variable? Explain why.

3 a You might have noticed that different items of electrical equipment in the house use different diameters of wire. You want to find out why. You use some accepted theory to try to answer the question.

Explain what you understand by a hypothesis.

Figure 1 Different diameters of wire in a household

b The diameter of the wire can affect the resistance of the wire. This is a hypothesis. Use this to make a prediction.

c Suppose you have tested your prediction and have some data. What might this do for your hypothesis?

d Suppose the data does not support the hypothesis. What should you do to the theory that gave you the hypothesis?

4 a What do you understand by a fair test?

b Suppose you were carrying out an investigation into how changing the current in an electromagnet affects the magnetic field. You would need to carry out a trial. Describe what a trial would tell you about how to plan your method.

c How could you decide if your results showed precision?

d It is possible to calculate the theoretical magnetic field around a coil. How could you use this to check on the accuracy of your results?

5 Suppose you were watching a friend carry out an investigation using the equipment shown in Figure 2. You have to mark your friend on how accurately he is making his measurements. Make a list of points that you would be looking for.

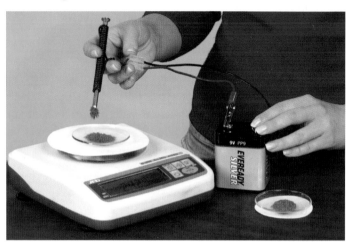

Figure 2 Student using an electromagnet to pick up iron filings

6 a How do you decide on the range of a set of data?

b How do you calculate the mean?

c When should you use a bar chart?

d When should you use a line graph?

7 a What should happen to anomalous results?

b What does a line of best fit allow you to do?

c When making a conclusion, what must you take into consideration?

d How can you check on the repeatability and reproducibility of your results?

8 a Why is it important when reporting science to 'tell the truth, the whole truth and nothing but the truth'?

b Why might some people be tempted not to be completely fair when reporting their opinions on scientific data?

9 a 'Science can advance technology and technology can advance science.' What do you think is meant by this statement?

b Who answers the questions that start with 'Should we ...'?

10 Wind turbines are an increasingly popular way of generating electricity. It is very important that they are sited in the best place to maximise energy output. Clearly they need to be where there is plenty of wind. Energy companies have to be confident that they get value for money. Therefore they must consider the most economic height at which to build them. Put them too high and they might not get enough extra energy to justify the extra cost of the turbine. Before deciding finally on a site they will carry out an investigation to decide the best height.

The prediction is that increasing the height will increase the power output of the wind turbine. A test platform was erected and the turbine placed on it. The lowest height that would allow the turbines to move was 32 metres. The correct weather conditions were waited for, and the turbine began turning and the power output was measured in kilowatts.

The results are in the table.

Height of turbine (m)	Power output Test 1 (kW)	Power output Test 2 (kW)
32	162	139
40	192	195
50	223	219
60	248	245
70	278	270
80	302	304
85	315	312

a What was the prediction for this test?

b What was the independent variable?

c What was the dependent variable?

d What is the range of the heights for the turbine?

e Suggest a control variable that should have been used.

f This is a fieldwork investigation. Is it possible to control all of the variables? If not, say what you think the scientist should have done to produce more accurate results.

g Is there any evidence for an anomalous result in this investigation? Explain your answer.

h What was the resolution of the power output measurement? Provide some evidence for your answer from the data in the table.

i Draw a graph of the results for the second test.

j Draw a line of best fit.

k Describe the pattern in these results.

l What conclusion can you reach?

m How might this data be of use to people who want to stop a wind farm being built?

n Who should carry out these tests for those who might object?

Figure 3 Wind turbines

P1 1.1

Infrared radiation

Learning objectives

- What is infrared radiation?
- Do all objects give off infrared radiation?
- How does infrared radiation depend on the temperature of an object?

Figure 1 Keeping watch in darkness

∞ links

For more information on infrared heaters, see P1 1.9 Heating and insulating buildings.

 Did you know ... ?

A **passive infrared (PIR) detector** in a burglar alarm circuit will 'trigger' the alarm if someone moves in front of the detector. The detector contains sensors that detect infrared radiation from different directions.

∞ links

For more information on electromagnetic waves, see P1 6.1 The electromagnetic spectrum.

Seeing in the dark

We can use special cameras to 'see' animals and people in the dark. These cameras detect **infrared radiation**. Every object gives out (**emits**) infrared radiation.

The hotter an object is, the more infrared radiation it emits in a given time.

Look at the photo in Figure 1. The rhinos are hotter than the ground.

- **a** Why is the ground darker than the rhinos?
- **b** Which part of each rhino is coldest?

Practical

Detecting infrared radiation

You can use a thermometer with a blackened bulb to detect infrared radiation. Figure 2 shows how to do this.

- The glass prism splits a narrow beam of white light into the colours of the spectrum.
- The thermometer reading rises when it is placed just beyond the red part of the spectrum. Some of the infrared radiation in the beam goes there. Our eyes cannot detect it but the thermometer can.
- Infrared radiation is beyond the red part of the visible spectrum.

What would happen to the thermometer reading if the thermometer were moved away from the screen?

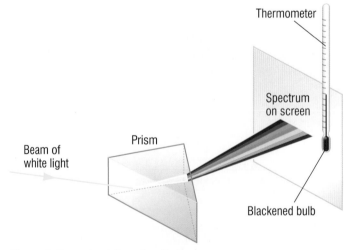

Figure 2 Detecting infrared radiation

The electromagnetic spectrum

Radio waves, **microwaves**, infrared radiation and **visible light** are parts of the electromagnetic spectrum. So too are ultraviolet rays and X-rays. Electromagnetic waves are electric and magnetic waves that travel through space.

Energy from the Sun

The Sun emits all types of electromagnetic radiation. Fortunately for us, the Earth's atmosphere blocks most of the radiation that would harm us. But it doesn't block infrared radiation from the Sun.

Figure 3 shows a solar furnace. This uses a giant reflector that focuses sunlight.

The temperature at the focus can reach thousands of degrees. That's almost as hot as the surface of the Sun, which is 5500 °C.

The greenhouse effect

The Earth's atmosphere acts like a greenhouse made of glass. In a greenhouse:

- short wavelength infrared radiation (and light) from the Sun can pass through the glass to warm the objects inside the greenhouse
- infrared radiation from these warm objects is trapped inside by the glass because the objects emit infrared radiation of longer wavelengths that can't pass through the glass.

So the greenhouse stays warm.

Gases in the atmosphere, such as water vapour, methane and carbon dioxide, trap infrared radiation from the Earth. This makes the Earth warmer than it would be if it had no atmosphere.

But the Earth is becoming too warm. If the polar ice caps melt, it will cause sea levels to rise. Reducing our use of fossil fuels will help to reduce the production of 'greenhouse gases'.

Figure 3 A solar furnace in the Eastern Pyrenees, France

How Science Works

A huddle test

Design an investigation to model the effect of penguins huddling together. You could use beakers of hot water to represent the penguins.

Figure 4 Penguins keeping warm

Summary questions

1 Copy and complete **a** and **b** using the words below. Each word can be used more than once.

temperature   radiation   waves

  a Infrared .... is energy transfer by electromagnetic ....

  b The higher the .... of an object is, the more .... it emits each second.

2 **a** Copy and complete the table to show if the object emits infrared radiation or light or both.

Object	Infrared	Light
A hot iron		
A light bulb		
A TV screen		
The Sun		

  b How can you tell if an electric iron is hot without touching it?

3 Explain why penguins huddle together to keep warm.

Key points

- Infrared radiation is energy transfer by electromagnetic waves.
- All objects emit infrared radiation.
- The hotter an object is, the more infrared radiation it emits in a given time.

P1 1.2   Surfaces and radiation

Learning objectives

- Which surfaces are the best emitters of infrared radiation?
- Which surfaces are the best absorbers of infrared radiation?
- Which surfaces are the best reflectors of infrared radiation?

Figure 1 A thermal blanket in use

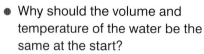

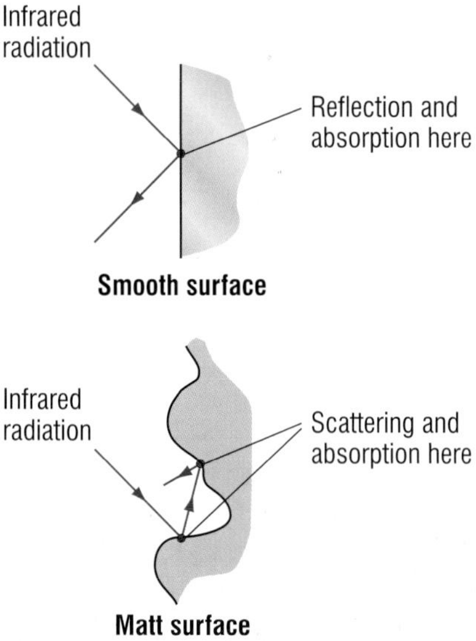

Smooth surface

Matt surface

Figure 3 Absorbing infrared radiation

⚭ links

Infrared heaters use light, shiny surfaces to reflect infrared radiation. See P1 3.4 Cost effectiveness matters.

Which surfaces are the best **emitters** of radiation?

Rescue teams use light shiny thermal blankets to keep accident survivors warm (see Figure 1). A light, shiny outer surface emits much less radiation than a dark, matt surface.

Practical

Testing radiation from different surfaces

To compare the radiation from two different surfaces, you can measure how fast two cans of hot water cool. The surface of one can is light and shiny and the other has a dark matt surface (see Figure 2). At the start, the volume and temperature of the water in each can must be the same.

- Why should the volume and temperature of the water be the same at the start?
- Which can will cool faster?

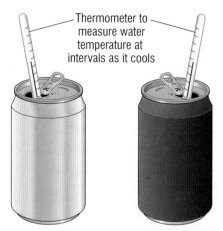

Thermometer to measure water temperature at intervals as it cools

Figure 2 Testing different surfaces

Your tests should show that:

- **Dark, matt surfaces are better at emitting radiation than light, shiny surfaces.**

Which surfaces are the best **absorbers** and **reflectors** of radiation?

When you use a photocopier, infrared radiation from a lamp dries the ink on the paper. Otherwise, the copies would be smudged. Black ink absorbs more infrared radiation than white paper.

A light, shiny surface absorbs less radiation than a dark, matt surface. A matt surface has lots of cavities, as shown in Figure 3.

- The radiation reflected from the matt surface hits the surface again.
- The radiation reflected from the shiny surface travels away from the surface.

So the shiny surface absorbs less and reflects more radiation than a matt surface.

In general:

- **Light, shiny surfaces absorb less radiation than dark, matt surfaces.**
- **Light, shiny surfaces reflect more radiation than dark, matt surfaces.**

   a   Why does ice on a road melt faster in sunshine if sand is sprinkled on it?
   b   Why are solar heating panels painted matt black?

Practical

Absorption tests

Figure 4 shows how we can compare absorption by two different surfaces.

● The front surfaces of the two metal plates are at the same distance from the heater.

● The back of each plate has a coin stuck on with wax. The coin drops off the plate when the wax melts.

● The coin at the back of the matt black surface drops off first. The matt black surface absorbs more radiation than the light shiny surface.

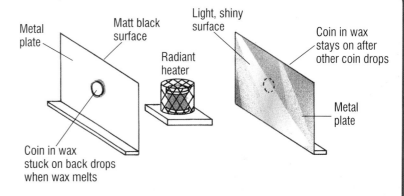

Figure 4 Testing different absorbers of infrared radiation

Summary questions

1 Copy and complete **a** and **b** using the words below:

absorber   emitter   reflector

   a A dark, matt surface is a better .... and a better .... of infrared radiation than a light, shiny surface.

   b A light, shiny surface is a better .... of infrared radiation than a dark, matt surface.

2 A black car and a metallic silver car are parked next to each other on a sunny day. Why does the temperature inside the black car rise more quickly than the temperature inside the silver car?

3 A metal cube filled with hot water was used to compare the infrared radiation emitted from its four vertical faces, A, B, C and D.

   An infrared sensor was placed opposite each face at the same distance, as shown in Figure 5. The sensors were connected to a computer. The results of the test are shown in the graph below.

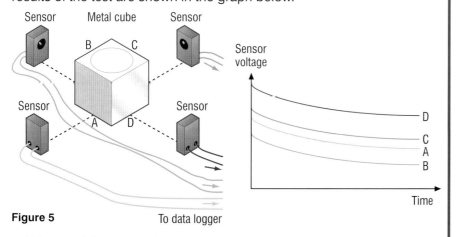

Figure 5

   a Why was it important for the distance from each sensor to the face to be the same?

   b One face was light and shiny, one was light and matt, one was dark and shiny, and one was dark and matt.
   Which face A, B, C or D emits the **i** least radiation, **ii** most radiation?

   c What are the advantages of using data logging equipment to collect the data in this investigation?

?¿? Did you know ... ?

Scientists are developing blacker and blacker materials. These new materials have very tiny pits in the surface to absorb almost all the light that hits them. They can be used to coat the insides of telescopes so that there are no reflections.

Key points

● Dark, matt surfaces emit more infrared radiation than light, shiny surfaces.

● Dark, matt surfaces absorb more infrared radiation than light, shiny surfaces.

● Light, shiny surfaces reflect more infrared radiation than dark matt surfaces.

P1 1.3

States of matter

Learning objectives

- How are solids, liquids and gases different?
- How are the particles in a solid, liquid and a gas arranged?
- Why is a gas much less dense than a solid or a liquid?

Everything around us is made of matter in one of three states – solid, liquid or gas. The table below summarises the main differences between the three **states of matter**.

	Flow	Shape	Volume	Density
Solid	no	fixed	fixed	much higher than a gas
Liquid	yes	fits container shape	fixed	much higher than a gas
Gas	yes	fills container	can be changed	low compared with a solid or liquid

a We can't see it and yet we can fill objects like balloons with it. What is it?

b When an ice cube melts, what happens to its shape?

Change of state

A substance can change from one state to another, as shown in Figure 2. We can make these changes by heating or cooling the substance. For example,

- when water in a kettle boils, the water turns to steam. Steam, also called water vapour, is water in its gaseous state
- when solid carbon dioxide or 'dry ice' warms up, the solid turns into gas directly
- when steam touches a cold surface, the steam condenses and turns to water.

Figure 1 Spot the three states of matter

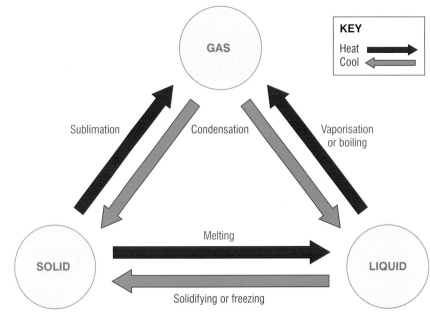

Figure 2 Change of state

c What change of state occurs when hailstones form?

Practical

Changing state

1 Heat some water in a beaker using a Bunsen burner, as shown in Figure 3. Notice that:
   - steam or 'vapour' leaves the water surface before the water boils
   - when the water boils, bubbles of vapour form inside the water and rise to the surface to release steam.
2 Switch the Bunsen burner off and hold a cold beaker or cold metal object above the boiling water. Observe condensation of steam from the boiling water on the cold object. Take care with boiling water.

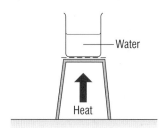

Figure 3 Changing state

The kinetic theory of matter

Solids, liquids and gases consist of particles. Figure 4 shows the arrangement of the particles in a solid, a liquid and a gas. When the temperature of the substance is increased, the particles move faster.

- The particles in a solid are held next to each other in fixed positions. They vibrate about their fixed positions so the solid keeps its own shape.
- The particles in a liquid are in contact with each other. They move about at random. So a liquid doesn't have its own shape and it can flow.
- The particles in a gas move about at random much faster. They are, on average, much further apart from each other than in a liquid. So the density of a gas is much less than that of a solid or liquid.
- The particles in solids, liquids and gases have different amounts of energy. In general, the particles in a gas have more energy than those in a liquid, which have more energy than those in a solid.

 **Did you know ... ?**

Random means unpredictable. Lottery numbers are chosen at random.

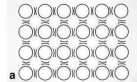  

Figure 4 The arrangement of particles in **a** a solid, **b** a liquid and **c** a gas

Summary questions

1 Copy and complete **a** to **d** using the words below. Each word can be used more than once.

   gas   liquid   solid

   a A .... has a fixed shape and volume.
   b A .... has a fixed volume but no shape.
   c A .... and a .... can flow.
   d A .... does not have a fixed volume.

2 State the scientific word for each of the following changes.
   a A mist appears on the inside of a window in a bus full of people.
   b Steam is produced from the surface of the water in a pan when the water is heated before it boils.
   c Ice cubes taken from a freezer thaw out.
   d Water put into a freezer gradually turns to ice.

3 Describe the changes that take place in the movement and arrangement of the particles in an ice cube when the ice melts.

Key points

- Flow, shape, volume and density are the properties used to describe each state of matter.

- The particles in a solid are held next to each other in fixed positions.

- The particles in a liquid move about at random and are in contact with each other.

- The particles in a gas move about randomly and are much further apart than particles in a solid or liquid.

P1 1.4

Conduction

Learning objectives

- What materials make the best conductors?
- What materials make the best insulators?
- Why are metals good conductors?
- Why are non-metals poor conductors?

Figure 1 At a barbecue – the steel cooking utensils have wooden or plastic handles

When you have a barbecue, you need to know which materials are good **conductors** and which are good **insulators**. If you can't remember, you are likely to burn your fingers!

Testing rods of different materials as conductors

The rods need to be the same width and length for a fair test. Each rod is coated with a thin layer of wax near one end. The uncoated ends are then heated together.

Look at Figure 2. The wax melts fastest on the rod that conducts best.

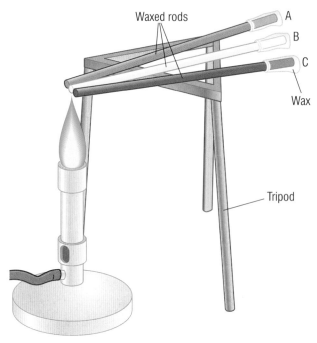

Figure 2 Comparing conductors

- Metals conduct energy better than non-metals.
- Copper is a better conductor than steel.
- Wood conducts better than glass.

a Why do steel pans have handles made of plastic or wood?
b Name the independent and the dependent variables investigated in Figure 2.

 links

For more information on independent and dependent variables, look back at H3 Starting an investigation.

Practical

Testing sheets of materials as insulators

Use different materials to insulate identical cans (or beakers) of hot water. The volume of water and its temperature at the start should be the same.

Use a thermometer to measure the water temperature after a fixed time. The results should tell you which insulator was best.

The table below gives the results of comparing two different materials using the method explained in the practical.

Material	Starting temperature (°C)	Temperature after 300 s (°C)
paper	40	32
felt	40	36

c Which material, felt or paper, was the better insulator?

d Which variable shown in the table was controlled to make this a fair test?

Conduction in metals

Metals contain lots of **free electrons**. These electrons move about at random inside the metal and hold the positive metal ions together. They collide with each other and with the positive ions. (Ions are charged particles.)

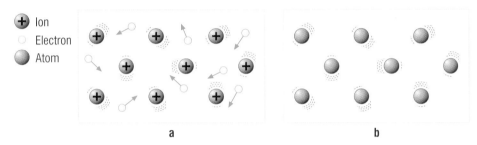

- ⊕ Ion
- ○ Electron
- ⚫ Atom

a   b

Figure 4 Energy transfer in **a** a metal, **b** a non-metal

Figure 3 Insulating a loft. The air trapped between fibres make fibreglass a good insulator.

Did you know ... ?

Materials like wool and fibreglass are good insulators. This is because they contain air trapped between the fibres. Trapped air is a good insulator. We use materials like fibreglass for loft insulation and for lagging water pipes.

When a metal rod is heated at one end, the free electrons at the hot end gain kinetic energy and move faster.

- These electrons **diffuse** (i.e. spread out) and collide with other free electrons and ions in the cooler parts of the metal.
- As a result, they transfer kinetic energy to these electrons and ions.

So energy is transferred from the hot end of the rod to the colder end.

In a non-metallic solid, all the electrons are held in the atoms. Energy transfer only takes place because the atoms vibrate and shake each other. This is much less effective than energy transfer by free electrons. This is why metals are much better conductors than non-metals.

Summary questions

1 Copy and complete **a** to **c** using the words below:

fibreglass plastic steel wood

   a A material called .... is used to insulate a house loft.
   b The handle of a frying pan is made of .... or ....
   c A radiator in a central heating system is made from ....

2 **a** Choose a material you would use to line a pair of winter boots. Explain your choice of material.
   b How could you carry out a test on three different lining materials?

3 Explain why metals are good conductors of energy.

Key points

- Metals are the best conductors of energy.
- Materials such as wool and fibreglass are the best insulators.
- Conduction of energy in a metal is due mainly to free electrons transferring energy inside the metal.
- Non-metals are poor conductors because they do not contain free electrons.

Convection

Learning objectives

- What is convection?
- Where can convection take place?
- Why does convection occur?

Figure 1 A natural glider – birds use convection currents to soar high above the ground

Gliders and birds use convection to stay in the air. **Convection currents** can keep them high above the ground for hours.

Convection happens whenever we heat **fluids**. A fluid is a gas or a liquid. Look at the diagram in Figure 2. It shows a simple demonstration of convection.

- The hot gases from the burning candle go straight up the chimney above the candle.
- Cold air is drawn down the other chimney to replace the air leaving the box.

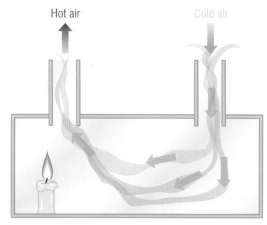

Figure 2 Convection

Using convection

Hot water at home

Many homes have a hot water tank. Hot water from the boiler rises and flows into the tank where it rises to the top. Figure 3 shows the system. When you use a hot water tap at home, you draw off hot water from the top of the tank.

a What would happen if we connected the hot taps to the bottom of the tank?

Sea breezes

Sea breezes keep you cool at the seaside. On a sunny day, the ground heats up faster than the sea. So the air above the ground warms up and rises. Cooler air from the sea flows in as a 'sea breeze' to take the place of the rising warm air (see Figure 4).

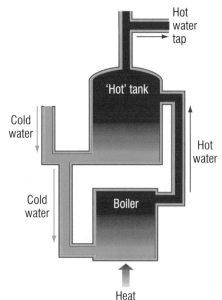

Figure 3 Hot water at home

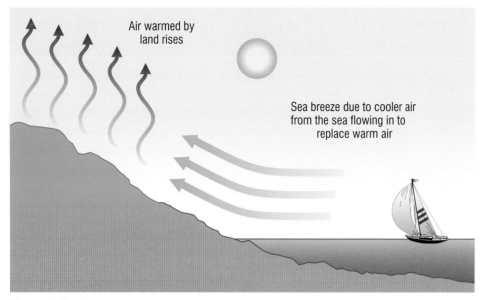

Figure 4 Sea breezes

How convection works

Convection takes place:

● only in fluids (liquids and gases)

● due to circulation (convection) currents within the fluid.

The circulation currents are caused because fluids rise where they are heated (as heating makes them less dense). Then they fall where they cool down (as cooling makes them more dense). Convection currents transfer energy from the hotter parts to the cooler parts.

So why do fluids rise when heated?

Most fluids expand when heated. This is because the particles move about more, taking up more space. Therefore the **density** decreases because the same mass of fluid now occupies a bigger volume. So heating part of a fluid makes that part less dense and therefore it rises.

AQA **Examiner's tip**

When you explain convection, remember it is the hot fluid that rises, NOT 'heat'.

Summary questions

1 Copy and complete **a** and **b** using the words below:

cools   falls   mixes   rises

   a When a fluid is heated, it _____ and _____ with the rest of the fluid.
   b The fluid circulates and _____ then it _____ .

2 Figure 5 shows a convector heater. It has an electric heating element inside and a metal grille on top.
   a What does the heater do to the air inside the heater?
   b Why is there a metal grille on top of the heater?
   c Where does air flow into the heater?

3 Describe how you could demonstrate convection currents in water using a strongly coloured crystal or a suitable dye. Explain in detail what you would see.

Hot air

Figure 5 A convector heater

Key points

● Convection is the circulation of a fluid (liquid or gas) caused by heating it.

● Convection takes place only in liquids and gases.

● Heating a liquid or a gas makes it less dense so it rises and causes circulation.

P1 1.6

Evaporation and condensation ⓚ

Learning objectives

- What is evaporation?
- What is condensation?
- How does evaporation cause cooling?
- What factors affect the rate of evaporation from a liquid?
- What factors affect the rate of condensation on a surface?

Drying off

If you hang wet clothes on a washing line in fine weather, they will gradually dry off. The water in the wet clothes **evaporates**. You can observe evaporation of water if you leave a saucer of water in a room. The water in the saucer gradually disappears. Water molecules escape from the surface of the water and enter the air in the room.

In a well-ventilated room, the water molecules in the air are not likely to re-enter the liquid. They continue to leave the liquid until all the water has evaporated.

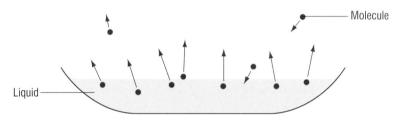

Figure 1 Water molecules escaping from a liquid

Condensation

In a steamy bathroom, a mirror is often covered by a film of water. There are lots of water molecules in the air. Some of them hit the mirror, cool down and stay there. We say water vapour in the air **condenses** on the mirror.

   a Why does opening a window in a steamy room clear the condensation?

Cooling by evaporation

If you have an injection, the doctor or nurse might 'numb' your skin by dabbing it with a liquid that easily evaporates. As the liquid evaporates, your skin becomes too cold to feel any pain.

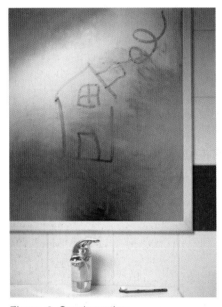

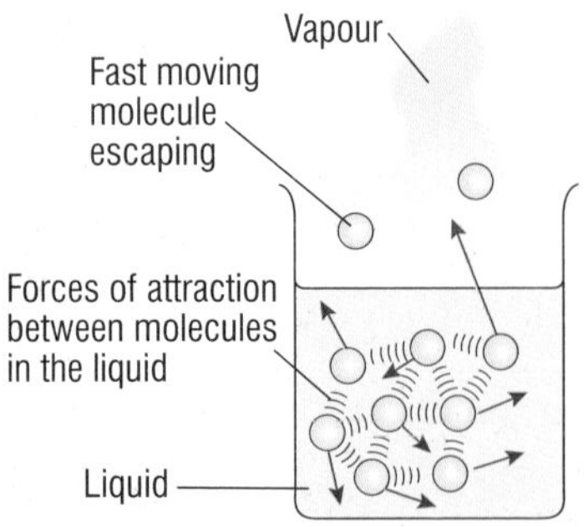

Figure 2 Condensation

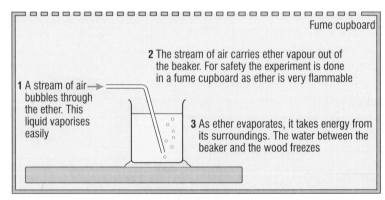

Figure 3 Explanation of cooling by evaporation

Demonstration

Cooling by evaporation

Watch your teacher carry out this experiment in a fume cupboard.

- Why is ether used in this experiment?

Fume cupboard

2 The stream of air carries ether vapour out of the beaker. For safety the experiment is done in a fume cupboard as ether is very flammable

1 A stream of air bubbles through the ether. This liquid vaporises easily

3 As ether evaporates, it takes energy from its surroundings. The water between the beaker and the wood freezes

Figure 4 A demonstration of cooling by evaporation

Figure 3 explains why evaporation causes this cooling effect.

- Weak attractive forces exist between the molecules in the liquid.
- The faster molecules, which have more kinetic energy, break away from the attraction of the other molecules and escape from the liquid.
- After they leave, the liquid is cooler because the average kinetic energy of the remaining molecules in the liquid has decreased.

Factors affecting the rate of evaporation

Clothes dry faster on a washing line:

- if each item of wet clothing is spread out when it is hung on the line. This increases the area of the wet clothing that is in contact with dry air.
- if the washing line is in sunlight. Wet clothes dry faster the warmer they are.
- if there is a breeze to take away the molecules that escape from the water in the wet clothes.

The example above shows that the rate of evaporation from a liquid is increased by:

- increasing the surface area of the liquid
- increasing the temperature of the liquid
- creating a draught of air across the liquid's surface.

Factors affecting the rate of condensation

In a steamy kitchen, water can often be seen trickling down a window pane. The glass pane is a cold surface so water vapour condenses on it. The air in the room is moist or 'humid'. The bigger the area of the window pane, or the colder it is, the greater the rate of condensation. This example shows that the rate of condensation of a vapour on a surface is increased by:

- increasing the surface area
- reducing the surface temperature.

b Why does washing on a line take longer to dry on a damp day?

Summary questions

1 Copy and complete **a** to **c** using the words below. Each word can be used more than once.

condenses   cools   evaporates

a A liquid .... when its molecules escape into the surrounding air.
b When water .... on glass, water molecules in the air form a liquid on the glass.
c When a liquid ...., it loses its faster-moving molecules and it ....

2 Why do the windows on a bus become misty when there are lots of people on the bus?

3 Explain the following statements.
a Wet clothes on a washing line dry faster on a hot day than on a cold day.
b A person wearing wet clothes on a cold windy day is likely to feel much colder than someone wearing dry clothes.

Did you know ...?

Air conditioning

An **air conditioning unit** in a room transfers energy from inside the room to the outside. The unit contains a 'coolant' liquid that easily evaporates. The coolant is pumped round a sealed circuit of pipes that go through the unit and the outside.

- The liquid coolant evaporates in the pipes in the room and cools the room.
- The evaporated coolant condenses in the pipes outside and transfers energy to the surroundings.

Figure 5 An air conditioning unit

Key points

- Evaporation is when a liquid turns into a gas.
- Condensation is when a gas turns into a liquid.
- Cooling by evaporation of a liquid is due to the faster-moving molecules escaping from the liquid.
- Evaporation can be increased by increasing the surface area of the liquid, by increasing the liquid's temperature, or by creating a draught of air across the liquid's surface.
- Condensation on a surface can be increased by increasing the area of the surface or reducing the temperature of the surface.

P1 1.7

Energy transfer by design

Learning objectives

- What design factors affect the rate at which a hot object transfers energy?

- What can we do to control the rate of energy transfer to or from an object?

Figure 1 A car radiator

Figure 2 Motorcycle engine fins

??? Did you know ...?

Some electronic components get warm when they are working, but if they become too hot they stop working. Such components are often fixed to a metal plate to keep them cool. The metal plate increases the effective surface area of the component. We call the metal plate a **heat sink**.

Figure 3 A heat sink in a computer

⚙⚙⚙ How Science Works

Cooling by design

Lots of things can go wrong if we don't control energy transfer. For example, a car engine that overheats can go up in flames.

- The cooling system of a car engine transfers energy from the engine to a radiator. The radiator is shaped so it has a large surface area. This increases the rate of energy transfer through convection in the air and through radiation.

- A motorcycle engine is shaped with **fins** on its outside surface. The fins increase the surface area of the engine in contact with air so the engine transfers energy to its surroundings faster than if it had no fins.

- Most cars also have a cooling fan that switches on when the engine is too hot. This increases the flow of air over the surface of the radiator.

  a Why do car radiators have a large surface area?
  b What happens to the rate of energy transfer when the cooling fan switches on?

The vacuum flask

If you are outdoors in cold weather, a hot drink from a vacuum flask keeps you warm. In the summer the same vacuum flask keeps your drinks cold.

In Figure 4, the liquid you drink is in the double-walled glass container.

- The vacuum between the two walls of the container cuts out energy transfer by conduction and convection between the walls.

- Glass is a poor conductor so there is little energy transfer by conduction through the glass.

- The glass surfaces are silvery to reduce radiation from the outer wall.

- The spring supporting the double-walled container is made of plastic which is a good insulator.

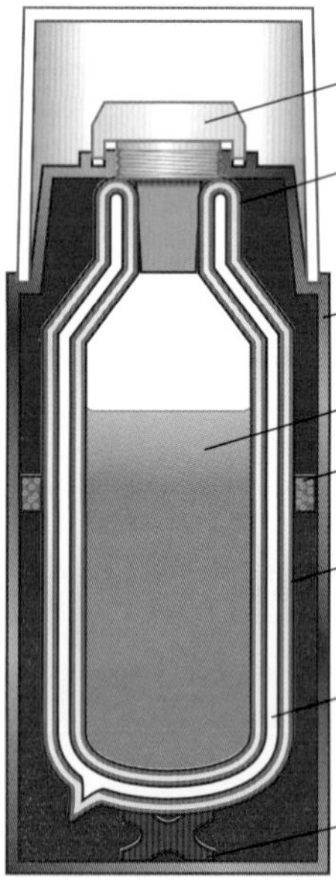

- Plastic cap
- Double-walled glass (or plastic) container
- Plastic protective cover
- Hot or cold liquid
- Sponge pad (for protection)
- Inside surfaces silvered to stop radiation
- Vacuum prevents conduction and convection
- Plastic spring for support

Figure 4 A vacuum flask

- The plastic cap stops cooling by evaporation as it stops vapour loss from the flask. In addition, energy transfer by conduction is cut down because the cap is made from plastic.

So why does the liquid in the flask eventually cool down?

The above features cut down but do not totally stop the transfer of energy from the liquid. Energy transfer occurs at a very low rate due to radiation from the silvery glass surface and conduction through the cap, spring and glass walls. The liquid transfers energy slowly to its surroundings so it eventually cools.

  c List the other parts of the flask that are good insulators.
     What would happen if they weren't good insulators?

Factors affecting the rate of energy transfer

The bigger the **temperature difference** between an object and its surroundings, the faster the rate at which energy is transferred. In addition, the above examples show that the rate at which an object transfers energy depends on its design. The design factors that matter are:

- the materials the object is in contact with
- the object's shape
- the object's surface area.

In addition, the object's mass and the material it is made from are important. That is because they affect how quickly its temperature changes (and therefore the rate of transfer of energy to or from it) when it loses or gains energy.

How Science Works

Foxy survivors

A desert fox has much larger ears than an arctic fox. Blood flowing through the ears transfers energy from inside the body to the surface of the ears. Big ears have a much larger surface area than little ears so they transfer energy to the surroundings more quickly than little ears.

- A desert fox has big ears so it keeps cool by transferring energy quickly to its surroundings.
- An arctic fox has little ears so it transfers energy more slowly to its surroundings. This helps keep it warm.

⟳ links

For more information on factors affecting energy transfer, see P1 1.8 Specific heat capacity.

Practical

Investigating the rate of energy transfer

You can plan an investigation using different beakers and hot water to find out what affects the rate of cooling.

- Write a question that you could investigate.
- Identify the independent, dependent and control variables in your investigation.

Figure 5 Fox ears **a** A desert fox **b** An arctic fox

Summary questions

1 Hot water is pumped through a radiator like the one in Figure 6.
   Copy and complete **a** to **c** using the words below:

   conduction   radiation   convection

   a Energy transfer through the walls of the radiator is due to ....

   b Hot air in contact with the radiator causes energy transfer to the room by ....

   c Energy transfer to the room takes place directly due to ....

Figure 6 A central heating radiator

2 An electronic component in a computer is attached to a heat sink.
   a i Explain why the heat sink is necessary.
   ii Why is a metal plate used as the heat sink?
   b Plan a test to show that double glazing is more effective at preventing energy transfer than single glazing.

3 Describe, in detail, how the design of a vacuum flask reduces the rate of energy transfer.

Key points

- The rate of energy transferred to or from an object depends on:
  - the shape, size and type of material of the object
  - the materials the object is in contact with
  - the temperature difference between the object and its surroundings.

P1 1.8 Specific heat capacity ⓚ

Learning objectives

- How does the mass of a substance affect how quickly its temperature changes when it is heated?

- What else affects how quickly the temperature of a substance changes when it is heated?

- How do storage heaters work?

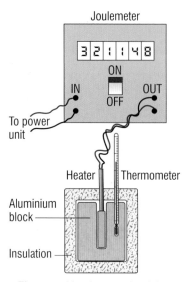

Figure 1 Heating an aluminium block

⁇⁇⁇ Did you know ... ?

Coastal towns are usually cooler in summer and warmer in winter than towns far inland. This is because water has a very high specific heat capacity. Energy from the Sun (or lack of energy) affects the temperature of the sea much less than the land.

A car in strong sunlight can become very hot. A concrete block of equal **mass** would not become as hot. Metal heats up more easily than concrete. Investigations show that when a substance is heated, its temperature rise depends on:

- the amount of energy supplied to it
- the mass of the substance
- what the substance is.

⚙⚙ Practical

Investigating heating

Figure 1 shows how we can use a low voltage electric heater to heat an aluminium block.

Energy is measured in units called joules (J).

Use the energy meter (or joulemeter) to measure the energy supplied to the block. Use the thermometer to measure its temperature rise.

Replace the block with an equal mass of water in a suitable container. Measure the temperature rise of the water when the same amount of energy is supplied to it by the heater.

Your results should show that aluminium heats up more than water.

The following results were obtained using two different amounts of water. They show that:

- 1600 J was used to heat 0.1 kg of water by 4 °C
- 3200 J was used to heat 0.2 kg of water by 4 °C.

Using these results we can say that:

- 16 000 J of energy would have been needed to heat 1.0 kg of water by 4 °C
- 4000 J of energy is needed to heat 1.0 kg of water by 1 °C.

More accurate measurements would give 4200 J per kg per °C for water. This is its **specific heat capacity**.

The specific heat capacity of a substance is the energy needed or energy transferred to 1 kg of the substance to raise its temperature by 1 °C.

The unit of specific heat capacity is the joule per kilogram per °C.

For a known change of temperature of a known mass of a substance:

$$E = m \times c \times \theta$$

Where:

E is the energy transferred in joules, J; m is the mass in kilograms, kg; c is the specific heat capacity, J/kg °C; θ is the temperature change in degrees Celsius, °C

To find the specific heat capacity you need to rearrange the above equation:

$$c = \frac{E}{m \times \theta}$$

a How much energy is needed to heat 5.0 kg of water from 20 °C to 60 °C?

Practical

Measuring the specific heat capacity of a metal

Use the arrangement shown in Figure 1 to heat a metal block of known mass. Here are some measurements using an aluminium block of mass 1.0 kg.

Starting temperature = 14 °C
Final temperature = 22 °C
Energy supplied = 7200 J

To find the specific heat capacity of aluminium, the measurements above give:
E = energy transferred = energy supplied = 7200 J
θ = temperature change = 22 °C − 14 °C = 8 °C

Inserting these values into the rearranged equation gives:

$$c = \frac{E}{m \times \theta} = \frac{7200\,J}{1.0\,kg \times 8\,°C} = 900\,J/kg\,°C$$

The table below shows the values for some other substances.

Substance	water	oil	aluminium	iron	copper	lead	concrete
Specific heat capacity (joules per kg per °C)	4200	2100	900	390	490	130	850

Storage heaters

A storage heater uses electricity at night (off-peak) to heat special bricks or concrete blocks in the heater. Energy transfer from the bricks keeps the room warm. The bricks have a high specific heat capacity so they store lots of energy. They warm up slowly when the heater element is on and cool down slowly when it is off.

Electricity consumed at off-peak times is sometimes charged for at a cheaper rate, so storage heaters are designed to be cost effective.

Figure 2 A storage heater

b How would the temperature of the room change if the bricks cooled quickly?

Key points

- The greater the mass of an object, the more slowly its temperature increases when it is heated.

- The rate of temperature change of a substance when it is heated depends on:
  – the energy supplied to it
  – its mass
  – its specific heat capacity.

- Storage heaters use off-peak electricity to store energy in special bricks.

Summary questions

1 A small bucket of water and a large bucket of water are left in strong sunlight. Which one warms up faster? Give a reason for your answer.

2 Use the information in the table above to answer this question.
   a Explain why a mass of lead heats up more quickly than an equal mass of aluminium.
   b Calculate the energy needed
      i to raise the temperature of 0.20 kg of aluminium from 15 °C to 40 °C
      ii to raise the temperature of 0.40 kg of water from 15 °C to 40 °C.

3 State two ways in which a storage heater differs from a radiant heater.

P1 1.9 Heating and insulating buildings

Learning objectives

- How can we reduce the rate of energy transfer from our homes?
- What are U-values?
- Is solar heating free?

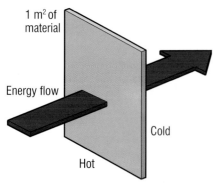

1 m² of material

Energy flow

Cold

Hot

U-value of the material = energy/s passing per m² for 1°C temperature difference

Figure 2 U-values

How Science Works

Reducing the rate of energy transfers at home

Home heating bills can be expensive. Figure 1 shows how we can reduce the rate of **energy transfer** at home and reduce our home heating bills.

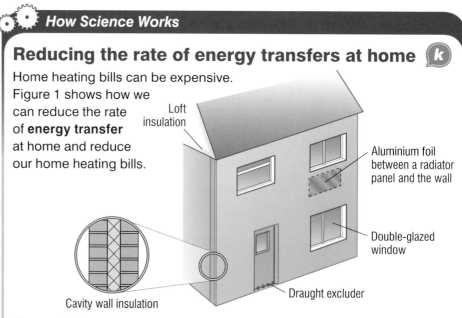

Loft insulation

Aluminium foil between a radiator panel and the wall

Double-glazed window

Draught excluder

Cavity wall insulation

Figure 1 Saving money

- **Loft insulation** such as fibreglass reduces the rate of energy transfer through the roof. Fibreglass is a good insulator. The air between the fibres also helps to reduce the rate of energy transfer by conduction.
- **Cavity wall insulation** reduces energy loss through the outer walls of the house. The 'cavity' of an outer wall is the space between the two layers of brick that make up the wall. The insulation is pumped into the cavity. It is a better insulator than the air it replaces. It traps the air in small pockets, reducing convection currents.
- **Aluminium foil** between a radiator panel and the wall reflects radiation away from the wall.
- **Double-glazed windows** have two glass panes with dry air or a vacuum between the panes. Dry air is a good insulator so it reduces the rate of energy transfer by conduction. A vacuum cuts out energy transfer by convection as well.

a Why is cavity wall insulation better than air in the cavity between the walls of a house?

U-values

We can compare different insulating materials if we know their U-values. This is the energy per second that passes through one square metre of material when the temperature difference across it is 1 °C.

The lower the U-value, the more effective the material is as an insulator.

For example, replacing a single-glazed window with a double-glazed window that has a U-value four times smaller would make the energy loss through the window four times smaller.

b The U-value of 'MoneySaver' loft insulation is twice that of 'Staywarm'. Which type is more effective as an insulator?

Solar heating panels

Heating water at home using electricity or gas can be expensive. A **solar heating panel** uses solar energy to heat water. The panel is usually fitted on a roof that faces south, making the most of the Sun's energy. Figure 3 shows the design of one type of solar heating panel.

The panel is a flat box containing liquid-filled copper pipes on a matt black metal plate. The pipes are connected to a heat exchanger in a water storage tank in the house.

A transparent cover on the top of the panel allows solar radiation through to heat the metal plate. Insulating material under the plate stops energy being transferred through the back of the panel.

On a sunny day, the metal plate and the copper pipes in the box become hot. Liquid pumped through the pipes is heated when it passes through the panel. The liquid may be water or a solution containing antifreeze. The hot liquid passes through the heat exchanger and transfers energy to the water in the storage tank.

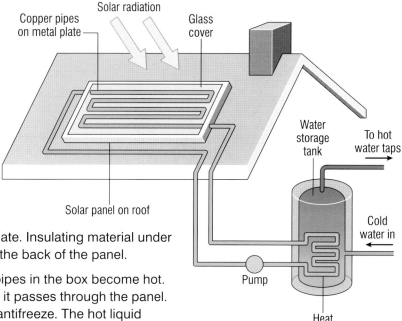

Figure 3 A solar heating panel

How Science Works

Payback time

Solar heating panels save money because no fuel is needed to heat the water. But they are expensive to buy and to install.

Suppose you pay £2000 to buy and install a solar panel and you save £100 each year on your fuel bills. After 20 years you would have saved £2000. In other words, the **payback time** for the solar panel is 20 years. This is the time taken to recover the up-front costs from the savings on fuel bills.

∞ **links**

For more information on payback times, see P1 3.4 Cost effectiveness matters.

Summary questions

1 Copy and complete **a** to **c** using the words below. Each word can be used more than once.

conduction   convection   radiation

  a Cavity wall insulation reduces the rate of energy transfer due to ....
  b Aluminium foil behind a radiator reduces the rate of energy transfer due to ....
  c Closing the curtains in winter reduces the rate of energy transfer due to .... and ....

2 Some double-glazed windows have a plastic frame and a vacuum between the panes.
  a Why is a plastic frame better than a metal frame?
  b Why is a vacuum between the panes better than air?

3 A manufacturer of loft insulation claimed that each roll of loft insulation would save £10 per year on fuel bills. A householder bought 6 rolls of the loft insulation at £15 per roll and paid £90 to have the insulation fitted in her loft.
  a How much did it cost to buy and install the loft insulation?
  b What would be the saving each year on fuel bills?
  c Calculate the payback time.

Key points

● Energy transfer from our homes can be reduced by fitting:
  – loft insulation
  – cavity wall insulation
  – double glazing
  – draught proofing
  – aluminium foil behind radiators.

● U-values tell us how much energy per second passes through different materials.

● Solar heating panels do not use fuel to heat water but they are expensive to buy and install.

Summary questions (k)

1 a Why does a matt surface in sunshine get hotter than a shiny surface?

b What type of surface is better for a flat roof – a matt dark surface or a smooth shiny surface? Explain your answer.

c A solar heating panel is used to heat water. Why is the top surface of the metal plate inside the panel painted matt black?

d Why is a car radiator painted matt black?

2 Copy and complete **a** and **b** using the words below:

collide   electrons   atoms   vibrate

a Energy transfer in a metal is due to particles called .... moving about freely inside the metal. They transfer energy when they .... with each other.

b Energy transfer in a non-metallic solid is due to particles called .... inside the non-metal. They transfer energy because they ....

3 A heat sink is a metal plate or clip fixed to an electronic component to stop it overheating.

Figure 1 A heat sink

a When the component becomes hot, how does energy transfer from where it is in contact with the plate to the rest of the plate?

b Why does the plate have a large surface area?

4 Copy and complete **a** to **d** using the words below. Each word can be used more than once.

conduction   convection   radiation

a .... cannot happen in a solid or through a vacuum.

b Energy transfer from the Sun is due to ....

c When a metal rod is heated at one end, energy transfer due to .... takes place in the rod.

d .... is energy transfer by electromagnetic waves.

5 a In winter, why do gloves keep your hands warm outdoors?

b Why do your ears get cold outdoors in winter if they are not covered?

6 Energy transfer takes place in each of the following examples. In each case, state where the energy transfer occurs and if the energy transfer is due to conduction, convection or radiation.

a The metal case of an electric motor becomes warm due to friction when the motor is in use.

b A central heating radiator warms up first at the top when hot water is pumped through it.

c A slice of bread is toasted under a red-hot electric grill.

7 A glass tube containing water with a small ice cube floating at the top was heated at its lower end. The time taken for the ice cube to melt was measured. The test was repeated with a similar ice cube weighted down at the bottom of the tube of water. The water in this tube was heated near the top of the tube. The time taken for the ice cube to melt was much longer than in the first test.

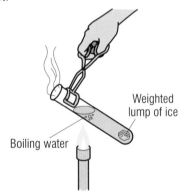

Weighted lump of ice

Boiling water

Figure 2 Energy transfer in water

a Energy transfer in the tube is due to conduction or convection or both.
  i Why was convection the main cause of energy transfer to the ice cube in the first test?
  ii Why was conduction the only cause of energy transfer in the second test?

b Which of the following conclusions about these tests is true?
  1 Energy transfer due to conduction does not take place in water.
  2 Energy transfer in water is mainly due to convection.
  3 Energy transfer in water is mainly due to conduction.

AQA Examination-style questions

1 Convection takes place in fluids.

Use words from the list to complete each sentence. Each word can be used once, more than once or not at all.

contracts   expands   rises   sinks   transfers

When a fluid is heated it ...., becomes less dense, and .... The warm fluid is replaced by cooler, denser, fluid. The resulting convection current .... energy throughout the fluid. (3)

2 There are three states of matter: solid, liquid and gas.

Complete each sentence.

a A solid has
a fixed shape and a fixed volume.
a fixed shape but not a fixed volume.
a fixed volume but not a fixed shape.
neither a fixed shape nor a fixed volume. (1)

b A liquid has
a fixed shape and a fixed volume.
a fixed shape but not a fixed volume.
a fixed volume but not a fixed shape.
neither a fixed shape nor a fixed volume. (1)

c A gas has
a fixed shape and a fixed volume.
a fixed shape but not a fixed volume.
a fixed volume but not a fixed shape.
neither a fixed shape nor a fixed volume. (1)

d Fluids are
solids or liquids.
solids or gases.
liquids or gases. (1)

e The particles in a solid
move about at random in contact with each other.
move about at random away from each other.
vibrate about fixed positions. (1)

3 In an experiment a block of copper is heated from 25 °C to 45 °C.

a Give the name of the process by which energy is transferred through the copper block. (1)

b The mass of the block is 1.3 kg.
Calculate the energy needed to increase the temperature of the copper from 25 °C to 45 °C.
Specific heat capacity of copper = 380 J/kg °C.
Show clearly how you work out your answer. (3)

4 The diagram shows some water being heated with a solar cooker.

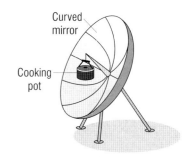

The curved mirror reflects the sunlight that falls on it. The sunlight can be focused on to the cooking pot. The energy from the sunlight is absorbed by the pot, heating up the water inside.

a Suggest **one** reason why a matt black pot has been used. (2)

b When the water has been heated, equal amounts of the water are poured into two metal pans. The pans are identical except one has a matt black surface and the other has a shiny metal surface.

Which pan will keep the water warm for the longer time? Explain your answer. (2)

5 The continuous movement of water from the oceans to the air and land and back to the oceans is called the water cycle.

a The Sun heats the surface of the oceans, which causes water to evaporate.
How does the rate of evaporation depend on
  i the wind speed (1)
  ii the temperature (1)
  iii the humidity? (1)

b Explain how evaporation causes a cooling effect. (3)

6 Double-glazed windows are used to reduce the rate of energy transfer from buildings. The diagrams show cross-sections of single-glazed and double-glazed windows.

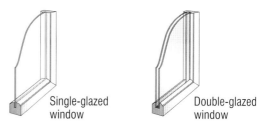

Single-glazed window   Double-glazed window

Give two reasons why a double-glazed window reduces conduction more effectively than a single-glazed window. (2)

7 *In this question you will be assessed on using good English, organising information clearly and using specialist terms where appropriate.*

Compare the similarities and differences between the process of conduction in metals and non-metals. (6)

P1 2.1

Forms of energy

Learning objectives

● What forms of energy are there?

● How can we describe energy changes?

● What energy changes take place when an object falls to the ground?

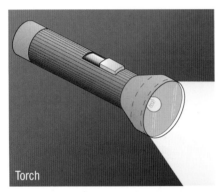

Torch

Skier

Microwave oven

Figure 2 Energy transfers

On the move (k)

Cars, buses, planes and ships all use energy from fuels. They carry their own fuel. Electric trains use energy from fuel in power stations. Electricity transfers energy from the power station to the train.

Figure 1 The French TGV (Train à grande vitesse) electric train can reach speeds of more than 500 km/hour

We describe energy stored or transferred in different ways as **forms of energy**.

Here are some examples of forms of energy:

● **Chemical energy** is energy stored in fuel (including food). This energy is released when chemical reactions take place.

● **Kinetic energy** is the energy of a moving object.

● **Gravitational potential energy** is the energy of an object due to its position.

● **Elastic potential energy** is the energy stored in a springy object when we stretch or squash it.

● **Electrical energy** is energy transferred by an electric current.

a What form of energy is supplied to the train in Figure 1?

Energy may be transferred from one form to another.

In the torch in Figure 2, the torch's battery pushes a current through the bulb. This makes the torch bulb emit light and it also gets hot. We can show the energy transfers using a flow diagram:

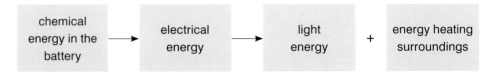

| chemical energy in the battery | → | electrical energy | → | light energy | + | energy heating surroundings |

b What happens to the energy of the torch bulb?

Practical

Energy transfers

When an object starts to fall freely, it gains kinetic energy because it speeds up as it falls. So its gravitational potential energy is transferred to kinetic energy as it falls.

Look at Figure 3. It shows a box that hits the floor with a thud. All of its kinetic energy is transferred by heating and to **sound** energy at the point of impact. The proportion of kinetic energy transferred to sound is much smaller than that transferred by heating.

● Draw an energy flow diagram to show the changes in Figure 3.

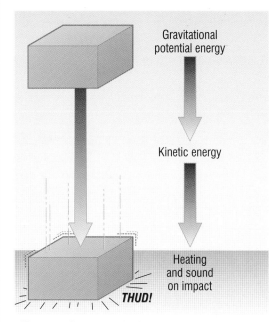

Figure 3 An energetic drop

Did you know ... ?

Tall buildings need firm foundations. Engineers make the foundations using a pile driver to hammer steel girders end-on into the ground. The pile driver lifts a heavy steel block above the top end of the girder. Then it lets the block crash down onto the girder. The engineers keep doing this until the bottom end of the girder reaches solid rock.

Figure 4 A pile driver in action

Summary questions

1 Copy and complete **a** and **b** using the words below:

   electrical   kinetic   gravitational potential

   a When a ball falls in air, it loses .... energy and gains .... energy.
   b When an electric heater is switched on, it transfers .... energy by heating.

2 **a** List two different objects you could use to light a room if you have a power cut. For each object, describe the energy transfers that happen when it lights up the room.
   b Which of the two objects in **a** is:
   i easier to obtain energy from?
   ii easier to use?

3 Read the 'Did you know?' box on this page about the pile driver.
   a What form of energy does the steel block have after it has been raised?
   b Draw an energy flow diagram for the steel block from the moment it is released to when it stops moving.

Key points

● Energy exists in different forms.

● Energy can change from one form into another form.

● When an object falls and gains speed, its gravitational potential energy decreases and its kinetic energy increases.

P1 2.2 Conservation of energy

Learning objectives

- What do we mean by 'conservation of energy'?
- Why is conservation of energy a very important idea?

At the funfair

Funfairs are very exciting places because lots of energy transfers happen quickly. A roller coaster gains gravitational potential energy when it climbs. This energy is then transferred as the roller coaster races downwards.

As it descends:

its gravitational potential energy → kinetic energy + sound + energy transfer by heating due to air resistance and friction

The energy transferred by heating is 'wasted' energy, which you will learn more about in P1 2.3.

a When a roller coaster gets to the bottom of a descent, what energy transfers happen if:
  i we apply the brakes to stop it
  ii it goes up and over a second 'hill'?

AQA Examiner's tip

Never use the term 'movement energy' in the exam; you will only gain marks for using 'kinetic energy'.

Figure 1 On a roller coaster – having fun with energy transfers!

Practical

Investigating energy changes

Pendulum swinging

When energy changes happen, does the total amount of energy stay the same? We can investigate this question with a simple pendulum.

Figure 2 shows a pendulum bob swinging from side to side.

As it moves towards the middle, its gravitational potential energy is transferred to kinetic energy.

As it moves away from the middle, its kinetic energy transfers back to gravitational potential energy. If the air resistance on the bob is very small, you should find that the bob reaches the same height on each side.

- What does this tell you about the energy of the bob when it goes from one side at maximum height to the other side at maximum height?
- Why is it difficult to mark the exact height the pendulum bob rises to? How could you make your judgement more accurate?

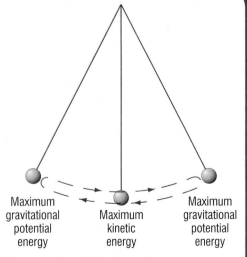

Maximum gravitational potential energy

Maximum kinetic energy

Maximum gravitational potential energy

Figure 2 A pendulum in motion

Conservation of energy

Scientists have done lots of tests to find out if the total energy after a transfer is the same as the energy before the transfer. All the tests so far show it is the same.

This important result is known as the **conservation of energy**.

It tells us that **energy cannot be created or destroyed**.

Bungee jumping

What energy transfers happen to a bungee jumper after jumping off the platform?

- When the rope is slack, some of the gravitational potential energy of the bungee jumper is transferred to kinetic energy as the jumper falls.

- Once the slack in the rope has been used up, the rope slows the bungee jumper's fall. Most of the gravitational potential energy and kinetic energy of the jumper is transferred into elastic strain energy.

- After reaching the bottom, the rope pulls the jumper back up. As the jumper rises, most of the elastic strain energy of the rope is transferred back to gravitational potential energy and kinetic energy of the jumper.

The bungee jumper doesn't return to the same height as at the start. This is because some of the initial gravitational potential energy has been transferred to the surroundings by heating as the rope stretched then shortened again.

b What happens to the gravitational potential energy lost by the bungee jumper?

c Draw a flow diagram to show the energy changes.

Figure 3 Bungee jumping

Practical

Bungee jumping

You can try out the ideas about bungee jumping using the experiment shown in Figure 4.

Summary questions

1 Copy and complete using the words below:

electrical   gravitational potential   kinetic

A person going up in a lift gains .... energy. The lift is driven by electric motors. Some of the .... energy supplied to the motors is wasted instead of being transferred to .... energy.

2 a A ball dropped onto a trampoline returns to almost the same height after it bounces. Describe the energy transfer of the ball from the point of release to the top of its bounce.

b What can you say about the energy of the ball at the point of release compared with at the top of its bounce?

c You could use the test in **a** above to see which of three trampolines was the bounciest.

   i Name the independent variable in this test.

   ii Is this variable categoric or continuous?

3 One exciting fairground ride acts like a giant catapult. The capsule, in which you are strapped, is fired high into the sky by the rubber bands of the catapult. Explain the energy transfers taking place in the ride.

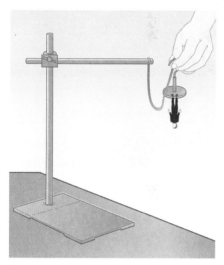

Figure 4 Testing a bungee jump

∞ links

For more information on variables, look back at H2 Fundamental ideas about how science works.

Key points

- Energy cannot be created or destroyed.

- Conservation of energy applies to all energy changes.

P1 2.3
Useful energy

Learning objectives

- What is 'useful' energy?
- What do we mean by 'wasted' energy?
- What eventually happens to wasted energy?
- Does energy become less useful after we use it?

Figure 1 Using energy

??? Did you know ...?

Lots of energy is transferred in a car crash. The faster the car travels the more kinetic energy it has and the more it has to transfer before stopping. In a crash, kinetic energy is quickly transferred to elastic strain energy, distorting the car's shape, and energy is transferred by heating the metal. There is usually quite a lot of sound energy too!

Energy for a purpose

Where would we be without **machines**? We use washing machines at home. We use machines in factories to make the goods we buy. We use them in the gym to keep fit and we use them to get us from place to place.

a What eventually happens to all the energy you use in a gym?

A machine transfers energy for a purpose. Friction between the moving parts of a machine causes the parts to warm up. So not all of the energy supplied to a machine is usefully transferred. Some energy is wasted.

- **Useful energy** is energy transferred to where it is wanted, in the form it is wanted.
- **Wasted energy** is energy that is not usefully transferred.

b What eventually happens to the kinetic energy of a machine when it stops?

Practical

Investigating friction

Friction in machines always causes energy to be wasted. Figure 2 shows two examples of friction in action. Try one of them out.

In **a**, friction acts between the drill bit and the wood. The bit becomes hot as it bores into the wood. Some of the electrical energy supplied to the bit heats up the drill bit (and the wood).

In **b**, when the brakes are applied, friction acts between the brake blocks and the wheel. This slows the bicycle and the cyclist down. Some of the kinetic energy of the bicycle and the cyclist is transferred to energy heating the brake blocks (and the bicycle wheel).

You can practise your skills in 'How Science Works' by investigating friction on different surfaces.

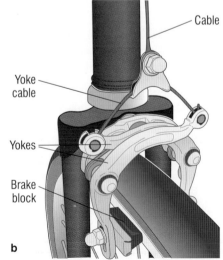

Cable

Yoke cable

Yokes

Brake block

b

Figure 2 Friction in action
a Using a drill **b** Braking on a bicycle

Disc brakes at work

The next time you are in a car slowing down at traffic lights, think about what is making the car stop. Figure 3 shows how the disc brakes of a car work. When the brakes are applied, the pads are pushed on to the disc in each wheel. Friction between the pads and each disc slows the wheel down. Some of the kinetic energy of the car is transferred to energy heating the disc pads and the discs. In Formula One racing cars you can sometimes see the discs glow red hot.

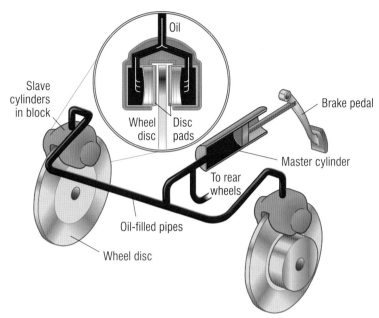

Figure 3 Disc brakes

Spreading out

- **Wasted energy is dissipated (spreads out) to the surroundings.**
  For example, the gears of a car get hot due to friction when the car is running. So energy transfers from the gear box to the surrounding air.

- **Useful energy eventually transfers to the surroundings too.**
  For example, the useful energy supplied to the wheels of a car is transferred to energy heating the tyres. This energy is then transferred to the road and the surrounding air.

- **Energy becomes less useful, the more it spreads out.**
  For example, the hot water from the cooling system of a CHP (combined heat and power) power station gets used to heat nearby buildings. The energy supplied to heat the buildings will eventually be transferred to the surroundings.

c The hot water from many power stations flows into rivers or lakes. Why is this wasteful?

Summary questions

1 Copy and complete the table below.

Energy transfer by	Useful energy output	Wasted energy output
a An electric heater		
b A television		
c An electric kettle		
d Headphones		

2 What would happen, in terms of energy transfer, to
  a a gear box that was insulated so it could not transfer energy heating the surroundings?
  b a jogger wearing running shoes, which are well-insulated?
  c a blunt electric drill if you use it to drill into hard wood?

3 a Describe the energy transfers of a pendulum as it swings from one side to the middle then to the opposite side.
  b Explain why a swinging pendulum eventually stops.

Key points

- Useful energy is energy in the place we want it and in the form we need it.

- Wasted energy is energy that is not useful energy.

- Useful energy and wasted energy both end up being transferred to the surroundings, which become warmer.

- As energy spreads out, it gets more and more difficult to use for further energy transfers.

P1 2.4 Energy and efficiency

Learning objectives

- What do we mean by efficiency?
- How efficient can a machine be?
- How can we make machines more efficient?

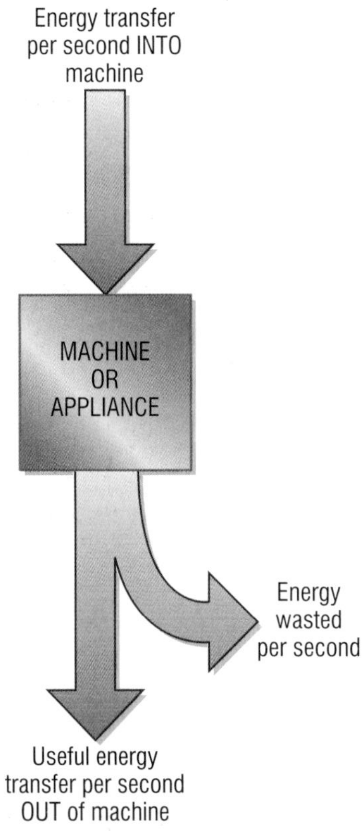

Energy transfer per second INTO machine

↓

MACHINE OR APPLIANCE

→ Energy wasted per second

↓

Useful energy transfer per second OUT of machine

Figure 1 A Sankey diagram

When you lift an object, the useful energy from your muscles goes to the object as gravitational potential energy. This depends on its weight and how high it is raised.

- Weight is measured in **newtons (N)**. The weight of a 1 kilogram object on the Earth's surface is about 10 N.
- Energy is measured in **joules (J)**. The energy needed to lift a weight of 1 N by a height of 1 metre is equal to 1 joule.

Your muscles get warm when you use them so they do waste some energy.

a Think about lowering a weight. What happens to its gravitational potential energy?

Sankey diagrams

Figure 1 represents the energy transfer through a device. It shows how we can represent any energy transfer where energy is wasted. This type of diagram is called a **Sankey diagram**.

Because energy cannot be created or destroyed:

Input energy (energy supplied) = useful energy delivered + energy wasted

For any device that transfers energy:

$$\text{Efficiency} = \frac{\text{useful energy transferred by the device}}{\text{total energy supplied to the device}} \ (\times 100\%)$$

Maths skills

Efficiency can be written as a number (which is never more than 1) or as a percentage.

For example, a light bulb with an efficiency of 0.15 would radiate 15 J of energy as light for every 100 J of electrical energy we supply to it.

- Its efficiency (as a number) $= \dfrac{15}{100} = 0.15$
- Its percentage efficiency $= 0.15 \times 100\% = 15\%$

b In the example above, how much energy is wasted for every 100 J of electrical energy supplied?

c What happens to the wasted energy?

Maths skills

Worked example

An electric motor is used to raise an object. The object gains 60 J of gravitational potential energy when the motor is supplied with 200 J of electrical energy. Calculate the percentage efficiency of the motor.

Solution

Total energy supplied to the device = 200 J

Useful energy transferred by the device = 60 J

Percentage efficiency of the motor

$= \dfrac{\text{useful energy transferred by the motor}}{\text{total energy supplied to the motor}} \times 100\%$

$= \dfrac{60\,\text{J}}{200\,\text{J}} \times 100\% = 0.30 \times 100\% = 30\%$

Efficiency limits

No machine can be more than 100% efficient because we can never get more energy from a machine than we put into it.

Practical

Investigating efficiency

Figure 2 shows how you can use an electric winch to raise a weight. You can use the joulemeter to measure the electrical energy supplied.

● If you double the weight for the same increase in height, do you need to supply twice as much electrical energy to do this task?

The gravitational potential energy gained by the weight = weight in newtons × height increase in metres.

● Use this equation and the joulemeter measurements to work out the percentage efficiency of the winch.

Safety: Protect the floor and your feet. Stop the winch before the masses wrap round the pulley.

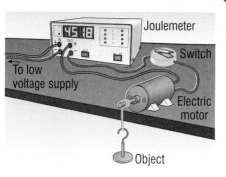

Figure 2 An electric winch

Improving efficiency

	Why machines waste energy	How to reduce the problem
1	Friction between the moving parts causes heating.	Lubricate the moving parts to reduce friction.
2	The resistance of a wire causes the wire to get hot when a current passes through it.	In circuits, use wires with as little electrical resistance as possible.
3	Air resistance causes energy transfer to the surroundings.	Streamline the shapes of moving objects to reduce air resistance.
4	Sound created by machinery causes energy transfer to the surroundings.	Cut out noise (e.g. tighten loose parts to reduce vibration).

d Which of the above solutions would hardly reduce the energy supplied?

AQA Examiner's tip

● The greater the percentage of the energy that is usefully transferred in a device, the more efficient the device is.

● Efficiency and percentage efficiency are numbers without units. The maximum efficiency is 1 or 100%, so if a calculation produces a number greater than this it must be wrong.

Summary questions

1 Copy and complete **a** to **c** using the words below. Each term can be used more than once.

supplied to   wasted by

a The useful energy from a machine is always less than the total energy .... it.

b Friction between the moving parts of a machine causes energy to be .... the machine.

c Because energy is conserved, the energy .... a machine is the sum of the useful energy from the machine and the energy .... the machine.

2 An electric motor is used to raise a weight. When you supply 60 J of electrical energy to the motor, the weight gains 24 J of gravitational potential energy. Work out:

a the energy wasted by the motor

b the efficiency of the motor.

3 A machine is 25% efficient. If the total energy supplied to the machine is 3200 J, how much useful energy can be transferred?

Key points

● The efficiency of a device = useful energy transferred by the device ÷ total energy supplied to the device (× 100%).

● No machine can be more than 100% efficient.

● Measures to make machines more efficient include reducing friction, air resistance, electrical resistance and noise due to vibrations.

Summary questions ⓚ

1 The devices listed below transfer energy in different ways.

1 Car engine 2 Electric bell 3 Electric light bulb

The next list gives the useful form of energy the devices are designed to produce.

Match words A, B and C with the devices numbered 1 to 3.

A Light B Kinetic energy C Sound

2 Copy and complete using the words below:

useful wasted light electrical

When a light bulb is switched on, ... energy is transferred into .... energy and energy that heats the surroundings. The energy that radiates from the light bulb is .... energy. The rest of the energy supplied to the light bulb is .... energy.

3 You can use an electric motor to raise a load. In a test, you supply the motor with 10 000 J of electrical energy and the load gains 1500 J of gravitational potential energy.

a Calculate its efficiency.

b How much energy is wasted?

c Copy and complete the Sankey diagram below for the motor.

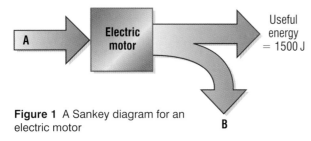

Figure 1 A Sankey diagram for an electric motor

4 A ball gains 4.0 J of gravitational potential energy when it is raised to a height of 2.0 m above the ground. When it is released, it falls to the ground and rebounds to a height of 1.5 m.

a How much kinetic energy did it have just before it hit the ground? Assume air resistance is negligible.

b How much gravitational potential energy did it transfer when it fell to the ground?

c The ball gained 3.0 J of gravitational potential energy when it moved from the ground to the top of the rebound. How much energy did it transfer in the impact at the ground?

d What happened to the energy it transferred on impact?

5 A low energy light bulb has an efficiency of 80%. Using an energy meter, a student found the light bulb used 1500 J of electrical energy in 100 seconds.

a How much useful energy did the light bulb transfer in this time?

b How much energy was wasted by the light bulb?

c Draw a Sankey diagram for the light bulb.

6 A bungee jumper jumps from a platform and transfers 12 000 J of gravitational potential energy before the rope attached to her becomes taut and starts to stretch. She then transfers a further 24 000 J of gravitational potential energy before she stops falling and begins to rise.

a Describe the energy changes:

  i after she jumps before the rope starts to stretch

  ii after the rope starts to stretch until she stops falling.

b What is the maximum kinetic energy she has during her descent?

7 On a building site, an electric winch and a pulley were used to lift bricks from the ground.

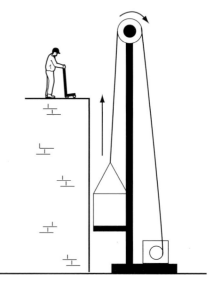

Figure 2 An electric winch and pulley

The winch transferred 12 000 J of electrical energy to raise a load through a height of 3.0 m. The load gained 1500 J of gravitational potential energy when it was raised.

a i How much useful energy was transferred by the motor ?

  ii Calculate the energy wasted.

  iii Calculate the percentage efficiency of the system.

b How could the efficiency of the winch be improved?

AQA Examination-style questions 🅚

1 A television transfers electrical energy.

Use words from the list to complete each sentence. Each word can be used once, more than once or not at all.

electrical   light   sound   warmer

A television is designed to transfer .... energy into light and .... energy. Some energy is transferred to the surroundings, which become .... (3)

2 A hairdryer contains an electrical heater and a fan driven by an electric motor. The hairdryer transfers electrical energy into other forms.

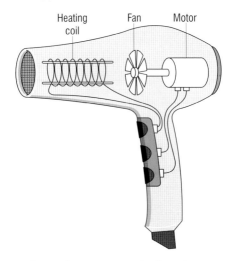

a Apart from energy by heating, name **two** of the other forms of energy. (2)

b Not all of the energy supplied to the fan is usefully transferred. Name **one** form of energy that is wasted by the fan. (1)

c Which of the following statements about the energy wasted by the fan is true?
   A It eventually becomes very concentrated.
   B It eventually makes the surroundings warmer.
   C It is eventually completely destroyed.
   D It is eventually transferred into electrical energy. (1)

d The fan in another hairdryer transfers useful energy at the same rate but wastes more of the energy supplied to it. What does this tell you about the efficiency of this hairdryer? (1)

3 In a hot water system water is heated by burning gas in a boiler. The hot water is then stored in a tank. For every 111 J of energy released from the gas, 100 J of energy is absorbed by the water in the boiler.

a Calculate the percentage efficiency of the boiler.

Write down the equation you use. Show clearly how you work out your answer. (4)

b The energy released from the gas but **not** absorbed by the boiler is 'wasted'. Explain why this energy is of little use for further energy transfers. (1)

c The tank in the hot water system is surrounded by a layer of insulation. Explain the effect of the insulation on the efficiency of the hot water system. (3)

4 A chairlift carries skiers to the top of a mountain. The chairlift is powered by an electric motor.

a What type of energy have the skiers gained when they reach the top of the mountain? (1)

b The energy required to lift two skiers to the top of the mountain is 240 000 J.
The electric motor has an efficiency of 40%.

Calculate the energy wasted in the motor.

Write down the equation you use. Show clearly how you work out your answer and give the unit. **[H]** (4)

c Explain why some energy is wasted in the motor. (2)

5 A light bulb transfers electrical energy into useful light energy and wasted energy to the surroundings. For every 100 J of energy supplied to the bulb, 5 J of energy is transferred into light.

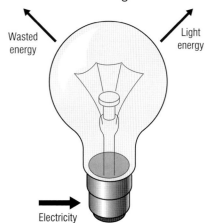

Draw and label a Sankey diagram for the light bulb. (3)

6 *In this question you will be assessed on using good English, organising information clearly and using specialist terms where appropriate.*

Explain why an electric heater is the only appliance that can possibly be 100% efficient. (6)

P1 3.1 Electrical appliances

Learning objectives

- Why are electrical appliances so useful?

- What do we use most everyday electrical appliances for?

- How do we choose an electrical appliance for a particular job?

Practical

Energy transfers

Carry out a survey of electrical appliances you find at school or at home.

Record the useful and wasted energy transfers of each appliance.

Everyday electrical appliances

We use **electrical appliances** every day. They transfer electrical energy into useful energy at the flick of a switch. Some of the electrical energy we supply to them is wasted.

Figure 1 Electrical appliances – how many can you see in this photo?

Table 1

Appliance	Useful energy	Energy wasted
Light bulb	Light from the glowing filament.	Energy transfer from the filament heating surroundings.
Electric heater	Energy heating the surroundings.	Light from the glowing element.
Electric toaster	Energy heating bread.	Energy heating the toaster case and the air around it.
Electric kettle	Energy heating water.	Energy heating the kettle itself.
Hairdryer	Kinetic energy of the air driven by the fan. Energy heating air flowing past the heater filament.	Sound of fan motor (energy heating the motor heats the air going past it, so is not wasted). Energy heating the hairdryer itself.
Electric motor	Kinetic energy of object driven by the motor. Potential energy of objects lifted by the motor.	Energy heating the motor and sound energy of the motor.
Computer disc drive	Energy stored in magnetic dots on the disc.	Energy heating the motor that drives the disc.

a What energy transfers happen in an electric toothbrush?

Did you know …?

Unlike high voltage electrical injuries, people do not get many burns when they are struck by lightning. Damage is usually to the nervous system. The brain is frequently damaged as the skull is the most likely place to be struck. Lightning that strikes near the head can enter the body through the eyes, ears and mouth and flow internally through the body.

How Science Works

Clockwork radio

People without electricity supplies can now listen to radio programmes – thanks to the British inventor Trevor Baylis. In the early 1990s, he invented and patented the clockwork radio. When you turn a handle on the radio, you wind up a clockwork spring in the radio. When the spring unwinds, it turns a small electric generator in the radio. It doesn't need batteries or mains electricity. So people in remote areas where there is no mains electricity can listen to their radios without having to walk miles for a replacement battery. But they do have to wind up the spring every time it runs out of energy.

Figure 2 Clockwork radios are now mass-produced and sold all over the world

Choosing an electrical appliance

We use electrical appliances for many purposes. Each appliance is designed for a particular purpose and it should waste as little energy as possible. Suppose you were a rock musician at a concert. You would need appliances that transfer sound energy into electrical energy and then back into sound energy. But you wouldn't want them to produce lots of energy heating the appliance itself and its surroundings. See if you can spot some of these appliances in Figure 3.

Figure 3 On stage

b What electrical appliance transfers:
  i sound energy into electrical energy?
  ii electrical energy into sound energy?
c What other electrical appliance would you need at a concert?

Summary questions

1 Copy and complete using the words below:

electrical   light   heating

When a battery is connected to a light bulb, ... energy is transferred from the battery to the light bulb. The filament of the light bulb becomes hot and so energy transfers to its surroundings by .... and as .... energy.

2 Match each electrical appliance in the list below with the energy transfer A or B it is designed to bring about.
  1 Electric drill
  2 Food mixer
  3 Electric bell

Energy transfer A Electrical energy → sound energy
                  B Electrical energy → kinetic energy

3 a Why does a clockwork radio need to be wound up before it can be used?
  b What energy transfers take place in a clockwork radio when it is wound up then switched on?
  c Give an advantage and a disadvantage of a clockwork radio compared with a battery-operated radio.

Key points

- Electrical appliances can transfer electrical energy into useful energy at the flick of a switch.

- Uses of everyday electrical appliances include heating, lighting, making objects move (using an electric motor) and creating sound and visual images.

- An electrical appliance is designed for a particular purpose and should waste as little energy as possible.

P1 3.2 | Electrical power

Learning objectives

- What do we mean by power?
- How can we calculate the power of an appliance?
- How can we calculate the efficiency of an appliance in terms of power?

Figure 1 A lift motor

When you use a lift to go up, a powerful electric motor pulls you and the lift upwards. The lift motor transfers energy from electrical energy to gravitational potential energy when the lift goes up at a steady speed. We also get electrical energy transferred to wasted energy heating the motor and the surroundings, and sound energy.

- The energy we supply per second to the motor is the **power** supplied to it.
- The more powerful the lift motor is, the faster it moves a particular load.

In general, we can say that:

the more powerful an appliance, the faster the rate at which it transfers energy.

We measure the power of an appliance in watts (W) or kilowatts (kW).

1 **watt** is a rate of transfer of energy of 1 joule per second (J/s).

1 **kilowatt** is equal to 1000 watts (i.e. 1000 joules per second or 1 kJ/s).

You can calculate power using:

$$P = \frac{E}{t}$$

Where:

P is the power in watts, W

E is the energy transferred to the appliance in joules, J

t is the time taken for the energy to be transferred in seconds, s.

 Maths skills

Worked example

A motor transfers 10 000 J of energy in 25 s. What is its power?

Solution

$$P = \frac{E}{t}$$

$$P = \frac{10\ 000\ \text{J}}{25\ \text{s}} = 400\,\text{W}$$

a What is the power of a lift motor that transfers 50 000 J of energy from the electricity supply in 10 s?

Power ratings

Here are some typical values of power ratings for different energy transfers:

Appliance	Power rating
A torch	1 W
An electric light bulb	100 W
An electric cooker	10 000 W = 10 kW (where 1 kW = 1000 watts)
A railway engine	1 000 000 W = 1 megawatt (MW) = 1 million watts
A Saturn V rocket	100 MW
A very large power station	10 000 MW
World demand for power	10 000 000 MW
A star like the Sun	100 000 000 000 000 000 000 000 MW

Figure 2 Rocket power

b How many 100 W electric light bulbs would use the same amount of power as a 10 kW electric cooker?

Muscle power

How powerful is a weightlifter?

A 30 kg dumbbell has a weight of 300 N. Raising it by 1 m would give it 300 J of gravitational potential energy. A weightlifter could lift it in about 0.5 seconds. The rate of energy transfer would be 600 J/s (= 300 J ÷ 0.5 s). So the weightlifter's power output would be about 600 W in total!

c An inventor has designed an exercise machine that can also generate 100 W of electrical power. Do you think people would buy this machine in case of a power cut?

Efficiency and power

For any appliance

- its useful power out (or output power) is the useful energy **per second** transferred by it.
- its total power in (or input power) is the energy **per second** supplied to it.

In P1 2.4 Energy and efficiency, we saw that the efficiency of an appliance

$$= \frac{\text{useful energy transferred by the device}}{\text{total energy supplied to it}} \text{ (} \times 100\% \text{)}$$

Because power = energy **per second** transferred or supplied, we can write the efficiency equation as:

$$\text{Efficiency} = \frac{\text{useful power out}}{\text{total power in}} \text{ (} \times 100\% \text{)}$$

For example, suppose the useful power out of an electric motor is 20 W and the total power in is 80 W, the percentage efficiency of the motor is:

$$\frac{\text{useful power out}}{\text{total power in}} \times 100\% = \frac{20\ W}{80\ W} \times 100\% = 25\%$$

Figure 3 Muscle power

Summary questions

1 a Which is more powerful?
   i A torch bulb or a mains filament bulb.
   ii A 3 kW electric kettle or a 10 000 W electric cooker.
   b There are about 20 million occupied homes in England. If a 3 kW electric kettle was switched on in 1 in 10 homes at the same time, how much power would need to be supplied?

2 The total power supplied to a lift motor is 5000 W. In a test, it transfers 12 000 J of electrical energy to gravitational potential energy in 20 seconds.
   a How much electrical energy is supplied to the motor in 20 s?
   b What is its efficiency in the test?

3 A machine has an input power rating of 100 kW. If the useful energy transferred by the machine in 50 seconds is 1500 kJ, calculate
   a its output power in kilowatts
   b its percentage efficiency.

Key points

- Power is rate of transfer of energy.
- $P = \dfrac{E}{t}$
- Efficiency = $\dfrac{\text{useful power out}}{\text{total power in}}$ (× 100%)

Electrical energy

Using electrical energy

Learning objectives

- What is the kilowatt-hour?
- How can we work out the energy used by a mains appliance?
- How is the cost of mains electricity worked out?

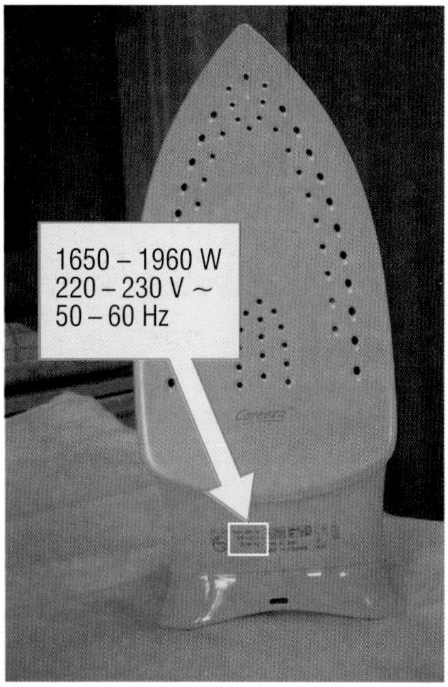

1650 – 1960 W
220 – 230 V ~
50 – 60 Hz

Figure 1 Mains power

When you use an electric heater, how much electrical energy is transferred from the mains? You can work this out if you know its power and how long you use it for.

For any appliance, the energy supplied to it depends on:

- how long it is used for
- the power supplied to it.

A 1 kilowatt heater uses the same amount of electrical energy in 1 hour as a 2 kilowatt heater would use in half an hour. For ease, we say that:

the energy supplied to a 1 kW appliance in 1 hour is 1 **kilowatt-hour (kWh)**.

We use the kilowatt-hour as the unit of energy supplied by mains electricity. You can use this equation to work out the energy, in kilowatt-hours, transferred by a mains appliance in a certain time:

$$E = P \times t$$

Where:

E is the energy transferred in kilowatt hours, kWh

P is the power in kilowatts, kW

t is the time taken for the energy to be transferred in hours, h.

Maths skills

Worked example

You have used this equation before in P1 3.2 to calculate the power of an appliance. It is the same equation, just rearranged and with different units.

$$E = P \times t$$

Divide both sides by t $\quad \dfrac{E}{t} = P$

This is the same as $\quad P = \dfrac{E}{t}$

For example:

- a 1 kW heater switched on for 1 hour uses 1 kWh of electrical energy (= 1 kW × 1 hour)
- a 1 kW heater switched on for 10 hours uses 10 kWh of electrical energy (= 1 kW × 10 hours)
- a 0.5 kW or 500 W heater switched on for 6 hours uses 3 kWh of electrical energy (= 0.5 kW × 6 hours).

If we want to calculate the energy transferred in joules, we can use the equation:

$$E = P \times t$$

Where:

E is the energy transferred in joules, J

P is the power in watts, W

t is the time taken for the energy to be transferred in seconds, s.

a How many kWh of energy are used by a 100 W lamp in 24 hours?

b How many joules of energy are used by a 5 W torch lamp in 3000 seconds (= 50 minutes)?

Did you know ...?

One kilowatt-hour is the amount of electrical energy supplied to a 1 kilowatt appliance in 1 hour.

So **1 kilowatt-hour**
= 1000 joules per second × 60 × 60 seconds
= 3 600 000 J
= **3.6 million joules**.

Paying for electrical energy

The **electricity meter** in your home measures how much electrical energy your family uses. It records the total energy supplied, no matter how many appliances you all use. It gives us a reading of the number of kilowatt-hours (kWh) of energy supplied by the mains.

In most houses, somebody reads the meter every three months. Look at the electricity bill in Figure 2.

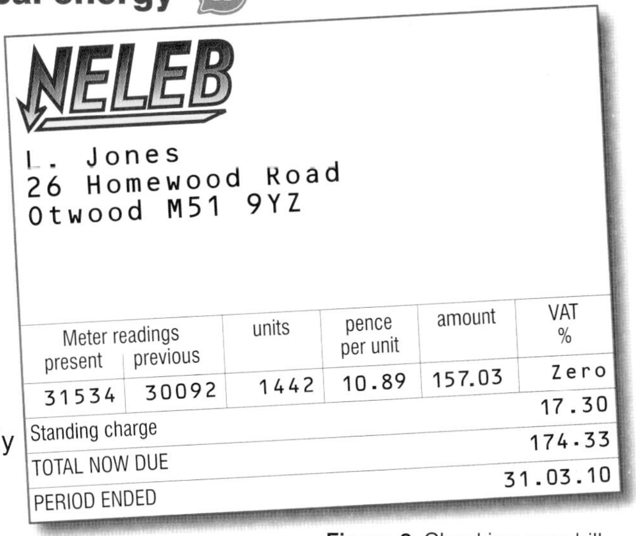

NELEB

L. Jones
26 Homewood Road
Otwood M51 9YZ

Meter readings		units	pence per unit	amount	VAT %
present	previous				
31534	30092	1442	10.89	157.03	Zero
					17.30
Standing charge					174.33
TOTAL NOW DUE					
PERIOD ENDED					31.03.10

Figure 2 Checking your bill

The difference between the two readings is the number of kilowatt-hours supplied since the last bill.

c Check for yourself that 1442 kWh of electrical energy is supplied in the bill shown.

We use the kilowatt-hour to work out the cost of electricity. For example, a cost of 12p per kWh means that each kilowatt-hour of electrical energy costs 12p. Therefore:

total cost = number of kWh used × cost per kWh

d Work out the cost of 1442 kWh at 12p per kWh.

Figure 3 An electricity meter

AQA **Examiner's tip**

Remember that a kilowatt-hour (kWh) is a unit of energy.

Summary questions

1 Copy and complete **a** to **c** using the words below. Each word can be used more than once.

hours   kilowatt   kilowatt-hours

a The .... is a unit of power.

b Electricity meters record the mains electrical energy transferred in units of ....

c Two .... is the energy transferred by a 1 .... appliance in 2 ....

2 a Work out the number of kWh transferred in each case below.

   i A 3 kilowatt electric kettle is used 6 times for 5 minutes each time.

   ii A 1000 watt microwave oven is used for 30 minutes.

   iii A 100 watt electric light is used for 8 hours.

b Calculate the total cost of the electricity used in part **a** if the cost of electricity is 12p per kWh.

3 An electric heater is left on for 3 hours. During this time it uses 12 kWh of electrical energy.

a What is the power of the heater?

b How many joules are supplied?

Key points

- The kilowatt-hour is the energy supplied to a 1 kW appliance in 1 hour.

- $E = P \times t$

- Total cost = number of kWh used × cost per kWh

P1 3.4 Cost effectiveness matters

Costs **k**

When we compare the effectiveness of different energy-saving appliances that do the same job, we need to make sure we get value for money. In other words, we need to make sure the appliance we choose is **cost effective**.

To compare the cost effectiveness of different cost-cutting measures, we need to consider:

- the capital costs such as buying and installing equipment
- the running costs, including fuel and maintenance
- environmental costs, for example
  - removal or disposal of old equipment (e.g. refrigerators, used batteries)
  - tax charges such as carbon taxes of fossil fuels
- other costs such as interest on loans.

Payback time again!

A householder wants to cut her fuel bills by reducing energy losses from her home. This would save fuel and reduce fuel bills. She is comparing loft insulation with cavity wall insulation in terms of payback time.

- The loft insulation costs £200 (including gloves and a safety mask) and she would fit the insulation herself. This could save £100 per year on the fuel bill. So the payback time would be 2 years.

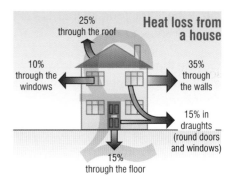

Figure 1 Heat loss from a house

- The cavity wall insulation for a house costs £500 and an additional £100 to fit the insulation. This could save £200 per year on the fuel bill. It would pay for itself after 3 years.

a For each type of insulation, how much would the householder have saved after 5 years?

b A double-glazed window costs £200. It saves £10 per year on the fuel bill. How long is the payback time?

links

For more information on payback time, look back at P1 1.9 Heating and insulating buildings.

Activity

Buying a heater

An artist wants to buy an electric heater to provide instant heating in his workshop when he starts work on a cold morning. He can't decide between a fan heater, a radiant heater and a tubular heater.

Table 1 shows how each type of heater works and its main drawback.

Assuming the heaters cost the same to buy, write a short report advising the artist which type of heater would be most suitable for him.

Table 1

Heater type	Input power	How the heater works	Drawbacks
Fan heater	2.0 kW	blows warm air from the hot element round the room	energy needed to run the fan motor
Radiant heater	1.0 kW	uses a reflector to direct radiation from the glowing element	the radiation only heats the air and objects in front of the heater
Tubular heater	0.5 kW	the heater element is inside a metal tube which heats the room	provides background heat gradually

Lighting costs

Low-energy bulbs use much less electrical energy than filament bulbs. This is why the UK government has banned the sale of filament bulbs. Table 2 gives some data about different types of mains bulbs.

Table 2

Type	Power in watts	Efficiency	Lifetime in hours	Cost of bulb	Typical use	Drawbacks
Filament bulb	100 W	20%	1000	50p	room lighting	inefficient, gets hot
Halogen bulb	100 W	25%	2500	£2.00	spotlight	inefficient, gets hot
Low-energy compact fluorescent bulb	25 W	80%	15 000	£2.50	room lighting	takes a few minutes for full brightness, disposal must be in a sealed bag due to mercury (which is toxic) in it
Low-energy light-emitting diode	2 W	90%	30 000	£7.00	spotlight	expensive to buy, brightness of one halogen bulb needs several LEDs

c Which bulb has the greatest output in terms of useful energy?
d Which type of spotlight wastes the least energy by heating?

Key points

- Cost effectiveness means getting the best value for money.
- To compare the cost effectiveness of different appliances, we need to take account of costs to buy it, running costs and other costs such as environmental costs.

Summary questions

1 State with a reason which type of heater from Table 1 you would choose to keep a bedroom warm at night in winter.
2 Use the information in Table 2 to answer the following questions.
   a State one advantage and one disadvantage of a CFL bulb compared with a filament bulb.
   b State one advantage and one disadvantage of an LED compared with a halogen bulb.

Summary questions (k)

1 a Name an appliance that transfers electrical energy into:
  i light and sound energy
  ii kinetic energy.

b Complete the sentences below.
  i In an electric bell, electrical energy is transferred into useful energy in the form of .... energy, and .... energy.
  ii In a dentist's drill, electrical energy is transferred into useful energy in the form of .... energy and sometimes as .... energy.

2 a Which two words in the list below are units that can be used to measure energy?

  joule   kilowatt   kilowatt-hour   watt

b Rank the electrical appliances below in terms of energy used from highest to lowest.
  A a 0.5 kW heater used for 4 hours
  B a 100 W lamp left on for 24 hours
  C a 3 kW electric kettle used 6 times for 10 minutes each time
  D a 750 W microwave oven used for 10 minutes.

3 a The readings of an electricity meter at the start and the end of a month are shown below.

0	9	3	7	2

0	9	6	1	5

  i Which is the reading at the end of the month?
  ii How many kilowatt-hours of electricity were used during the month?
  iii How much would this electricity cost at 12p per kWh?

b A pay meter in a holiday home supplies electricity at a cost of 12p per kWh.
  i How many kWh would be supplied for £1.20?
  ii How long could a 2 kW heater be used for after £1.20 is put in the meter slot? **[H]**

4 An escalator in a shopping centre is powered by a 50 kW electric motor. The escalator is in use for a total time of 10 hours every day.

a How much electrical energy in kWh is supplied to the motor each day?

b The electricity supplied to the motor costs 12p per kWh. What is the daily cost of the electricity supplied to the motor?

c How much would be saved each day if the motor was replaced by a more efficient 40 kW motor?

5 The data below show the electrical appliances used in a house in one evening.
  A a 1.0 kW heater for 4 hours
  B a 0.5 kW television for 2 hours
  C a 3 kW electric kettle three times for 10 minutes each time.

a Which appliance uses most energy?

b How many kWh of electrical energy is used by each appliance?

c Each kWh costs 12p. How much did it cost to use the three appliances?

6 The battery of a laptop computer is capable of supplying 60 watts to the computer circuits for 2 hours before it needs to be recharged.

a Calculate the electrical energy the battery can supply in two hours in:
  i kilowatt-hours
  ii joules.

b Describe the energy transfers that take place when the computer is being used.

c A mains charging unit can be connected to the computer when in use to keep its battery fully charged. Would the computer use less energy with the charging unit connected than without it connected?

7 A student has an HD television at home that uses 120 watts of electrical power when it is switched on. He monitors its usage for a week and finds it is switched on for 30 hours.

Figure 1 An HD TV in use

a How many kilowatt-hours of electrical energy are supplied to it in this time?

b Calculate the cost of this electrical energy at 12p per kilowatt-hour.

AQA Examination-style questions

1 The pictures show six different household appliances.

Fan heater

Vacuum cleaner

Washing machine

Iron

Kettle

Blender

Name the **four** appliances in which electrical energy is usefully transferred into kinetic energy. (4)

2 An electric motor is used to lift a load. The useful power output of the motor is 30 W. The total input power to the motor is 75 W.

Calculate the efficiency of the motor.

Write down the equation you use. Show clearly how you work out your answer. (3)

3 Which **two** of the following units are units of energy?

a J

b J/s

c kWh

d W (1)

4 The diagram shows the readings on a household electricity meter at the beginning and end of one week.

5	2	3	4	0

5	2	5	5	5

Beginning of the week   **End of the week**

a How many kWh of electricity were used during the week? (1)

b On one day 35 kWh of electricity were used. The total cost of this electricity was £5.25.

Calculate how much the electricity cost per kWh.

Write down the equation you use. Show clearly how you work out your answer and give the unit. [H] (3)

c During the week a 2.4 kW kettle was used for 2 hours.

Calculate how much energy was transferred by the kettle.

Write down the equation you use. Show clearly how you work out your answer and give the unit. (3)

5 A student uses some hair straighteners.

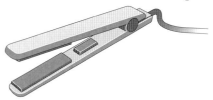

a The hair straighteners have a power of 90 W.

What is meant by *a power of 90 W*? (2)

b Calculate how many kilowatt-hours of electricity are used when the straighteners are used for 15 minutes.

Write down the equation you use. Show clearly how you work out your answer and give the unit. (3)

c The electricity supplier is charging 14p per kWh.

Calculate how much it will cost to use the straighteners for 15 minutes a day for one year.

Write down the equation you use. Show clearly how you work out your answer and give the unit. (2)

6 Filament bulbs are being replaced by compact fluorescent bulbs.

A compact fluorescent bulb costs £12, a filament bulb costs 50p.

A 25 W compact fluorescent bulb gives out as much light as a 100 W filament bulb.

A filament bulb lasts for about 1000 hours; a compact fluorescent bulb lasts for about 8000 hours, although this time is significantly shorter if the bulb is turned on and off very frequently.

A compact fluorescent bulb contains a small amount of poisonous mercury vapour.

a Explain how a 25 W compact fluorescent bulb provides the same amount of light as a 100 W filament bulb but use less electricity. (2)

b *In this question you will be assessed on using good English, organising information clearly and using specialist terms where appropriate.*

Compare the advantages and disadvantages of buying compact fluorescent bulbs rather than filament bulbs. (6)

P1 4.1 Fuel for electricity

Learning objectives

- How is electricity generated in a power station?
- Which fossil fuels do we burn in power stations?
- How do we use nuclear fuels in power stations?
- What other fuels can be used to generate electricity?

Inside a power station

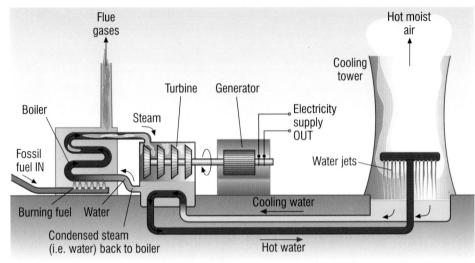

Figure 1 Inside a fossil fuel power station

Figure 2 Inside a gas-fired power station

Practical

Turbines

See how we can use water to drive round the blades of a turbine.

- Why is steam better than water?

Figure 3 Using biofuel to generate electricity

Almost all the electricity you use is generated in power stations.

- In **coal-** or **oil-fired power stations**, and in most **gas-fired power stations**, the burning fuel heats water in a boiler. This produces steam. The steam drives a **turbine** that turns an electricity **generator**. Coal, oil and gas are fossil fuels, which are fuels obtained from long-dead biological material.

  a What happens to the steam after it has been used?
  b What happens to the energy of the steam after it has been used?

- In some gas-fired power stations, we burn natural gas directly in a gas turbine engine. This heats the air drawn into the engine. It produces a powerful jet of hot gases and air that drives the turbine. A gas-fired turbine can be switched on very quickly.

Biofuels

We can get methane gas from cows or animal manure and from sewage works, decaying rubbish and other sources. It can be used in small-scale gas-fired power stations. Methane is an example of a **biofuel**.

A biofuel is any fuel obtained from living or recently living organisms such as animal waste or woodchip. Other biofuels include ethanol (from fermented sugar cane), straw, nutshells and woodchip.

A biofuel is:

- **renewable** because its biological source continues to exist and never dies out as a species
- **carbon-neutral** because, in theory, the carbon it takes in from the atmosphere as carbon dioxide can 'balance' the amount released when it is burned.

Nuclear power

Figure 4 shows you that every atom contains a positively charged nucleus surrounded by electrons. The **atomic nucleus** is composed of two types of particles: neutrons and protons. Atoms of the same element can have different numbers of neutrons in the nucleus.

How is electricity obtained from a nuclear power station?

The fuel in a nuclear power station is uranium (or plutonium). The uranium fuel is in sealed cans in the core of the reactor. The nucleus of a uranium atom is unstable and can split in two. Energy is released when this happens. We call this process **nuclear fission**. Because there are lots of uranium atoms in the core, it becomes very hot.

The energy of the core is transferred by a fluid (called the 'coolant') that is pumped through the core.

● The coolant is very hot when it leaves the core. It flows through a pipe to a 'heat exchanger', then back to the reactor core.

● The energy of the coolant is used to turn water into steam in the heat exchanger. The steam drives turbines that turn electricity generators.

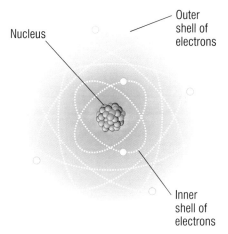

Figure 4 The structure of the atom

⚙ *How Science Works*

Comparing nuclear power and fossil fuel power

	Nuclear power station	Fossil fuel power station
Fuel	Uranium or plutonium	Coal, oil or gas
Energy released per kg of fuel	1 000 000 kWh (= about 10 000 × energy released per kg of fossil fuel)	100 kWh
Waste	Radioactive waste that needs to be stored for many years	Non-radioactive waste
Greenhouse gases	No – because uranium releases energy without burning	Yes – because fossil fuels produce gases such as carbon dioxide when they burn

---

Summary questions

1 Copy and complete **a** to **c** using the words below:

coal   gas   oil   uranium

a The fuel that is not a fossil fuel is ....

b Power stations that use .... as the fuel can be switched on very quickly.

c Greenhouse gases are produced in a power station that uses coal, gas or .... as fuel.

2 **a** State one advantage and one disadvantage of:

  i an oil-fired power station compared with a nuclear power station

  ii a gas-fired power station compared with a coal-fired power station.

 b Look at the table above.
  How many kilograms of fossil fuel would give the same amount of energy as 1 kilogram of uranium fuel?

3 **a** Explain why ethanol is described as a biofuel.

 b Ethanol is also described as carbon-neutral. What is a carbon-neutral fuel?

Key points

● Electricity generators in power stations are driven by turbines.

● Coal, oil and natural gas are burned in fossil fuel power stations.

● Uranium or plutonium are used as the fuel in a nuclear power station. Much more energy is released per kg from uranium or plutonium than from fossil fuel.

● Biofuels are renewable sources of energy. Biofuels such as methane and ethanol can be used to generate electricity.

P1 4.2 — Energy from wind and water

Learning objectives

- What does a wind turbine consist of?
- How do we use waves to generate electricity?
- What type of power station uses water running downhill to generate electricity?
- How can we use the tides to generate electricity?

Strong winds can cause lots of damage on a very stormy day. Even when the wind is much weaker, it can still turn a wind turbine. Energy from the wind and other natural sources such as **waves** and **tides** is called **renewable energy**. That's because such natural sources of energy can never be used up.

In addition, no fuel is needed to produce electricity from these natural sources so they are carbon-free to run.

Wind power

A wind turbine is an electricity generator at the top of a narrow tower. The force of the wind drives the turbine's blades around. This turns a generator. The power generated increases as the wind speed increases.

   a What happens if the wind stops blowing?

Wave power

A wave generator uses the waves to make a floating generator move up and down. This motion turns the generator so it generates electricity. A cable between the generator and the shore delivers electricity to the grid system.

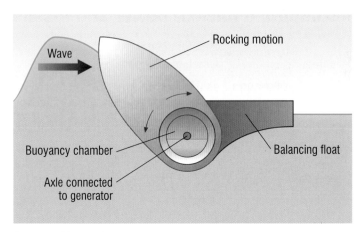

Figure 2 Energy from waves

Labels: Wave; Rocking motion; Buoyancy chamber; Axle connected to generator; Balancing float

Figure 1 A wind farm – why do some people oppose these developments?

Wave generators need to withstand storms and they don't produce a constant supply of electricity. Also, lots of cables (and buildings) are needed along the coast to connect the wave generators to the electricity grid. This can spoil areas of coastline. Tidal flow patterns might also change, affecting the habitats of marine life and birds.

   b What could happen if the waves get too high?

Hydroelectric power

We can generate hydroelectricity when rainwater collected in a reservoir (or water in a pumped storage scheme) flows downhill. The flowing water drives turbines that turn electricity generators at the foot of the hill.

   c Where does the energy for hydroelectricity come from?

How Science Works

When electricity demand is low, we can use electricity from wind turbines, wave generators and other electricity generators to pump water uphill into a reservoir. When demand is high, we can let the water run downhill through a hydroelectric generator.

Tidal power

A tidal power station traps water from each high tide behind a barrage. We can then release the high tide into the sea through turbines. The turbines drive generators in the barrage.

One of the most promising sites in Britain is the Severn estuary. This is because the estuary rapidly becomes narrower as you move up-river away from the open sea. So it funnels the incoming tide and makes it higher.

d Why is tidal power more reliable than wind power?

Figure 3 A hydroelectric scheme

Figure 4 A tidal power station

Summary questions

1 Copy and complete **a** to **d** using the words below:

hydroelectric   tidal   wave   wind

a .... power does not need water.
b .... power does not need energy from the Sun.
c .... power is obtained from water running downhill.
d .... power is obtained from water moving up and down.

2 a Use the table below for this question. The output of each source is given in millions of watts (MW).
 i How many wind turbines would give the same total power output as a tidal power station?
 ii How many kilometres of wave generators would give the same total output as a hydroelectric power station?
b Use the words below to fill in the location column in the table.

coastline   estuaries   hilly or coastal areas   mountain areas

	Output	Location	Total cost in £ per MW
Hydroelectric power station	500 MW per station		50
Tidal power station	2000 MW per station		300
Wave power generators	20 MW per kilometre of coastline		100
Wind turbines	2 MW per wind turbine		90

3 The last column of the table above shows an estimate of the total cost per MW of generating electricity using different renewable energy sources. The total cost for each includes its running costs and the capital costs to set it up.
a The capital cost per MW of a tidal power station is much higher than that of a hydroelectric power station. Give one reason for this difference.
b i Which energy resource has the lowest total cost per MW?
 ii Give two reasons why this resource might be unsuitable in many areas.

Key points

- A wind turbine is an electricity generator on top of a tall tower.
- Waves generate electricity by turning a floating generator.
- Hydroelectricity generators are turned by water running downhill.
- A tidal power station traps each high tide and uses it to turn generators.

P1 4.3 Power from the Sun and the Earth

Learning objectives

- What are solar cells and how do we use them?
- What is the difference between a panel of solar cells and a solar heating panel?
- What is geothermal energy?
- How can we use geothermal energy to generate electricity?

Figure 2 A solar-powered vehicle. Think of some advantages and disadvantages of this car.

∞ links

For more information on solar heating panels, look back at P1 1.9 Heating and insulating buildings.

Solar radiation transfers energy to you from the Sun. That can sometimes be more energy than you want if you get sunburnt. But we can use the Sun's energy to generate electricity using **solar cells**. We can also use the Sun's energy to heat water directly in solar heating panels.

a Which generates electricity – a solar cell or a solar heating panel?

Practical

Solar cells

Use a solar cell panel to drive a small electric motor.

- See what happens if you gradually cover the solar cells with a card.

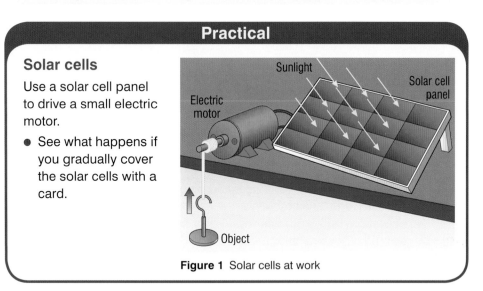

Figure 1 Solar cells at work

Solar cells at present convert less than 10% of the solar energy they absorb into electrical energy. We can connect them together to make solar cell panels.

- They are useful where we only need small amounts of electricity (e.g. in watches and calculators) or in remote places (e.g. on small islands in the middle of an ocean).
- They are very expensive to buy even though they cost nothing to run.
- We need lots of them – and plenty of sunshine – to generate enough power to be useful.

A **solar heating panel** heats water that flows through it. Even on a cloudy day in Britain, a solar heating panel on a house roof can supply plenty of hot water.

b If the water stopped flowing through a solar heating panel, what would happen?

A **solar power tower** uses thousands of flat mirrors to reflect sunlight on to a large water tank at the top of a tower. The mirrors on the ground surround the base of the tower.

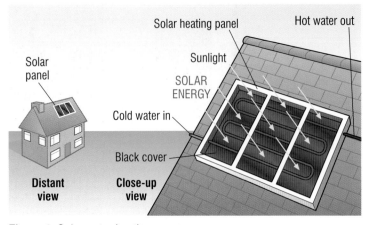

Figure 3 Solar water heating

- The water in the tank is turned to steam by the heating effect of the solar radiation directed at the water tank.
- The steam is piped down to ground level where it turns electricity generators.
- The mirrors are controlled by a computer so they track the Sun.

A solar power tower in a hot dry climate can generate more than 20 MW of electrical power.

c The solar furnace shown in Figure 3 in P1 1.1 uses 63 flat tracking mirrors to reflect solar radiation on to the giant reflector. Why does the solar power tower in Figure 4 opposite collect much more solar radiation than this solar furnace?

Geothermal energy

Geothermal energy comes from energy released by radioactive substances, deep within the Earth.

- The energy released by these radioactive substances heats the surrounding rock.
- As a result, energy is transferred by heating towards the Earth's surface.

We can build **geothermal power stations** in volcanic areas or where there are hot rocks deep below the surface. Water gets pumped down to these rocks to produce steam. Then the steam produced drives electricity turbines at ground level.

In some areas, we can heat buildings using geothermal energy directly. Heat flow from underground is called **ground heat.** It can be used to heat water in long lengths of underground pipes. The hot water is then pumped round the building. Ground heat is used as under-floor heating in some large 'eco-buildings'.

d Why do geothermal power stations not need energy from the Sun?

Figure 4 A solar power tower

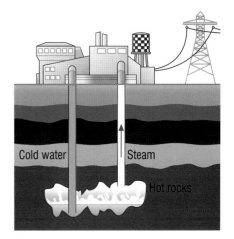

Cold water   Steam

Hot rocks

Figure 5 A geothermal power station

Summary questions

1 Copy and complete **a** to **c** using the words below:

geothermal   solar   radiation   radioactivity

  a A suitable energy resource for a calculator is .... energy.
  b .... inside the Earth releases .... energy.
  c .... from the Sun generates electricity in a solar cell.

2 A satellite in space uses a solar cell panel for electricity. The panel generates 300 W of electrical power and has an area of 10 m².
  a Each cell generates 0.2 W. How many cells are in the panel?
  b The satellite carries batteries that are charged by electricity from the solar cell panels. Why are batteries carried as well as solar cell panels?

3 A certain geothermal power station has a power output of 200 000 W.
  a How many kilowatt-hours of electrical energy does the power station generate in 24 hours?
  b State one advantage and one disadvantage of a geothermal power station compared with a wind turbine.

Key points

- Solar cells are flat solid cells that convert solar energy directly into electricity.
- Solar heating panels use the Sun's energy to heat water directly.
- Geothermal energy comes from the energy released by radioactive substances deep inside the Earth.
- Water pumped into hot rocks underground produces steam to drive turbines that generate electricity.

P1 4.4

Energy and the environment

Learning objectives

- What do fossil fuels do to our environment?
- Why are people concerned about nuclear power?
- How do renewable energy resources affect our environment?

Can we get energy without creating any problems? Look at the pie chart in Figure 1.

It shows the energy sources we use at present to generate electricity. What effect does each one have on our environment?

How Science Works

When a popular TV programme ends, lots of people decide to put the kettle on. The national demand for electricity leaps as a result. Engineers meet these surges in demand by switching gas turbine engines on in gas-fired power stations.

Fossil fuel problems

- When we burn coal, oil or gas, greenhouse gases such as carbon dioxide are released. We think that these gases cause global warming. We get some of our electricity from oil-fired power stations. We use much more oil to produce fuels for transport.
- Burning fossil fuels can also produce sulfur dioxide. This gas causes **acid rain**. We can remove the sulfur from a fuel before burning it to stop acid rain. For example, natural gas has its sulfur impurities removed before we use it.
- Fossil fuels are non-renewable. Sooner or later, we will have used up the Earth's reserves of fossil fuels. We will then have to find alternative sources of energy. But how soon? Oil and gas reserves could be used up within the next 50 years. Coal reserves will last much longer.
- **Carbon capture and storage** (CCS) could be used to stop carbon dioxide emissions into the atmosphere from fossil fuel power stations. Old oil and gas fields could be used for storage.

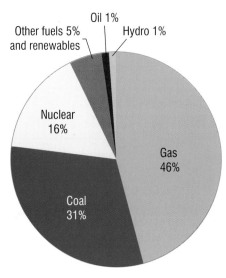

Figure 1 Energy sources for electricity

  a Burning fossil fuels in power stations pollutes our atmosphere. Which gas contributes towards:
   i global warming?
   ii acid rain?

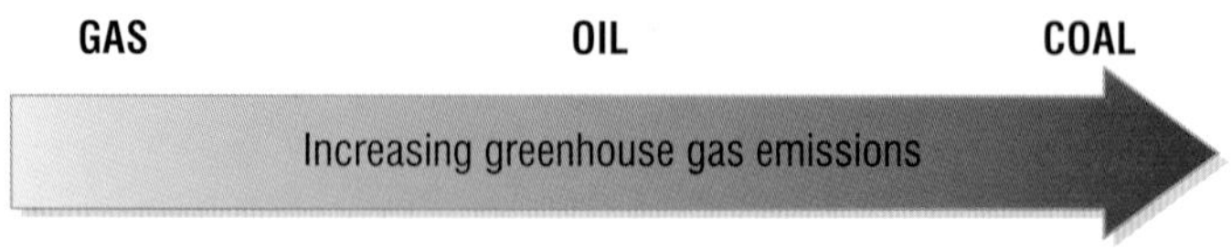

GAS   OIL   COAL

Increasing greenhouse gas emissions

Figure 2 Greenhouse gases from fossil fuels

Nuclear v. renewable

We need to cut back on our use of fossil fuels to stop global warming. Should we rely on nuclear power or on renewable energy in the future?

Nuclear power

Advantages

- No greenhouse gases (unlike fossil fuel).
- Much more energy from each kilogram of uranium (or plutonium) fuel than from fossil fuel.

Did you know ...?

The Gobi Desert is one of the most remote regions on Earth. Many areas do not have mains electricity. Yet people who live there can watch TV programmes – just as you can. All they need is a solar panel and satellite TV.

Disadvantages

- Used fuel rods contain radioactive waste, which has to be stored safely for centuries.
- Nuclear reactors are safe in normal operation. However, an explosion at one could release radioactive material over a wide area. This would affect these areas for many years.

b Why is nuclear fuel non-renewable?

Renewable energy sources and the environment

Advantages

- They will never run out.
- They do not produce greenhouse gases or acid rain.
- They do not create radioactive waste products.
- They can be used where connection to the National Grid is uneconomic. For example, solar cells can be used for road signs and hydroelectricity can be used in remote areas.

Disadvantages

- Wind turbines create a whining noise that can upset people nearby and some people consider them unsightly.
- Tidal barrages affect river estuaries and the habitats of creatures and plants there.
- Hydroelectric schemes need large reservoirs of water, which can affect nearby plant and animal life. Habitats are often flooded to create dams.
- Solar cells would need to cover large areas to generate large amounts of power.

c Do wind turbines affect plant and animal life?

Figure 4 The effects of acid rain

Summary questions

1 Copy and complete **a** to **c** using the words below:

acid rain fossil fuels greenhouse gas plant and animal life radioactive waste

  a Most of Britain's electricity is produced by power stations that burn ....

  b A gas-fired power station does not produce .... or much ....

  c A tidal power station does not produce .... as a nuclear power station does but it does affect .... locally.

2 Match each energy source with a problem it causes.

Energy source	Problem
i Coal	A Noise
ii Hydroelectricity	B Acid rain
iii Uranium	C Radioactive waste
iv Wind power	D Takes up land

3 **a** List three possible renewable energy resources that could be used to generate electricity for people on a remote flat island in a hot climate.

  b List three types of power stations that do not release greenhouse gases into the atmosphere.

Key points

- Fossil fuels produce increased levels of greenhouse gases which could cause global warming.
- Nuclear fuels produce radioactive waste.
- Renewable energy resources can affect plant and animal life.

P1 4.5

The National Grid (k)

Learning objectives

- What is the National Grid?
- What do the transformers do in the National Grid?
- Why do we use high voltages in the National Grid?

∞ links

For more information on transformers, see P3 3.4 Transformers and P3 3.5 Transformers in action.

Your electricity supply at home reaches you through the **National Grid**. This is a network of cables that distributes electricity from power stations to homes and other buildings. The network also contains **transformers**. Step-up transformers are used at power stations. Step-down transformers are used at substations near homes.

The National Grid's voltage is 132 000 V or more. This is because transmitting electricity at a high voltage reduces power loss, making the system more efficient.

Power stations produce electricity at a voltage of 25 000 V.

- We use **step-up transformers** to step this voltage up to the grid voltage.
- We use **step-down transformers** at local substations to step the grid voltage down to 230 V for use in homes and offices.

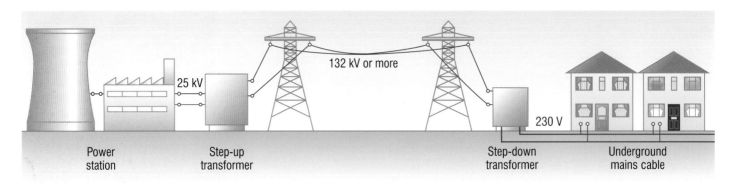

Figure 1 The National Grid

Figure 2 Electricity pylons carry the high voltage cables of the National Grid

Demonstration

Modelling the National Grid

Watch a demonstration of the effect of a transformer using this apparatus.

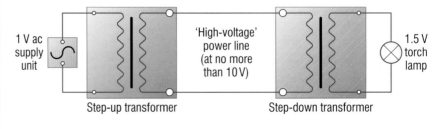

Figure 3 A model power line

AQA Examiner's tip

Remember that step-up transformers are used at power stations and step-down transformers are used at sub-stations near homes.

??? Did you know ...?

The National Grid was set up in 1926. The UK government decided electricity would be supplied to homes at 240 V. This was lowered to 230 V in 1994.

Power and the grid voltage

The electrical power supplied to any appliance depends on the appliance's current and its voltage. To supply a certain amount of power, we can lower the current if we raise the voltage. This is what a step-up transformer does in the grid system.

A step-up transformer raises the voltage, so less current is needed to transfer the same amount of power. A lower current passes through the grid cables. So energy losses due to the heating effect of the current are reduced to almost zero. But we need to lower the voltage at the end of the grid cables before we can use mains electricity at home.

a What difference would it make if we didn't step-up the grid voltage?

How Science Works

Underground or overhead?

Lots of people object to electricity pylons. They say they spoil the landscape or they affect their health. Electric currents produce electric and magnetic fields that might affect people.

Why don't we bury all cables underground?

Underground cables would be much more expensive, much more difficult to repair, and difficult to bury where they cross canals, rivers and roads.

What's more, overhead cables are high above the ground. Underground cables could affect people more because the cables wouldn't be very deep.

b Suggest two reasons why underground cables are more difficult to repair than overhead ones.

Summary questions

1 Copy and complete **a** and **b** using the words below:
higher   down   lower   up
  a Power stations are connected to the National Grid using step-............ transformers. This type of transformer makes the voltage ....
  b Homes are connected to the National Grid using step-............ transformers. This type of transformer makes the voltage ....

2 a Why is electrical energy transferred through the National Grid at a much higher voltage than the voltage generated in a power station?
  b Why are transformers needed to connect local substations to the National Grid?

3 A step-up transformer connects a power station to the cables of the National Grid.
  a What does the transformer do to
    i the voltage
    ii the current?
  b Why are step-down transformers used between the end of the grid cables and the mains cables that supply mains electricity to our homes?

?? Did you know …?

The insulators used on electricity pylons need to be very effective or else the electricity would short-circuit to the ground. In winter, ice on the cables can cause them to snap. Teams of electrical engineers are always on standby to deal with sudden emergencies.

Figure 4 Engineers at work on the Grid. They certainly need a head for heights!

Key points

● The National Grid is a network of cables and transformers that distributes electricity to our homes from distant power stations and renewable energy generators.

● Step-up transformers are used to step up power station voltages to the grid voltage. Step-down transformers are used to step the grid voltage down for use in our homes.

● A high grid voltage reduces energy loss and makes the system more efficient.

P1 4.6 Big energy issues

Learning objectives

- How do we best use our electricity supplies to meet variations in demand?

- How do we best use our electricity supplies to meet base-load demand?

- Which energy resources need to be developed to meet our energy needs in future?

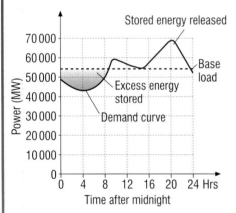

Figure 1 Example of electricity demand

Supply and demand

The demand for electricity varies during each day. It is also higher in winter than in summer. Our electricity generators need to match these changes in demand.

Power stations can't just 'start up' instantly. The **start-up time** depends on the type of power station.

NATURAL GAS	OIL	COAL	NUCLEAR
Shortest start-up time			Longest start-up time

a Which type of power station takes longest to start up?

Renewable energy resources are unreliable. The amount of electricity they generate depends on the conditions.

Table 1

Hydroelectric	Upland reservoir could run dry
Wind, waves	Wind and waves too weak on very calm days
Tidal	Height of tide varies both on a monthly and yearly cycle
Solar	No solar energy at night and variable during the day.

The variable demand for electricity is met by:

- using nuclear, coal- and oil-fired power stations to provide a constant amount of electricity (the **base load** demand)

- using gas-fired power stations and pumped-storage schemes to meet daily variations in demand and extra demand in winter

- using renewable energy sources when demand is high and renewables are in operation (e.g. use of wind turbines in winter when wind speeds are suitable)

- using renewable energy sources when demand is low to store energy in pumped storage schemes.

b Which type of power station can be used to satisfy sudden high demands for electricity which occur every day?

Figure 2 A nuclear power station

Activity

The big energy debate

A big energy debate is taking place at your school. Is it possible to generate enough electricity and to reduce the release of greenhouse gases? Your teacher will chair the debate.

Professor Jenny Jones has already spoken in favour of nuclear power and carbon capture. Here is a summary of what she said:

- About a quarter of Britain's electricity comes from nuclear power stations. Many of these stations are due to close by 2020. A new nuclear power station takes several years to build. We need to build more new nuclear power stations – or the lights will go out!

- We can't rely on wind power because when there is no wind, the wind turbines would not generate electricity. We can't rely on solar power at night or in winter. Nuclear power on its own won't give us enough electricity. We have to continue to burn fossil fuels but we can capture and store the greenhouse gases they produce in old oil or gas fields.

The leader of GoGreenUK, Peter Potts, has just finished speaking in favour of renewable energy and energy saving. Here is his summary:

- We need to reduce our greenhouse gas emissions so we have to stop burning fossil fuels. We need to develop renewable energy resources on a much larger scale. We think that we can get most of our electricity from renewable energy devices like wind turbines and solar panels fitted to buildings. We should use public transport more to cut down on how much oil we need.

- If we insulate our homes better and make domestic appliances like fridges more efficient, we wouldn't need as much electricity. We need to use energy more efficiently. Then we wouldn't need new nuclear power stations.

Debate

Now it's your turn to raise points and ask questions. Choose which side of the debate you are on – for, against or undecided!

Some possible points that could be raised are listed below. Add some more points if you think they are reasonable. Your teacher will invite people to ask questions.

- The cost of building and running a nuclear power station is very high. So is the cost of decommissioning it (i.e. taking it out of use).

- Radioactive waste products are dangerous. No one wants a nuclear reactor to be built where they live.

- The capital costs of setting up renewable energy resources are high because lots of expensive equipment is needed to 'collect' large quantities of renewable energy.

- Carbon capture and storage is a new technology and likely to be expensive.

- Most home owners are unlikely to buy energy-saving improvements until energy bills go up even more.

Summary questions

1 Copy and complete using the words below:

coal gas nuclear oil

A .... power station can be started faster than any other type of power station. A .... power station does not produce greenhouse gases. The reserves of .... are likely to last longer than any other fossil fuel reserves. More public transport would reduce our use of ....

2 We need to cut back on fossil fuels to reduce the production of greenhouse gases. What could happen if the only energy we used was:
   a renewable energy
   b nuclear power?

3 a Why are nuclear power stations unsuitable for meeting daily variations in the demand for electricity?
   b What are pumped storage schemes and why are they useful?

Key points

- Gas-fired power stations and pumped-storage stations can meet variations in demand.

- Nuclear, coal and oil power stations can meet base-load demand.

- Nuclear power stations, fossil-fuel power stations using carbon capture and renewable energy are all likely to contribute to future energy supplies.

Summary questions 🄚

1 Answer **a** to **e** using the list of fuels below:

coal   natural gas   oil   uranium   wood

  a Which fuels from the list are fossil fuels?

  b Which fuels from the list cause acid rain?

  c Which fuels release chemical energy when they are used?

  d Which fuel releases the most energy per kilogram?

  e Which fuel produces radioactive waste?

2 a Copy and complete **i** to **iv** using the words below:

hydroelectric   tidal   wave   wind

    i .... power stations trap sea water.
    ii .... power stations trap rain water.
    iii .... generators must be located along the coastline.
    iv .... turbines can be located on hills or offshore.

  b Which renewable energy resource transfers:
    i the kinetic energy of moving air to electrical energy
    ii the gravitational potential energy of water running downhill into electrical energy
    iii the kinetic energy of water moving up and down to electrical energy?

3 a Copy and complete **i** to **iv** using the words below:

coal-fired   geothermal   hydroelectric   nuclear

    i A .... power station does not produce greenhouse gases and uses energy which is from inside the Earth.
    ii A .... power station uses running water and does not produce greenhouse gases.
    iii A .... power station releases greenhouse gases.
    iv A .... power station does not release greenhouse gases but does produce waste products that need to be stored for many years.

  b Wood can be used as a fuel. State whether it is
    i renewable or non-renewable
    ii a fossil fuel or a non-fossil fuel.

4 a Figure 1 shows a landscape showing three different renewable energy resources, numbered 1 to 3. Match each type of energy resource with one of the labels below.

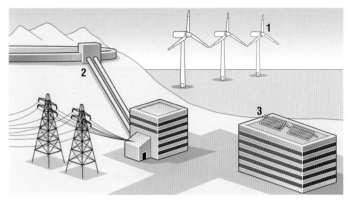

Figure 1 Renewable energy

Hydroelectricity   Solar energy   Wind energy

  b Which of the three resources shown is not likely to produce as much energy as the others if the area is
    i hot, dry and windy
    ii wet and windy?

5 Copy and complete **a** to **d** using the words below. Each word or phrase can be used more than once.

cheaper   more expensive   longer   shorter

  a Wind turbines are .... to build than nuclear power stations and .... to run.

  b Nuclear power stations take .... to decommission than fossil fuel power stations.

  c Solar cells are .... to install than solar heating panels.

  d A gas-fired power station has a .... start-up time compared to a nuclear power station.

6 a i What are transformers used for in the National Grid?
    ii What type of transformer is connected between the generators in the power station and the cables of the grid system?

  b i What can you say about the voltage of the cables in the grid system compared with the voltages at the power station generator and at the mains cables into the home?
    ii What can you say about the current through the grid cables compared with the current from the power station generator?
    iii What is the reason for making the grid voltage different from the generator voltage?

AQA Examination-style questions ⓚ

1 Electricity may be generated in a coal-fired power station.

Copy and complete the following sentences using words from the list below. Each word can be used once, more than once or not at all.

electricity fuel generator steam turbine
water wood

In a coal-fired power station, ... is burned to heat .... This produces .... at high pressure which makes a .... spin round. This then drives a .... that produces .... (6)

2 Various power sources can be used to generate electricity.

Match the power sources in the list with the statements 1 to 4 in the table.
A falling water
B tides
C waves
D wind

	Statement
1	the source of hydroelectric power
2	used with a floating generator
3	very unpredictable and at times may stop altogether
4	will produce a predictable cycle of power generation during the day

(4)

3 A solar cell panel and a solar heating panel work in different ways.

Which statement below is correct?

A A solar cell produces light when it is supplied with electricity.
B A solar cell generates electricity when it is supplied with light.
C A solar heating panel produces heat when it is supplied with electricity.
D A solar heating panel produces electricity when it is supplied with heat. (1)

4 Gas-fired power stations have a shorter start-up time than other power stations. Give **one** reason why is it important to have power stations with a short start-up time. (1)

5 During the night, when demand for electricity is low, a wind farm may be generating a large amount of power. Explain how, by using another type of power station, this power could be stored and used when it is needed. (3)

6 Explain why step-up transformers are used in the National Grid. (2)

7 Palm oil can be used to make a biofuel called biodiesel. Biodiesel can be used instead of the normal type of diesel obtained by refining crude oil.

a Suggest **two** advantages of using biodiesel rather than normal diesel. (2)

b Suggest **two** disadvantages of using biodiesel rather than normal diesel. (2)

8 The pie chart shows the main sources of energy used in power stations in a country last year.

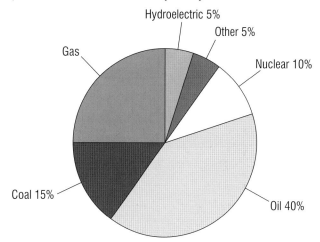

a What fraction of the energy used in power stations was obtained from gas? (2)

b Name **one** source of energy shown that is a fossil fuel. (1)

c Name **one** source of energy shown that is renewable. (1)

d Name **one** source of energy that could be included in the label 'other'. (1)

e Name **one** source of energy that does not cause carbon dioxide to be released when it is used. (1)

9 *In this question you will be assessed on using good English, organising information clearly and using specialist terms where appropriate.*

Power stations that burn fossil fuels produce waste gases that can cause pollution.

Describe the effect that these gases could have on the environment and what could be done to reduce the amount of these gases emitted by power stations. (6)

P1 5.1 The nature of waves

Learning objectives

- What can we use waves for?
- What are transverse waves?
- What are longitudinal waves?
- Which types of waves are transverse and which are longitudinal?

Figure 1 Big waves

∞ **links**

For more information on electromagnetic waves, see P1 6.1 The electromagnetic spectrum.

AQA *Examiner's tip*

You are **not** required to recall the value of the speed of electromagnetic waves through a vacuum. If you need it to answer a question, it will be provided for you.

We use waves to transfer information and we can use them to transfer energy. We can use information transferred by waves in communications, for example when you use a mobile phone or listen to the radio.

There are different types of waves. These include:

- sound waves, water waves, waves on springs and ropes and seismic waves produced by earthquakes. These are examples of **mechanical waves**, which are vibrations that travel through a medium (substance).
- light waves, radio waves and microwaves. These are examples of **electromagnetic waves** which can all travel through a vacuum at the same speed of 300 000 kilometres per second. No medium is needed.

Practical

Observing mechanical waves

Figure 2 shows how we can make waves on a rope by moving one end up and down.

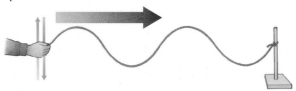

Figure 2 Transverse waves

Tie a ribbon to the middle of the rope. Move one end of the rope up and down. You will see that the waves move along the rope but the ribbon doesn't move along the rope – it just moves up and down. This type of wave is known as a **transverse wave.** We say the ribbon **vibrates** or **oscillates.** This means it moves repeatedly between two positions. When the ribbon is at the top of a wave, we say it is at the **peak** (or crest) of the wave.

Repeat the test with the slinky. You should observe the same effects if you move one end of the slinky up and down.

However, if you push and pull the end of the slinky as shown in Figure 3, you will see a different type of wave, known as a **longitudinal wave.** Notice that there are areas of **compression** (coils squashed together) and areas of **rarefaction** (coils spread further apart) moving along the slinky.

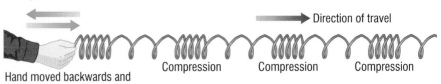

Hand moved backwards and
forwards along the line of the slinky

Figure 3 Making longitudinal waves on a slinky

- How does the ribbon move when you send **longitudinal** waves along the slinky?

Transverse waves

Imagine we send waves along a rope which has a white spot painted on it. The spot would be seen to move up and down without moving along the rope. In other words, the spot would vibrate **perpendicular** (at right angles) to the direction which the waves are moving. The waves on a rope are called **transverse waves** because the vibrations are up and down or from side to side. All electromagnetic waves are transverse waves.

The vibrations of a transverse wave are perpendicular to the direction in which the waves transfer energy.

a State one type of wave that is mechanical and transverse.

Longitudinal waves

The slinky spring in Figure 3 is useful to demonstrate how sound waves travel. When one end of the slinky is pushed in and out repeatedly, vibrations travel along the spring. The vibrations are parallel to the direction in which the waves transfer energy along the spring. Waves that travel in this way are called **longitudinal waves**.

Sound waves are longitudinal waves. When an object vibrates in air, it makes the air around it vibrate as it pushes and pulls on the air. The vibrations (**compressions** and **rarefactions**) which travel through the air are sound waves. The vibrations are along the direction in which the wave travels.

The vibrations of a longitudinal wave are parallel to the direction in which the waves are travelling.

Therefore mechanical waves can be transverse or longitudinal.

b When a sound wave passes through air, what happens to the air particles at a compression?

⚭ links

For more information on sound, see P1 5.5 Wave properties: diffraction, and P1 5.6 Sound.

AQA Examiner's tip

Make sure that you understand the difference between transverse waves and longitudinal waves.

?!? Did you know ...?

When we pluck a guitar string, it vibrates because we send transverse waves along the string. The vibrating string sends sound waves into the surrounding air. The sound waves are longitudinal.

Summary questions

1 Copy and complete **a** to **d** using the words below:
longitudinal   parallel   perpendicular   transverse
   a Sound waves are .... waves.
   b Light waves are .... waves.
   c Transverse waves vibrate .... to the direction of energy transfer of the waves.
   d Longitudinal waves vibrate .... to the direction of energy transfer of the waves.

2 A long rope with a knot tied in the middle lies straight along a smooth floor. A student picks up one end of the rope. This sends waves along the rope.
   a Are the waves on the rope transverse or longitudinal waves?
   b What can you say about:
      i the direction of energy transfer along the rope?
      ii the movement of the knot?

3 Describe how to use a slinky spring to demonstrate to a friend the difference between longitudinal waves and transverse waves.

Key points

- We use waves to transfer energy and transfer information.

- Transverse waves vibrate at right angles to the direction of energy transfer of the waves. All electromagnetic waves are transverse waves.

- Longitudinal waves vibrate parallel to the direction of energy transfer of the waves. A sound wave is an example of a longitudinal wave.

- Mechanical waves, which need a medium (substance) to travel through, may be transverse or longitudinal waves.

P1 5.2

Measuring waves

Learning objectives

- What do we mean by the amplitude of a wave?
- What do we mean by the frequency of a wave?
- What do we mean by the wavelength of a wave?
- What is the relationship between the speed, wavelength and frequency of a wave?

We need to measure waves if we want to find out how much energy or information they carry. Figure 1 shows a snapshot of waves on a rope. The **crests** or peaks are at the top of the wave. The **troughs** are at the bottom. They are equally spaced.

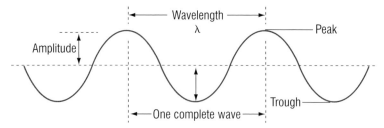

Figure 1 Waves on a rope

- The **amplitude** of the waves is the height of the wave crest or the depth of the wave trough from the middle, which is the position of the rope at rest. **The bigger the amplitude of the waves, the more energy the waves carry.**
- The **wavelength** of the waves is the distance from one wave crest to the next crest.

a Use a millimetre rule to measure the amplitude and the wavelength of the waves in Figure 1.

Frequency

If we made a video of the waves on the rope, we would see the waves moving steadily across the screen. The number of wavecrests passing a fixed point every second is the **frequency** of the waves.

The unit of frequency is the **hertz** (Hz). One wave crest passing each second is a frequency of 1 Hz.

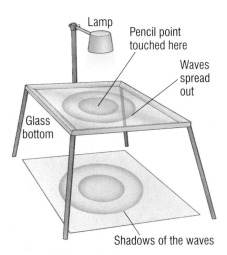

Figure 2 The ripple tank

Wave speed

Figure 2 shows a ripple tank, which is used to study water waves in controlled conditions. We can make straight waves by moving a ruler up and down on the water surface in a ripple tank. Straight waves are called **plane** waves. The waves all move at the same speed and keep the same distance apart.

The **speed** of the waves is the distance travelled by a wave crest or a wave trough every second.

For example, sound waves in air travel at a speed of 340 m/s. In 5 seconds, sound waves travel a distance of 1700 m (= 340 m/s × 5 s).

For waves of constant frequency, the speed of the waves depends on the frequency and the wavelength as follows:

$$\text{wave speed} = \text{frequency} \times \text{wavelength}$$
$$\text{(metre/second, m/s)} \quad \text{(hertz, Hz)} \quad \text{(metre, m)}$$

Maths skills

We can write the wave speed equation as $v = f \times \lambda$

where v = speed, f = frequency, λ = wavelength.

Note: λ is pronounced 'lambda'.

Practical

Making straight (plane) waves

To measure the speed of the waves:

Use a stopwatch to measure the time it takes for a wave to travel from the ruler to the side of the ripple tank.

Measure the distance the waves travel in this time.

Use the equation speed $= \dfrac{\text{distance}}{\text{time}}$ to calculate the speed of the waves.

Observe the effect on the waves of moving the ruler up and down faster. More waves are produced every second and they are closer together.

- Find out if the speed of the waves has changed.

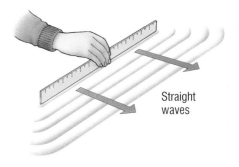

Straight waves

Figure 3 Making water waves

To understand what the wave speed equation means, look at Figure 4. The surfer is riding on the crest of some unusually fast waves.

Suppose the frequency of the waves is 3 Hz and the wavelength of the waves is 4.0 m.

- At this frequency, 3 wave crests pass a fixed point once every second (because the frequency is 3 Hz).
- The surfer therefore moves forward a distance of 3 wavelengths every second or 12 m (= 3 × 4.0 m).

The speed of the surfer is therefore 12 m/s.

This speed is equal to the frequency × the wavelength of the waves: $v = f \times \lambda$.

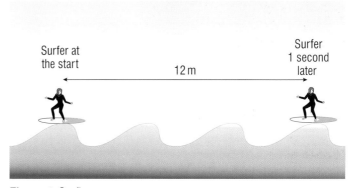

Surfer at the start

12 m

Surfer 1 second later

Figure 4 Surfing

Summary questions

1 Copy and complete **a** to **d** using the words below. Each word can be used more than once.

amplitude   frequency   speed   wavelength

  a The hertz is the unit of ....
  b The distance from one wave crest to the next is the .... of a wave.
  c For water waves, the height of a wave crest above the undisturbed water surface is the .... of the wave.
  d .... × frequency = ...

2 Figure 5 shows a snapshot of a wave travelling from left to right along a rope.
  a Copy Figure 5 and mark on your diagram
    i one wavelength
    ii the amplitude of the waves.
  b Describe the motion of point P on the rope when the wave crest at P moves along by a distance of one wavelength.

Figure 5 A wave on a rope

3 **a** A speedboat on a lake sends waves travelling across a lake at a frequency of 2.0 Hz and a wavelength of 3.0 m. Calculate the speed of the waves.
  b If the waves had been produced at a frequency of 1.0 Hz and travelled at the speed calculated in **a**, what would be their wavelength?   [H]

Key points

- For any wave, its amplitude is the height of the wave crest or the depth of the wave trough from the position at rest.

- For any wave, its frequency is the number of wave crests passing a point in one second.

- For any wave, its wavelength is the distance from one wave crest to the next wave crest. This is the same as the distance from one wave trough to the next wave trough.

- $v = f \times \lambda$

P1 5.3

Wave properties: reflection

If you visit a Hall of Mirrors at a funfair, you will see some strange images of yourself. A tall, thin image or a short, broad image of yourself means you are looking into a mirror that is curved. If you want to see a normal image of yourself, look in a **plane mirror**. Such a mirror is perfectly flat. You see an exact mirror **image** of yourself.

Figure 1 A good image

Learning objectives

- What is the normal in a diagram showing light rays?
- What is an angle of incidence?
- What can we say about the reflection of a light ray at a plane mirror?
- How is an image formed by a plane mirror?

Investigating the reflection of waves using a ripple tank

Light consists of waves. Figure 2 shows how we can investigate the reflection of waves using a ripple tank. The investigations show that when plane (straight) waves reflect from a flat reflector, the reflected waves are at the same angle to the reflector as the incident waves.

The law of reflection

We use light rays to show us the direction light waves are moving in. Figure 3 shows how we can investigate the reflection of a light ray from a ray box using a plane mirror.

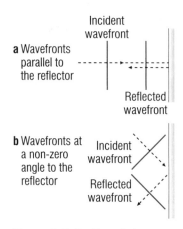

a Wavefronts parallel to the reflector

b Wavefronts at a non-zero angle to the reflector

Figure 2 Reflection of plane waves

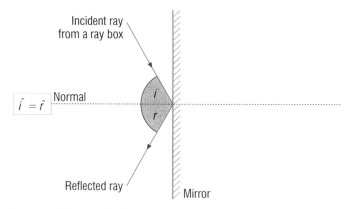

Figure 3 The law of reflection

- The perpendicular line to the mirror is called the **normal**.
- The **angle of incidence** is the angle between the incident ray and the normal.
- The **angle of reflection** is the angle between the reflected ray and the normal.

Measurements show that for any light ray reflected by a plane mirror:

the angle of incidence = the angle of reflection

a If the angle of reflection of a light ray from a plane mirror is 20° what is:
  i the angle of incidence?
  ii the angle between the incident ray and the reflected ray?

Practical

A reflection test

Use a ray box and a plane mirror as shown in Figure 3 to test the law of reflection for different angles of incidence.

Image formation by a plane mirror

Figure 4 shows how an image is formed by a plane mirror. This ray diagram shows the path of two light rays from a point object that reflect off the mirror. The image and the object in Figure 4 are at equal distances from the mirror.

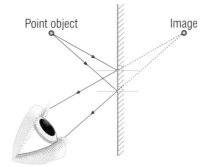

Figure 4 Image formation by a plane mirror

Real and virtual images

The image formed by a plane mirror is virtual, upright (the same way up as the object) and laterally inverted (back to front but not upside down). A **virtual image** can't be projected on to a screen like the movie images that you see at a cinema. An image on a screen is described as a **real image** because it is formed by focusing light rays on to the screen.

b When you use a mirror, is the image real or virtual?

Summary questions

1 Copy and complete **a** to **c** using the words below. Each word can be used once, more than once, or not at all.

equal to   greater than   less than

a The angle of incidence of a light ray at a plane mirror is always .... 90 degrees.

b The angle between the normal and the mirror is always .... 90 degrees.

c The angle of incidence of a light ray at a plane mirror is always .... the angle of reflection of the light ray.

2 A point object O is placed in front of a plane mirror, as shown.

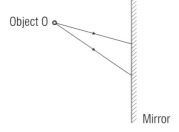

Figure 6

a Complete the path of the two rays shown from O after they have reflected off the mirror.

b i Use the reflected rays to locate the image of O.

ii Show that the image and the object are the same distance from the mirror.

3 Two plane mirrors are placed perpendicular to each other. Draw a ray diagram to show the path of a light ray at an angle of incidence of 60° that reflects off both mirrors.

Key points

● The normal at a point on a mirror is a line drawn perpendicular to the mirror.

● For a light ray reflected by a plane mirror:
  1 The angle of incidence is the angle between the incident ray and the normal.
  2 The angle of reflection is the angle between the reflected ray and the normal.

● The law of reflection states that:
  the angle of incidence = the angle of reflection.

P1 5.4

Wave properties: refraction

When you have your eyes tested, the optician might test different lenses in front of each of your eyes. Each lens changes the direction of light passing through it. This change of direction is known as **refraction**.

Refraction is a property of all forms of waves including light and sound. Figure 1 shows how we can see refraction of waves in a ripple tank.

A glass plate is submerged in a ripple tank. The water above the glass plate is shallower than the water in the rest of the tank. The waves are slower in shallow water than in deep water. If the waves are not parallel to the **boundary**, they change direction when they cross the boundary:

- towards the normal when they cross from deep to shallow water

- away from the normal when they cross from shallow to deep water.

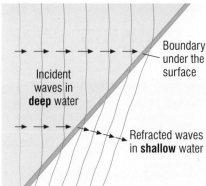

Figure 1 Refraction of water waves

Practical

Investigating refraction of light

Figure 2 shows how you can use a ray box and a rectangular glass block to investigate the refraction of a light ray when it enters glass. The ray changes direction at the boundary between air and glass (unless it is along the normal).

- At the point where the light ray enters the glass, compare the angle of refraction (the angle between the refracted ray and the normal) with the angle of incidence.

You should find that the angle of refraction at the point of entry is always less than the angle of incidence.

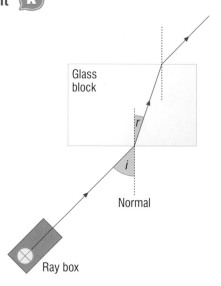

Figure 2 Refraction of light

Refraction rules

Your investigation should show that a light ray:

- changes direction towards the normal when it travels from air into glass. The angle of refraction (r) is smaller than the angle of incidence (i).

- changes direction away from the normal when it travels from glass into air. The angle of refraction (r) is greater than the angle of incidence (i).

a If a light ray enters a rectangular glass block along the normal, does it leave the block along the normal?

AQA Examiner's tip

Remember that angles of incidence, reflection and refraction are always measured between the ray and the normal.

Refraction by a prism

Figure 3 shows what happens when a narrow beam of white light passes through a triangular glass prism. The beam comes out of the prism in a different direction to the incident ray and is split into the colours of the spectrum.

White light contains all the colours of the spectrum. Each colour of light is refracted slightly differently. So the prism splits the light into colours.

Figure 3 Refraction by a prism

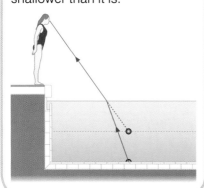

Summary questions

1 Copy and complete **a** to **d** using the words below:

away from greater than less than towards

  a When a light ray travels from air into glass, it refracts .... the normal.

  b When a light ray travels from glass into air, it refracts .... the normal.

  c When a light ray travels from air into glass, the angle of refraction is .... the angle of incidence.

  d When a light ray travels from glass into air, the angle of refraction is .... the angle of incidence.

2 Copy and complete the path of the light ray through each glass object below.

Figure 4

3 A light ray from the bottom of a swimming pool refracts at the water surface. Its angle of incidence is 40 degrees and its angle of refraction is 75 degrees.

  a Draw a diagram to show the path of this light ray from the bottom of the swimming pool into the air above the pool.

  b Use your diagram to explain why the swimming pool appears shallower than it really is when viewed from above.

Key points

● Refraction is the change of direction of waves when they travel across a boundary.

● When a light ray refracts as it travels from air into glass, the angle of refraction is less than the angle of incidence.

● When a light ray refracts as it travels from glass into air, the angle of refraction is more than the angle of incidence.

P1 5.5 | Wave properties: diffraction ⓚ

Learning objectives

- What do we mean by diffraction?
- What is the effect of gap width on the diffraction of waves?
- Why is radio and TV reception often poor in hilly areas?

Diffraction is the spreading of waves when they pass through a gap or move past an obstacle. The waves that pass through the gap or past the edges of the obstacle can spread out. Figure 1 shows waves in a ripple tank spreading out after they pass through two gaps. The effect is most noticeable if the wavelength of the waves is similar to the width of the gap. You can see from Figure 1 that

- the narrower the gap, the more the waves spread out
- the wider the gap, the less the waves spread out.

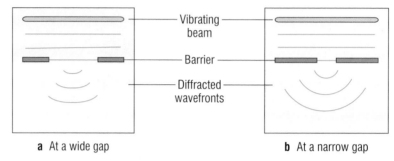

a At a wide gap   b At a narrow gap

Figure 1 Diffraction of waves by a gap: **a** A wide gap **b** A narrow gap

??? Did you know ...?

Sea waves entering a harbour through a narrow entrance spread out after passing through the entrance. Look out for this diffraction effect the next time you visit a harbour.

Practical

Investigating diffraction

Use a ripple tank as in Figure 1 to direct plane waves continuously at a gap between two metal barriers. Notice that the waves spread out after they pass through the gap. In other words, they are diffracted by the gap.

Change the gap spacing and observe the effect on the diffraction of the waves that pass through the gap. You should find that the diffraction of the waves increases as the gap is made narrower, as shown in Figure 1.

Diffraction details

Diffraction of light is important in any optical instrument. The Hubble Space Telescope in its orbit above the Earth has provided amazing images of objects far away in space. Its focusing mirror is 2.4 m in diameter. When it is used, astronomers can see separate images of objects which are far too close to be seen separately using a narrower telescope. Little diffraction occurs when light passes through the Hubble Space Telescope because it is so wide. So its images are very clear and very detailed.

Figure 2 Image of two colliding galaxies taken by the Hubble Space Telescope

Diffraction of ultrasonic waves is an important factor in the design of an ultrasonic scanner. Ultrasonic waves are sound waves at frequencies above the range of the human ear. An ultrasonic scan can be made of a baby in the womb. The ultrasonic waves spread out from a hand-held transmitter and then reflect from the tissue boundaries inside the womb. If the transmitter is too narrow, the waves spread out too much and the image is not very clear.

a The two examples of diffraction above show that both transverse and longitudinal waves can be diffracted. Which is which?

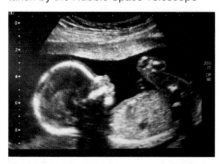

Figure 3 An ultrasonic scan of a baby in the womb

Demonstration

Tests using microwaves

A microwave transmitter and a detector can be used to demonstrate diffraction of microwaves. The transmitter produces microwaves of wavelength 3.0 cm.

1 Place a metal plate between the transmitter and the detector across the path of the microwaves. Microwaves can still be detected behind the metal plate. This is because some microwaves diffract round the edge of the plate.

● Why do the microwaves not go through the metal plates?

2 Place two metal plates separated by a gap across the path of the microwaves, as shown in Figure 4. The microwaves pass through the gap but not through the plates. When the detector is moved along an arc centred on the gap, it detects microwaves that have spread out from the gap.

When the gap is made wider, the microwaves passing through the gap spread out less. The detector needs to be nearer the centre of the arc to detect the microwaves.

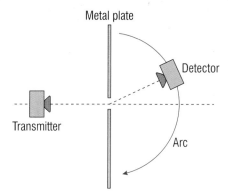

Figure 4 Using microwaves (top view)

Signal problems

People in hilly areas often have poor TV reception. The signal from a TV transmitter mast is carried by radio waves. If there are hills between a TV receiver and the transmitter mast, the signal may not reach the receiver. The radio waves passing the top of a hill are diffracted by the hill but they do not spread enough behind the hill.

b What type of waves carry TV signals?

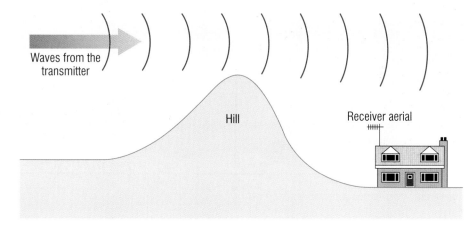

Figure 5 Poor reception

Summary questions

1 Copy and complete **a** and **b** using the words below. Each term can be used once, twice, or not at all.

more than   less than   the same as

a Diffracted waves spread out from a narrow gap .... they would from a wider gap.

b When waves pass through a gap, their wavelength is .... it was before it passed through the gap.

2 a State what is meant by diffraction.

b Explain why the TV reception from a transmitter mast can be poor in hilly areas.

3 A small portable radio inside a room can be heard all along a corridor that runs past the room when its door is open. Explain why it can be heard by someone in the corridor who is not near the door.

Key points

● Diffraction is the spreading out of waves when they pass through a gap or round the edge of an obstacle.

● The narrower a gap is, the greater the diffraction is.

● If radio waves do not diffract enough when they go over hills, radio and TV reception will be poor.

P1 5.6 · Sound

Learning objectives

- What range of frequencies can be detected by the human ear?
- What are sound waves?
- What are echoes?

Figure 1 Making sound waves

??? Did you know ... ?

When you blow a round whistle, you force a small ball inside the whistle to go round and round inside. Each time it goes round, its movement draws air in then pushes it out. Sound waves are produced as a result.

∞ links

For more information on alternating current, see P2 5.1 Alternating current.

Sound waves at frequencies above 20 000 Hz are called ultrasound. For more information on the use of ultrasound in medicine, see P3 1.2 Ultrasound.

Investigating sound waves

Sound waves are easy to produce. Your vocal cords vibrate and produce sound waves every time you speak. Any object vibrating in air makes the layers of air near the object vibrate. These layers make the layers of air further away vibrate. The vibrating object pushes and pulls repeatedly on the air. This sends out the vibrations of the air in waves of compressions and rarefactions. When the waves reach your ears, they make your eardrums vibrate in and out so you hear sound as a result.

The vibrations travelling through the air are sound waves. The waves are longitudinal because the air particles vibrate along the direction in which the waves transfer energy.

Practical

Investigating sound waves

You can use a loudspeaker to produce sound waves by passing alternating current through it. Figure 2 shows how to do this using a signal generator. This is an alternating current supply unit with a variable frequency dial.

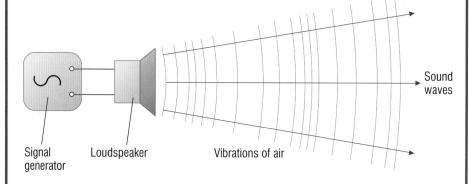

Figure 2 Using a loudspeaker

- If you observe the loudspeaker closely, you can see it vibrating. It produces sound waves as it pushes the surrounding air backwards and forwards.
- If you alter the frequency dial of the signal generator, you can change the frequency of the sound waves.

Find out the lowest and the highest frequency you can hear. Young people can usually hear sound frequencies from about 20 Hz to about 20 000 Hz. Older people in general can't hear frequencies at the higher end of this range.

a Which animal produces sound waves at a higher frequency, an elephant or a mouse?

Sound waves cannot travel through a vacuum. You can test this by listening to an electric bell in a bell jar. As the air is pumped out of the bell jar, the ringing sound fades away.

b What would you notice if the air is let back into the bell jar?

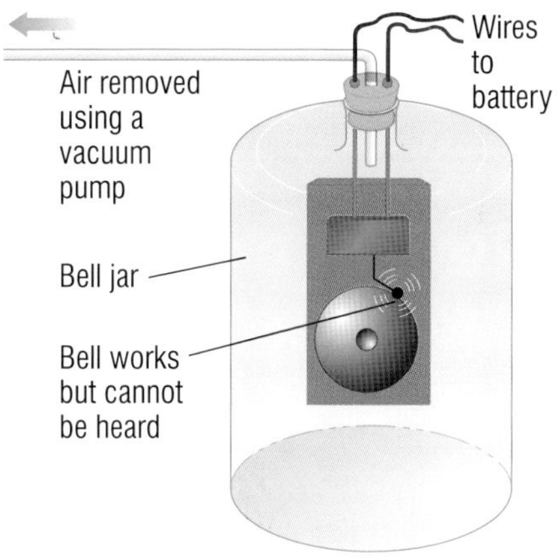

Reflection of sound

Have you ever created an echo? An **echo** is an example of reflection of sound. Echoes can be heard in a large hall or gallery which has bare, smooth walls.

- If the walls are covered in soft fabric, the fabric will absorb sound instead of reflecting it. No echoes will be heard.
- If the wall surface is uneven (not smooth), echoes will not be heard because the reflected sound is 'broken up' and scattered.

c What happens to the energy of the sound waves when they are absorbed by a fabric?

Figure 3 A sound test

Refraction of sound

Sound travels through air at a speed of about 340 m/s. The warmer the air is, the greater the speed of sound. At night you can hear sound a long way from its source. This is because sound waves refract back to the ground instead of travelling away from the ground. Refraction takes place at the boundaries between layers of air at different temperatures. In the daytime, sound refracts upwards, not downwards, because the air near the ground is warmer than air higher up.

Figure 4 Refraction of sound

Summary questions

1 Copy and complete **a** and **b** using the words below:

absorbed   reflected   scattered

a An echo is heard when sound is .... from a bare, smooth wall.
b Sound waves are .... by a rough wall and .... by soft fabric.

2 a What is the highest frequency of sound the human ear can hear?
b Why does a round whistle produce sound at a constant frequency when you blow steadily into it?

3 a A boat is at sea in a mist. The captain wants to know if the boat is near any cliffs so he sounds the horn and listens for an echo. Why would hearing an echo tell him he is near the cliffs?
b Explain why someone in a large cavern can sometimes hear more than one echo of a sound.

Key points

- The frequency range of the normal human ear is from about 20 Hz to about 20 000 Hz.

- Sound waves are vibrations that travel through a medium (substance). They cannot travel through a vacuum (as in space).

- Echoes are due to sound waves reflected from a smooth, hard surface.

P1 5.7 Musical sounds ⓚ

Learning objectives

- What determines the pitch of a note?
- What happens to the loudness of a note as the amplitude increases?
- How are sound waves created by musical instruments?

Figure 1 Making music

What type of music do you like? Whatever your taste in music is, when you listen to it you usually hear sounds produced by specially-designed instruments. Even your voice is produced by a biological organ that has the job of producing sound.

- Musical notes are easy to listen to because they are rhythmic. The sound waves change smoothly and the wave pattern repeats itself regularly.
- Noise consists of sound waves that vary in frequency without any pattern.

  a Name four different vehicles that produce sound through a loudspeaker or a siren.

Practical

Investigating different sounds

Use a microphone connected to an oscilloscope to display the waveforms of different sounds.

Figure 2 Investigating different sound waves

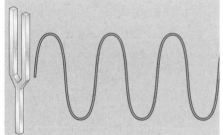

Figure 3 Tuning fork waves

1 Test a tuning fork to see the waveform of a sound of constant frequency.
2 Compare the pure waveform of a tuning fork with the sound you produce when you talk or sing or whistle. You may be able to produce a pure waveform when you whistle or sing but not when you talk.
3 Use a signal generator connected to a loudspeaker to produce sound waves. The waveform on the oscilloscope screen should be a pure waveform.

  b What can you say about the waveform of a sound when you make the sound quieter?

Your investigations should show you that:

- **increasing the loudness** of a sound increases the **amplitude** of the waves. So the waves on the screen become taller.
- **increasing the frequency of a sound** (the number of waves per second) increases its **pitch**. This makes more waves appear on the screen.

Figure 4 shows the waveforms for different sounds from the loudspeaker.

  c How would the waveform in Figure 4a change if the loudness and the pitch are both reduced?

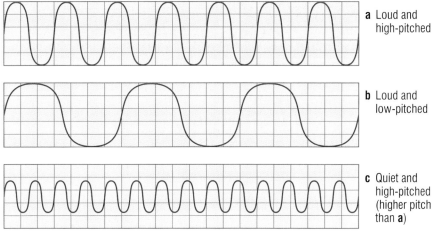

a Loud and high-pitched

b Loud and low-pitched

c Quiet and high-pitched (higher pitch than **a**)

Figure 4 Investigating sounds

Musical instruments

When you play a musical instrument, you create sound waves by making the instrument and the air inside it vibrate. Each new cycle of vibrations makes the vibrations stronger at certain frequencies. We say the instrument **resonates** at these frequencies. Because the instrument and the air inside it vibrate strongly at these frequencies when it is played, we hear recognisable notes of sound from the instrument.

- A wind instrument such as a flute is designed so that the air inside resonates when it is played. You can make the air in an empty bottle resonate by blowing across the top gently.
- A string instrument such as a guitar produces sound when the strings vibrate. The vibrating strings make the surfaces of the instrument vibrate and produce sound waves in the air. In an acoustic guitar, the air inside the hollow body of the guitar (the sound box) vibrates too.
- A percussion instrument such as a drum vibrates and produces sound when it is struck.

Practical

Musical instruments

Investigate the waveform produced by a musical instrument, such as a flute.

You should find its waveform changes smoothly, like the one in Figure 5 – but only if you can play it correctly. The waveform is a mixture of frequencies rather than a single frequency waveform like Figure 3.

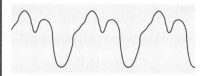

Figure 5 Flute wave pattern

Summary questions

1 Copy and complete **a** to **c** using the words below:

amplitude   frequency   vibrations

a When a drum is struck, sound waves are created by the .... of the drumskin.
b The loudness of a sound is increased by increasing the .... of the sound waves.
c The pitch of a sound is increased by increasing the .... of the sound waves.

2 A microphone and an oscilloscope are used to investigate sound from a loudspeaker connected to a signal generator. What change would you expect to see on the oscilloscope screen if the sound is:
a made louder at the same frequency
b made lower in frequency at the same loudness?

3 **a** How does the note produced by a guitar string change if the string is
i shortened   **ii** tightened?
b Compare the sound produced by a violin with the sound produced by a drum.

Key points

- The pitch of a note increases if the frequency of the sound waves increases.
- The loudness of a note increases if the amplitude of the sound waves increases.
- Vibrations created in an instrument when it is played produce sound waves.

Summary questions ⓚ

1 Figure 1 shows an incomplete ray diagram of image formation by a plane mirror.

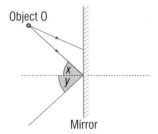

Object O

Mirror

Figure 1

a What can you say about the angles *x* and *y* in the diagram?

b Complete the ray diagram to locate the image.

c What can you say about the distance from the image to the mirror compared with the distance from the object to the mirror?

2 a Figure 2 shows a light ray directed into a glass block.

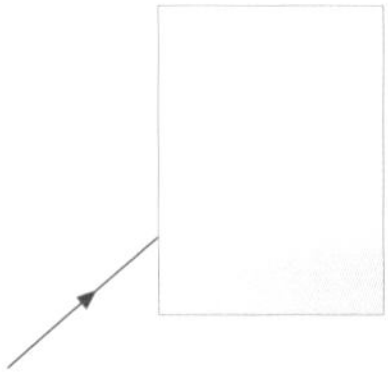

Figure 2

i Sketch the path of the light ray through the block.

ii Describe how the direction of the light ray changes as it passes into and out of the block.

b Copy and complete **i** to **iii** using the words below:

diffraction   reflection   refraction

i The change of the direction of a light ray when it enters a glass block from air is an example of ....

ii The spreading of waves when they pass through a gap is an example of ....

iii The image of an object seen in a mirror is formed because the mirror causes light from the object to undergo ....

3 Copy and complete **a** to **c** using the words below. Each word can be used more than once.

light   radio   sound

a .... waves and .... waves travel at the same speed through air.

b .... waves are longitudinal waves.

c .... waves cannot travel through a vacuum.

4 Waves travel a distance of 30 m across a pond in 10 seconds. The waves have a wavelength of 1.5 m.

a Calculate the speed of the waves.

b Show that the frequency of the waves is 2.0 Hz.

5 a A loudspeaker is used to produce sound waves. In terms of the amplitude of the sound waves, explain why the sound is fainter further away from the loudspeaker.

b A microphone is connected to an oscilloscope. Figure 3 shows the display on the screen of the oscilloscope when the microphone detects sound waves from a loudspeaker.

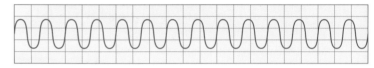

Figure 3

Describe how the waveform displayed on the oscilloscope screen changes if the sound from the loudspeaker is

i made louder

ii reduced in pitch.

6 Copy and complete **a** to **c** using the words below.

absorbed   reflected   scattered   smooth   soft   rough

a An echo is due to sound waves that are .... from a .... wall.

b When sound waves are directed at a .... surface, they are broken up and ....

c When sound waves are directed at a wall covered with a .... material, they are .... and not reflected.

7 a What is the highest frequency the human ear can hear?

b A sound meter is used to measure the loudness of the sound reflected from an object. Describe how you would use the meter and the arrangement shown in Figure 4 to test if more sound is reflected from a board than from a cushion in place of the board. The control knob and a frequency dial can be used to change the loudness and the frequency of the sound from the loudspeaker. List the variables that you would need to keep constant in your test.

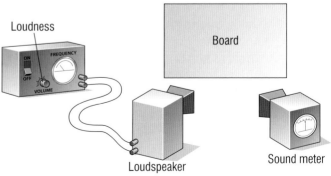

Loudness

Board

Loudspeaker

Sound meter

Figure 4

AQA/Examination-style questions

1 Draw labelled diagrams to explain what is meant by

  a a transverse wave (2)

  b a longitudinal wave. (2)

2 Match the words in the list with the descriptions **1** to **4** in the table.

  A amplitude

  B frequency

  C wave speed

  D wavelength

	Description
1	The distance travelled by a wave crest every second.
2	The distance from one crest to the next.
3	The height of the wave crest from the rest position.
4	The number of crests passing a fixed point every second.

  (4)

3 Which of the following is a correct description of the image in a plane mirror?

  A It is a virtual image

  B It can be focused on to a screen

  C It is on the surface of the mirror

  D It is upside down (1)

4 When a ray of light passes from air into glass it usually changes direction.

  a What is the name given to this effect? (1)

  b Which diagram correctly shows what happens to a ray of light as it passes through a glass block?

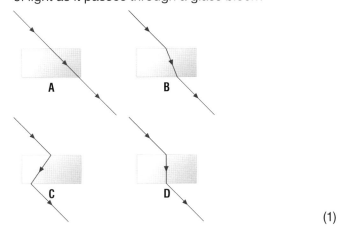

  (1)

5 The diagram represents some water waves passing through a narrow gap.

Give the name of the effect being shown by the waves. When is it most significant? (2)

6 Give one similarity and one difference between a sound wave and a light wave. (2)

7 A sound wave in air has a frequency of 256 Hz. The wavelength of the wave is 1.3 m.

Calculate the speed of sound in air. Write down the equation you use. Show clearly how you work out your answer and give the unit. (2)

8 a Give **one** example of each of the following from everyday life.

    i reflection of light (1)

    ii reflection of sound (1)

    iii refraction of light (1)

    iv diffraction of sound (1)

  b We do not normally see diffraction of light in everyday life.

    Suggest a reason for this. (2)

9 Electromagnetic waves travel at a speed of 300 000 000 m/s.

BBC Radio 4 is transmitted using a wavelength of 1500 metres.

Calculate the frequency of these waves.

Write down the equation you use. Show clearly how you work out your answer and give the unit. **[H]** (3)

10 *In this question you will be assessed on using good English, organising information clearly and using specialist terms where appropriate.*

The diagram shows an oscilloscope trace of the sound wave produced by a musical instrument.

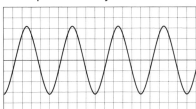

Explain, in detail, how the waveform would change if the instrument produced a sound which was louder and at a higher pitch. (6)

P1 6.1

The electromagnetic spectrum

Learning objectives

- What are the parts of the electromagnetic spectrum?

- How can we calculate the frequency or wavelength of electromagnetic waves?

∞ links

For more information on the use of X-rays in medicine, see P3 1.1 X-rays.

We all use waves from different parts of the **electromagnetic spectrum**. Figure 1 shows the spectrum and some of its uses.

Electromagnetic waves are electric and magnetic disturbances that transfer energy from one place to another.

Electromagnetic waves do not transfer matter. The energy they transfer depends on the **wavelength** of the waves. This is why waves of different wavelengths have different effects. Figure 1 shows some of the uses of each part of the electromagnetic spectrum.

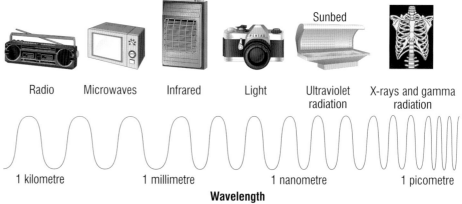

Radio   Microwaves   Infrared   Light   Ultraviolet radiation   X-rays and gamma radiation

1 kilometre   1 millimetre   1 nanometre   1 picometre

Wavelength

(1 nanometre = 0.000 001 millimetres, 1 picometre = 0.001 nanometres)

Figure 1 The spectrum is continuous. The frequencies and wavelengths at the boundaries are approximate as the different parts of the spectrum are not precisely defined.

Waves from different parts of the electromagnetic spectrum have different wavelengths.

- Long-wave radio waves have wavelengths as long as 10 km.

- X-rays and gamma rays have wavelengths as short as a millionth of a millionth of a millimetre (= 0.000 000 000 001 mm).

  a Where in the electromagnetic spectrum would you find waves of wavelength 10 millimetres?

The speed of electromagnetic waves

All electromagnetic waves travel at a speed of 300 million m/s through space or in a vacuum. This is the distance the waves travel each second.

We can link the speed of the waves to their frequency and wavelength using the **wave speed** equation:

$$v = f \times \lambda$$

Where:
v = wave speed in metres per second, m/s
f = frequency in hertz, Hz
λ = wavelength in metres, m

AQA *Examiner's tip*

The spectrum of visible light covers just a very tiny part of the electromagnetic spectrum. The wavelength decreases from radio waves to gamma rays.

∞ links

For more information on the wave speed equation, look back at P1 5.2 Measuring waves.

b Work out the wavelength of electromagnetic waves of frequency 200 million Hz.

c Work out the frequency of electromagnetic waves of wavelength 1500 m.

Maths skills

We can work out the wavelength if we know the frequency and the wave speed. To do this, we rearrange the equation into:

$$\lambda = \frac{v}{f}$$

We can work out the frequency if we know the wavelength and the wave speed. To do this, we rearrange the equation into:

$$f = \frac{v}{\lambda}$$

Where:

v = speed in metres per second, m/s

f = frequency in hertz, Hz

λ = wavelength in metres, m.

Worked example

A mobile phone gives out electromagnetic waves of frequency 900 million Hz. Calculate the wavelength of these waves. The speed of electromagnetic waves in air = 300 million m/s.

Solution

$$\text{wavelength } \lambda \text{ (in metres)} = \frac{\text{wave speed } v \text{ (in m/s)}}{\text{frequency } f \text{ (in Hz)}} =$$

$$\frac{300\,000\,000\,\text{m/s}}{900\,000\,000\,\text{Hz}} = 0.33\,\text{m}$$

Energy and frequency

The wave speed equation shows us that the shorter the wavelength of the waves, the higher their frequency is. The energy of the waves increases as the frequency increases. The energy and frequency of the waves therefore increases from radio waves to gamma rays as the wavelength decreases.

Summary questions

1 Copy and complete **a** to **c** using the words below:

greater than   smaller than   the same as

  a The wavelength of light waves is .... the wavelength of radio waves.

  b The speed of radio waves in a vacuum is .... the speed of gamma rays.

  c The frequency of X-rays is .... the frequency of infrared radiation.

2 Fill in the missing parts of the electromagnetic spectrum in the list below.

    radio   ...a...   infrared   visible   ...b...   X-rays   ...c...

3 Electromagnetic waves travel through space at a speed of 300 million metres per second. Calculate:

  a the wavelength of radio waves of frequency 600 million Hz

  b the frequency of microwaves of wavelength 0.30 m.

4 A distant star explodes and emits light and gamma rays simultaneously. Explain why the gamma rays and the light waves reach the Earth at the same time.

Key points

● The electromagnetic spectrum (in order of decreasing wavelength, increasing frequency and energy) is:

  – radio waves

  – microwaves

  – infrared radiation

  – light

  – ultraviolet radiation

  – gamma radiation and X-rays.

● The wave speed equation is used to calculate the frequency or wavelength of electromagnetic waves.

P1 6.2

Light, infrared, microwaves and radio waves

Learning objectives

- What is white light?
- What do we use infrared radiation, microwaves and radio waves for?
- What are the hazards of these types of electromagnetic radiation?

Light and colour

Light from ordinary lamps and from the Sun is called **white light**. This is because it has all the colours of the visible spectrum in it. The wavelength increases across the spectrum as you go from violet to red.

You see the colours of the spectrum when you look at a rainbow. You can also see them if you use a glass prism to split a beam of white light.

Photographers need to know how shades and colours of light affect the photographs they take.

1 **In a film camera**, the light is focused by the camera lens on to a light-sensitive film. The film then needs to be developed to see the image of the objects that were photographed.

2 **In a digital camera**, the light is focused by the lens on to a sensor. This consists of thousands of tiny light-sensitive cells called **pixels**. Each pixel gives a dot of the image. The image can be seen on a small screen at the back of the camera. When a photograph is taken, the image is stored electronically on a memory card.

a Why is a 10 million pixel camera better than a 2 million pixel camera?

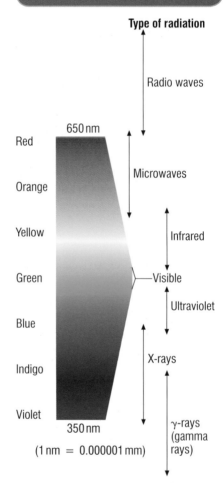

Figure 1 The electromagnetic spectrum with an expanded view of the visible range

∞ links

For more information on the uses of light in the camera and other optical instruments, see P3 1.5 Lenses and P3 1.6 Using lenses.

For more information on infrared radiation, look back at P1 1.1 Infrared radiation.

Infrared radiation

All objects emit infrared radiation.

- The hotter an object is, the more infrared radiation it emits.
- Infrared radiation is absorbed by the skin. It damages or kills skin cells because it heats up the cells.

b Where does infrared radiation lie in the electromagnetic spectrum?

Infrared devices

- **Optical fibres** in communications systems use infrared radiation instead of light. This is because infrared radiation is absorbed less than light in the glass fibres.
- **Remote control handsets** for TV and video equipment transmit signals carried by infrared radiation. When you press a button on the handset, it sends out a sequence of infrared pulses.
- **Infrared scanners** are used in medicine to detect 'hot spots' on the body surface. These hot areas can mean the underlying tissue is unhealthy.
- You can use **infrared cameras** to see people and animals in darkness.

c Does infrared radiation pass through a thin sheet of paper?

Microwaves

Microwaves lie between radio waves and infrared radiation in the electromagnetic spectrum. They are called '**micro**waves' because they are shorter in wavelength than radio waves.

We use microwaves for communications, e.g. **satellite TV**, because they can pass through the atmosphere and reach satellites above the Earth. We also use them to beam signals from one place to another. That's because microwaves don't spread out as much as radio waves. Microwaves (as well as radio waves) are used to carry **mobile phone** signals.

Radio waves

Radio wave frequencies range from about 300 000 Hz to 3000 million Hz (where microwave frequencies start). Radio waves are longer in wavelength and lower in frequency than microwaves.

As explained in P1 6.3, we use radio waves to carry **radio, TV and mobile phone** signals.

We can also use radio waves instead of cables to connect a computer to other devices such as a printer or a 'mouse'. For example, Bluetooth-enabled devices can communicate with each other over a range of about 10 metres. No cables are needed – just a Bluetooth radio in each device and the necessary software. Such wireless connections work at frequencies of about 2400 million hertz, and they operate at low power.

Bluetooth was set up by the electronics manufacturers. They realised the need to agree on the radio frequencies to be used for common software.

d If wireless-enabled devices operated at higher power, how would their range be affected?

Practical

Testing infrared radiation

Can infrared radiation pass through paper? Use a remote handset to find out.

Demonstration

Demonstrating microwaves

Look at the demonstration.

● What does this show?

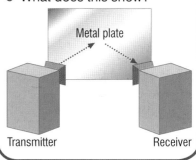

Transmitter   Receiver

Metal plate

Key points

● White light contains all the colours of the visible spectrum.

● Infrared radiation is used for carrying signals from remote handsets and inside optical fibres.
We use microwaves to carry satellite TV programmes and mobile phone calls.
Radio waves are used for radio and TV broadcasting, radio communications and mobile phone calls.

● Different types of electromagnetic radiation are hazardous in different ways. Microwaves and radio waves can cause internal heating. Infrared radiation can cause skin burns.

Summary questions

1 Copy and complete **a** and **b** using the words below:

infrared radiation   visible light   microwaves   radio waves

  a In a TV set, the aerial detects .... and the screen emits ....

  b A satellite TV receiver detects ...., which pass through the atmosphere, unlike ...., which have a shorter wavelength.

2 Mobile phones use electromagnetic waves in a wavelength range that includes short-wave radio waves and microwaves.

  a What would be the effect on mobile phone users if remote control handsets operated in this range as well?

  b Why do our emergency services use radio waves in a wavelength range that no else is allowed to use?

3 The four devices listed below each emit a different type of electromagnetic radiation. State the type of radiation each one emits.

  a A TV transmitter mast.

  b A TV satellite.

  c A TV remote handset.

  d A TV receiver.

P1 6.3

Communications

Learning objectives

- Why do we use radio waves of different frequencies for different purposes?
- Which waves do we use for satellite TV?
- How can we evaluate whether or not mobile phones are safe to use?
- What are optical fibres?

Figure 1 Sending microwave signals to a satellite

??? Did you know ... ?

Satellite TV signals are carried by microwaves. We can detect the signals on the ground because they pass straight through a layer of ionised gas in the upper atmosphere. This layer reflects lower-frequency radio waves.

Figure 2 A mobile phone mast

Radio communications

Radio waves are emitted from an aerial when we apply an alternating voltage to the aerial. The frequency of the radio waves produced is the same as the frequency of the alternating voltage.

When the radio waves pass across a receiver aerial, they cause a tiny alternating voltage in the aerial. The frequency of the alternating voltage is the same as the frequency of the radio waves received. The aerial is connected to a loudspeaker. The alternating voltage from the aerial is used to make the loudspeaker send out sound waves.

The radio and microwave spectrum is divided into **bands** of different wavelength ranges. This is because the shorter the wavelength of the waves:

- the more information they can carry
- the shorter their range (due to increasing absorption by the atmosphere)
- the less they spread out (because they diffract less).

Radio wavelengths

Microwaves and radio waves of different wavelengths are used for different communications purposes. Examples are given below.

- **Microwaves** are used for satellite phone and TV links and satellite TV broadcasting. This is because microwaves can travel between satellites in space and the ground. Also, they spread out less than radio waves do so the signal doesn't weaken as much.
- **Radio waves of wavelengths less than about 1 metre** are used for TV broadcasting from TV masts because they can carry more information than longer radio waves.
- **Radio waves of wavelengths from about 1 metre up to about 100 m** are used by local radio stations (and for the emergency services) because their range is limited to the area round the transmitter.
- **Radio waves of wavelengths greater than 100 m** are used by national and international radio stations because they have a much longer range than shorter wavelength radio waves.

   a Why do microwaves spread out less than radio waves do?

Mobile phone radiation k

A mobile phone sends a radio signal from your phone. The signal is picked up by a local mobile phone mast and is sent through the phone network to the other phone. The 'return' signal goes through the phone network back to the mobile phone mast near you and then on to you. The signals to and from your local mast are carried by radio waves of different frequencies.

The radio waves to and from a mobile phone have a wavelength of about 30 cm. Radio waves at this wavelength are not quite in the microwave range but they do have a similar heating effect to microwaves. So they are usually referred to as microwaves.

   b Why should signals to and from a mobile phone be at different frequencies?

How Science Works

Is mobile phone radiation dangerous?

The radiation is much weaker than the microwave radiation in an oven. But when you use a mobile phone, it is very close to your brain. Some scientists think the radiation might affect the brain. As children have thinner skulls than adults, their brains might be more affected by mobile phone radiation. A UK government report published in May 2000 recommended that the use of mobile phones by children should be limited.

Mobile phone hazards

Here are some findings by different groups of scientists:

The short-term memory of volunteers using a mobile phone was found to be unaffected by whether the phone was on or off.

The brains of rats exposed to microwaves were found to respond less to electrical impulses than the brains of unexposed rats.

Mice exposed to microwaves by some scientists developed more cancers than unexposed mice. Other scientists were unable to confirm this effect.

A survey of mobile phone users in Norway and Sweden found they experienced headaches and fatigue. No control group of people who did not use a mobile phone was surveyed.

- What conclusions do you draw from the evidence above?
- Suggest how researchers could improve the validity of any conclusions we can draw.

Optical fibre communications

Optical fibres are very thin glass fibres. We use them to transmit signals carried by light or infrared radiation. The light rays can't escape from the fibre. When they reach the surface of the fibre, they are reflected back into the fibre.

In comparison with radio waves and microwaves:

- optical fibres can carry much more information – this is because light has a much smaller wavelength than radio waves so can carry more pulses of waves
- optical fibres are more secure because the signals stay in the fibre.

  c Why are signals in an optical fibre more secure than radio signals?

Summary questions

1 Copy and complete **a** to **c** using the words below. Each term can be used more than once.

  infrared radiation   microwaves   radio waves

  a Mobile phone signals are carried by ....
  b Optical fibre signals are carried by ....
  c A beam of .... can travel from the ground to a satellite but a beam of .... cannot if its frequency is below 30 MHz.

2 **a** Why could children be more affected by mobile phone radiation than adults?

  b Why can light waves carry more information than radio waves?

3 Explain why microwaves are used for satellite TV and radio waves for terrestrial TV.

⚭ links

For more information on how optical fibres are used in the endoscope to see inside the body, see P3 1.4 The endoscope.

Demonstration

Demonstrating an optical fibre

Observe light shone into an optical fibre. You should see the reflection of light inside an optical fibre. This is known as total internal reflection.

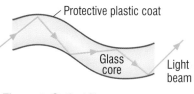

Figure 3 Optical fibres

Key points

- Radio waves of different frequencies are used for different purposes because the wavelength (and therefore frequency) of waves affects:
  - how far they can go
  - how much they spread
  - how much information they can carry.

- Microwaves are used for satellite TV signals.

- Further research is needed to evaluate whether or not mobile phones are safe to use.

- Optical fibres are very thin transparent fibres that are used to transmit signals by light and infrared radiation.

P1 6.4

The expanding universe

Figure 1 Galaxies

The Doppler effect

The **Doppler effect** is the change in the observed wavelength (and frequency) of waves due to the motion of the source of the waves. Christian Doppler discovered the effect in 1842 using sound waves. He demonstrated it by using an open railway carriage filled with trumpeters. The spectators had to listen to the pitch of the trumpets as they sped past. Another example, explained below, is the red-shift of the light from a distant galaxy moving away from us.

Red-shift

We live on the third rock out from a middle-aged star on the outskirts of a big galaxy we call the Milky Way. The galaxy contains about 100 000 million stars. Its size is about 100 000 light years across. This means that light takes 100 000 years to travel across it. But it's just one of billions of galaxies in the universe. The furthest galaxies are about 13 000 million light years away!

> **a** Why do stars appear as points of light?

We can find out lots of things about stars and galaxies by studying the light from them. We can use a prism to split the light into a spectrum. The wavelength of light increases across the spectrum from blue to red. We can tell from its spectrum if a star or galaxy is moving towards us or away from us. This is because:

- the light waves are stretched out if the star or galaxy is moving away from us. The wavelength of the waves is increased. We call this a **red-shift** because the spectrum of light is shifted towards the red part of the spectrum.
- the light waves are squashed together if the star or galaxy is moving towards us. The wavelength of the waves is reduced. We call this a **blue-shift** because the spectrum of light is shifted towards the blue part of the spectrum.

The dark spectral lines shown in Figure 2 are caused by absorption of light by certain atoms such as hydrogen that make up a star or galaxy. The position of these lines tells us if there is a shift and if so, whether it is a red-shift or a blue-shift.

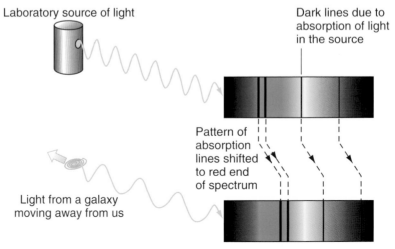

Laboratory source of light

Dark lines due to absorption of light in the source

Pattern of absorption lines shifted to red end of spectrum

Light from a galaxy moving away from us

Figure 2 Red-shift

The bigger the shift, the more the waves are squashed together or stretched out. So the faster the star or galaxy must be moving towards or away from us. In other words:

the faster a star or galaxy is moving (relative to us), the bigger the shift is.

b What do you think happens to the wavelength of the light from a star that is moving towards us?

Expanding universe

In 1929, Edwin Hubble discovered that:

1 the light from distant galaxies was red-shifted
2 the further a galaxy is from us, the bigger its red-shift is.

He concluded that:

● the distant galaxies are moving away from us (i.e. receding)
● the greater the distance a galaxy is from us, the greater the speed is at which it is moving away from us (its speed of recession).

Why should the distant galaxies be moving away from us? We have no special place in the universe, so all the distant galaxies must be moving away from each other. In other words, **the whole universe is expanding**.

c Galaxy X is 2000 million light years away. Galaxy Y is 4000 million light years away. Which galaxy, X or Y, has the bigger red-shift?

Summary questions

1 Copy and complete **a** to **d** using the words below:

approaching   expanding   orbiting   receding

a The Earth is .... the Sun.
b The universe is ....
c The distant galaxies are ....
d A blue-shift in the light from a star would tell us it is ....

2 **a** Put these objects in order of increasing size:

Andromeda galaxy   Earth   Sun   universe

b Copy and complete **i** and **ii** using the words below:

galaxy   star   red-shift   planet

i The Earth is a .... in orbit round a .... called the Sun.
ii There is a .... in the light from a distant ....

3 Galaxy X has a larger red-shift than galaxy Y.
a Which galaxy, X or Y, is
i nearer to us
ii moving away faster?
b The light from the Andromeda galaxy is not red-shifted. What does this tell you about Andromeda?

Key points

● The red-shift of a distant galaxy is the shift to longer wavelengths of the light from it because the galaxy is moving away from us.

● The faster a distant galaxy is moving away from us, the greater its red-shift is.

● All the distant galaxies show a red-shift. The further away a distant galaxy is from us, the greater its red-shift is.

● The distant galaxies are all moving away from us because the universe is expanding.

P1 6.5

The Big Bang ⓚ

The universe is expanding, but what is making it expand? The **Big Bang theory** was put forward to explain the expansion. This states that:

- the universe is expanding after exploding suddenly in a Big Bang from a very small initial point
- space, time and matter were created in the Big Bang.

Many scientists disagreed with the Big Bang theory. They put forward an alternative theory, the Steady State theory. The scientists said that the galaxies are being pushed apart. They thought that this is caused by matter entering the universe through 'white holes' (the opposite of black holes).

Which theory is weirder – everything starting from a Big Bang or matter leaking into the universe from outside? Until 1965, most people supported the Steady State theory.

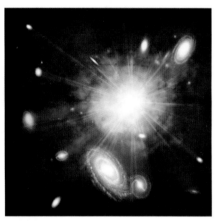

Figure 1 The Big Bang

⁇ Did you know ...?

You can use an analogue TV to detect background microwave radiation very easily – just disconnect your TV aerial. The radiation causes lots of fuzzy spots on the screen.

⚙ *How Science Works*

Evidence for the Big Bang

Scientists had two conflicting theories about the evolution of the universe: it was in a Steady State or it began at some point in the past with a Big Bang. Both theories could explain why the galaxies are moving apart, so scientists needed to find some way of selecting which theory was correct. They worked out that if the universe began in a Big Bang then there should have been high-energy electromagnetic radiation produced. This radiation would have 'stretched' as the universe expanded and become lower-energy radiation. Experiments were devised to look for this trace energy as extra evidence for the Big Bang model.

It was in 1965 that scientists first detected microwaves coming from every direction in space. The existence of this **cosmic microwave background radiation** can only be explained by the Big Bang theory.

The cosmic microwave background radiation is not as perfectly evenly spread as scientists thought it should be. Their model of the early universe needs to be developed further by gathering evidence and producing theories to explain this 'unevenness' in the early universe.

a How do scientists decide between two conflicting theories?

Cosmic microwave background radiation

- It was created as high-energy gamma radiation just after the Big Bang.
- It has been travelling through space since then.
- As the universe has expanded, it stretched out to longer and longer wavelengths and is now microwave radiation.
- It has been mapped out using microwave detectors on the ground and on satellites.

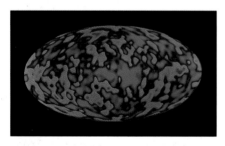

Figure 2 A microwave image of the universe from, the Cosmic Background Explorer satellite

b What will happen to cosmic microwave background radiation as the universe expands?

How Science Works

The future of the universe

Will the universe expand forever? Or will the force of gravity between the distant galaxies stop them from moving away from each other? The answer to this question depends on their total mass and how much space they take up – in other words, the density of the universe.

- If the density of the universe is less than a certain amount, it will expand forever. The stars will die out and so will everything else as the universe heads for a Big Yawn!

- If the density of the universe is more than a certain amount, it will stop expanding and go into reverse. Everything will head for a Big Crunch!

Recent observations by astronomers suggest that the distant galaxies are accelerating away from each other. These observations have been checked and confirmed by other astronomers. So astronomers have concluded that the expansion of the universe is accelerating. It could be we're in for a Big Ride followed by a Big Yawn.

The discovery that the distant galaxies are accelerating is puzzling astronomers. Scientists think some unknown source of energy, now called 'dark energy', must be causing this accelerating motion. The only known force on the distant galaxies, the force of gravity, can't be used to explain 'dark energy' as it is an attractive force and so acts against their outward motion away from each other.

c What could you say about the future of the universe if the galaxies were slowing down?
d i An object released above the ground accelerates as it falls. What makes it accelerate?
ii Why are scientists puzzled by the observation that the distant galaxies are accelerating?

links

For more information on how the chemical elements formed after the Big Bang, see P2 7.6 How the chemical elements formed.

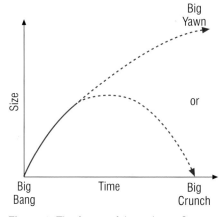

Figure 3 The future of the universe?

Summary questions

1 Copy and complete **a** to **d** using the words below:

created   detected   expanded   stretched

    a The universe was .... in an explosion called the Big Bang.
    b The universe .... suddenly in and after the Big Bang.
    c Microwave radiation from space can be .... from all directions.
    d Radiation created just after the Big Bang has been .... by the expansion of the universe and is now microwave radiation.

2 Put the following events A–D in the correct time sequence:
    A The distant galaxies were created.
    B Cosmic microwave background radiation was first detected.
    C The Big Bang happened.
    D The expansion of the universe began.

3 **a** Why do astronomers think that the expansion of the universe is accelerating?
    b What would have been the effect on the expansion of the universe if its density had been greater than a certain value?

Key points

- The universe started with the Big Bang, a massive explosion from a very small point.

- The universe has been expanding ever since the Big Bang.

- Cosmic microwave background radiation (CMBR) is electromagnetic radiation created just after the Big Bang.

- CMBR can only be explained by the Big Bang theory.

Summary questions (k)

1 a Place the four different types of electromagnetic waves listed below in order of increasing wavelength.
   A Infrared waves
   B Microwaves
   C Radio waves
   D Gamma rays

b The radio waves from a local radio station have a wavelength of 3.3 metres in air and a frequency of 91 million Hz.
   i Write down the equation that links frequency, wavelength and wave speed.
   ii Calculate the speed of the radio waves in air.

2 In P1 6.1 you will find the typical wavelengths of electromagnetic waves. Give the type of electromagnetic wave for each of the wavelengths given.
   A 0.0005 mm
   B 1 millionth of 1 millionth of 1 mm
   C 10 cm
   D 1000 m

3 Copy and complete **a** and **b** using the words below:

microwave   mobile phone   radio waves   TV

a A .... beam can travel from a ground transmitter to a satellite, but a beam of .... cannot if its frequency is below 30 MHz.

b .... signals and .... signals always come from a local transmitter.

4 Mobile phones send and receive signals using electromagnetic waves near or in the microwave part of the electromagnetic spectrum.

a Name the part of the electromagnetic spectrum which has longer wavelengths than microwaves have.

b Which two parts of the electromagnetic spectrum may be used to send information along optical fibres?

c New mobile phones are tested for radiation safety and given an SAR value before being sold. The SAR is a measure of the energy per second absorbed by the head while the phone is in use. For use in the UK, SAR values must be less than 2.0 W/kg. SAR values for two different mobile phones are given below.
   Phone A   0.2 W/kg
   Phone B   1.0 W/kg
   i What is the main reason why mobile phones are tested for radiation safety?
   ii Which phone, A or B, is safer? Give a reason for your answer.

5 Light from a distant galaxy has a change of wavelength due to the motion of the galaxy.

a Is this change of wavelength an increase or a decrease?

b What is the name for this change of wavelength?

c Which way is the galaxy moving?

d What would happen to the light it gives out if it were moving in the opposite direction?

6 a Galaxy A is further from us than galaxy B.
   i Which galaxy, A or B, produces light with a greater red-shift?
   ii Galaxy C gives a bigger red-shift than galaxy A. What can we say about the distance to galaxy C compared with galaxy A?

b All the distant galaxies are moving away from each other.
   i What does this tell us about the universe?
   ii What does it tell us about our place in the universe?

7 The diagram shows two galaxies X and Y, which have the same diameter.

a i Which galaxy, X or Y, is further from Earth? Give a reason for your answer.
   ii Which galaxy, X or Y, produces the larger red-shift?

b A third galaxy Z seen from Earth appears to be the same size as X but it has a larger red-shift than X.
   i What can you say about the speed at which Z is moving away from us, compared with the speed at which X is moving away?
   ii What can you deduce about the distance to Z compared with the distance to X? Give a reason for your answer.

AQA Examination-style questions

1 Electromagnetic waves can travel through the vacuum of space.

Copy and complete the following sentences using words from the list below. Each word can be used once, more than once or not at all.

energy   frequency   speed   wavelength

All electromagnetic waves travel at the same .... in a vacuum. They do not carry material, but they do carry .... Gamma waves have the greatest .... and the smallest ....   (4)

2 Different types of electromagnetic waves have different uses in communications.

Match the type of electromagnetic wave in the list with its use **1** to **4** in the table.

A infrared

B microwaves

C radio waves

D visible light

	These waves are used for
1	producing images in a video camera
2	mobile phone communication
3	television remote controls
4	carrying terrestrial television signals

                                                                                      (4)

3 Microwaves are used for communications. They can be used to send signals to other parts of the world by means of a satellite.

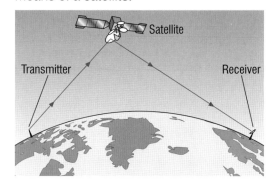

a Give **one** reason why the receiver shown in the diagram can only pick up the signal if a satellite is used.   (1)

b Explain why microwaves are used rather than:

  i long wave radio waves   (1)

  ii visible light   (2)

4 The diagram shows a ray passing through an optical fibre.

a Name **two** types of electromagnetic wave that can travel along an optical fibre.   (2)

b Suggest **two** advantages of sending signals along an optical fibre rather than using electrical signals in a metal wire.   (2)

5 Scientists have developed a theory about the universe called the 'Big Bang' theory. This theory is supported by evidence. Part of this evidence is the existence of cosmic background microwave radiation.

a What does the 'Big Bang' theory state?

  A The universe began with a massive explosion.

  B The universe will end with a massive explosion.

  C The universe began from a very small initial point.

  D The universe will end at a very small initial point. (1)

b Where does cosmic background microwave radiation come from?

  A people who use microwave ovens to heat food

  B gamma radiation created just after the Big Bang

  C mobile phone transmitters

  D radioactive rocks in the Earth's crust.   (1)

c If a scientist finds new evidence that does not support the Big Bang theory what should other scientists do?

  A Change the theory immediately.

  B Check the new evidence to make sure it is reproducible.

  C Ignore the new evidence.

  D Try to discredit the scientist who found the new evidence.   (1)

6 Red-shift from distant galaxies provides evidence for the Big Bang theory.

What is meant by red-shift?   (2)

7 *In this question you will be assessed on using good English, organising information clearly and using specialist terms where appropriate.*

Explain how red-shift provides evidence for the Big Bang theory.   (6)

1 The diagram shows a solar heating panel on the roof of a house.

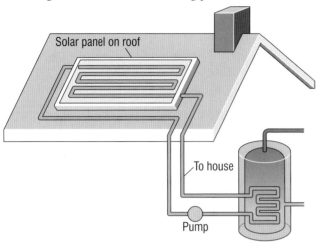

The solar heating panel consists of a flat box backed by a metal plate. The box contains copper pipes filled with a liquid. Liquid pumped through the pipes is heated as it passes through the panel.

Copy and complete the following sentences using words from the list below. Each word can be used once, more than once or not at all.

black   white   transparent   conduction   convection   insulation   radiation

There is a .... cover on top of the panel that allows infrared .... through to heat the metal plate. The metal plate is coloured .... for maximum absorption. There is a sheet of .... under the plate to stop heat loss by .... through the back of the panel.

(5)

2 **a** Microwaves are one type of electromagnetic wave.
   i   Which type of electromagnetic wave has a lower frequency than microwaves? (1)
   ii What do all types of electromagnetic wave transfer from one place to another?
   (1)

   b The picture shows a tennis coach using a speed gun to measure how fast the player serves the ball.

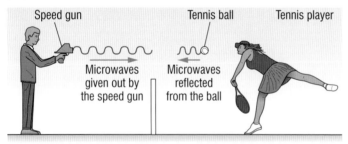

   i   The microwaves transmitted by the speed gun have a frequency of 2.4×10^{10} Hz and travel through the air at 3.0×10^8 m/s.
   Calculate the wavelength in metres of the microwaves emitted from the speed gun. Write down the equation you use. Show clearly how you work out your answer and give the unit. (2)
   ii Some of the microwaves transmitted by the speed gun are absorbed by the ball. What effect will the absorbed microwaves have on the ball? (2)
   iii Some of the microwaves transmitted by the speed gun are reflected from the moving ball back towards the speed gun. Describe how the wavelength and frequency of the microwaves change as they are reflected from the moving ball.
   (2)

   AQA, 2009

AQA Examiner's tip

Read through the whole passage first, to get the sense of it, before trying to put the words in.

AQA Examiner's tip

Make sure you have a way of remembering the order of the waves in the electromagnetic spectrum. For example, **g**ood **x**ylophones **u**pset **v**iolins **i**n **m**usical **r**ecitals for **g**amma, **X**-ray, **u**ltraviolet, **v**isible, **i**nfrared, **m**icrowave, **r**adio wave.

3 The diagram shows how electricity is distributed from power stations to consumers.

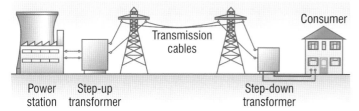

a i What name is given to the network of cables and transformers that links power stations to consumers? (1)

ii What does a step-up transformer do? (1)

iii Explain why step-up transformers are used in the electricity distribution system. (2)

b Most of the world's electricity is generated in power stations that burn fossil fuels. State **one** environmental problem that burning fossil fuels produces. (1)

c Electricity can be generated using energy from the wind. A company wants to build a new wind farm. Not everyone thinks that this is a good idea.

i What arguments could the company give to persuade people that a wind farm is a good idea? (2)

ii What reasons may be given by the people who think that wind farms are **not** a good idea? (2)

AQA, 2007

4 *In this question you will be assessed on using good English, organising information clearly and using specialist terms where appropriate.*

The diagram shows a vacuum flask. The flask can be used to keep hot liquids hot and cold liquids cold.

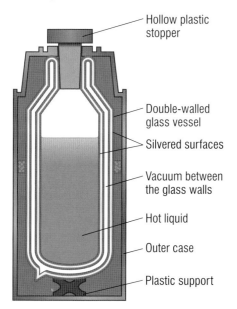

- Hollow plastic stopper
- Double-walled glass vessel
- Silvered surfaces
- Vacuum between the glass walls
- Hot liquid
- Outer case
- Plastic support

Explain how the flask reduces energy transfer by conduction, convection and radiation. (6)

> **AQA Examiner's tip**
>
> There are pros and cons to the use of any source for generating electricity, even the renewable ones. Make sure you know what they are.

> **AQA Examiner's tip**
>
> Make sure you understand that a vacuum is a completely empty space.

P2 1.1

Distance–time graphs

Learning objectives

- How can we tell from a distance–time graph if an object is stationary?

- How can we tell from a distance–time graph if an object is moving at constant speed?

- What does the gradient of a distance–time graph tell us?

- How do we calculate the speed of an object?

Figure 1 Capturing the land speed record

??? Did you know ...?

- Usain Bolt broke the 100 m sprint record in August 2009 in a time of 9.58 seconds – an average speed of 10.44 metres per second (100 ÷ 9.58). By the time you read this, there will probably be a new record.

- A cheetah is faster than any other animal. It can run about 30 metres every second – but only for about 20 seconds! This is nearly as fast as a vehicle travelling at 70 miles per hour (mph).

- The land speed record at present is 763 mph, which is more than Mach 1, the speed of sound. The Bloodhound Project is aiming to set a new record of 1000 mph.

Some motorways have marker posts every kilometre. If you are a passenger in a car on a motorway, you can use these posts to check the speed of the car. You need to time the car as it passes each post. The table below shows some measurements made on a car journey.

Distance (metres, m)	0	1000	2000	3000	4000	5000	6000
Time (seconds, s)	0	40	80	120	160	200	240

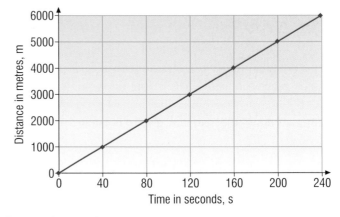

Figure 2 A distance–time graph

Look at the readings plotted on a graph of distance against time in Figure 2.

The graph shows that:

- the car took 40 s to go from each marker post to the next. So its speed was **constant** (or uniform).

- the car went a distance of 25 metres every second (= 1000 metres ÷ 40 seconds). So its speed was 25 metres per second.

If the car had travelled faster, it would have gone further than 1000 metres every 40 seconds. So the line on the graph would have been **steeper**. In other words, the **gradient** of the line would have been greater.

The gradient of a line on a distance–time graph represents speed.

> **a** What can you say about the gradient of the line if the car had travelled slower than 25 metres per second?

Speed

For an object moving at constant **speed**, we can calculate its speed using the formula:

$$\text{speed in metres per second, m/s} = \frac{\text{distance travelled in metres, m}}{\text{time taken in seconds, s}}$$

The scientific unit of speed is the metre per second, usually written as metre/second or m/s.

Speed in action

Long-distance vehicles are fitted with recorders called **tachographs.** These can check that their drivers don't drive for too long. Look at the distance–time graphs in Figure 3 for three lorries, X, Y and Z, on the same motorway.

- X went fastest because it travelled furthest in the same time.
- Y travelled more slowly than X. From the graph, you can see it travelled 30 000 metres in 1250 seconds. So its speed was:

$$\text{distance} \div \text{time} = 30\,000\,\text{m} \div 1250\,\text{s} = 24\,\text{m/s}.$$

b Calculate the speed of X.

- Z stopped for some of the time. Its speed was zero in this time.

c How long did Z stop for?
d Calculate the **average** speed of Z, using the total distance Z travels in its journey.

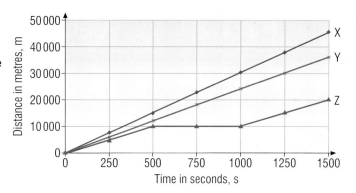

Figure 3 Comparing distance–time graphs

Practical

Be a distance recorder!

Take the measurements needed to plot distance–time graphs for a person:

- walking
- running
- riding a bike.

Remember that you must always label the graph axes, which includes units.

- Work out the average speeds.

Figure 4 Measuring distance

Maths skills

Rearranging the speed formula

If two of the three quantities are known, the third can be found. It may help to use the speed formula triangle below:

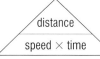

Cover up the unknown quantity and the triangle tells you how to use the other two known quantities.

Summary questions

1 Copy and complete sentences **a** to **c** using the words below:

distance   gradient   speed

a The unit of .... is the metre/second.
b An object moving at a constant speed travels the same .... every second.
c The steeper the .... of the line on a distance–time graph of a moving object, the greater its speed is.

2 A vehicle on a motorway travels 1800 m in 60 seconds. Calculate:
a the average speed of the vehicle in m/s.
b how far it would travel in 300 seconds if it continued travelling at this speed.

3 A car on a motorway travels 10 kilometres in six minutes. A coach takes seven minutes to travel the same distance. Which vehicle was travelling faster, the car or the coach? Give a reason for your answer.

Key points

- The distance–time graph for any object that is
  - stationary is a horizontal line
  - moving at constant speed is a straight line that slopes upwards.

- The gradient of a distance–time graph for an object represents the object's speed.

- Speed in metres per second, m/s =

$$\frac{\text{distance travelled in metres, m}}{\text{time taken in seconds, s}}$$

P2 1.2

Velocity and acceleration

Learning objectives

- What is the difference between speed and velocity?
- What is acceleration and what is its unit?
- How can we calculate the acceleration of an object?
- What is deceleration?

Figure 2 You experience plenty of changes in velocity on a corkscrew ride!

⬯ links

For more information on circular motion, see P3 2.6 Circular motion.

Figure 3 On a test circuit

When you visit a fairground, do you like the rides that throw you round? Your speed and your direction of motion keep changing. We use the word **velocity** for speed in a given direction. An exciting ride would be one that changes your velocity often and unexpectedly!

Velocity is speed in a given direction.

- An object moving steadily round in a circle has a constant speed. Its direction of motion changes continuously as it goes round so its velocity is not constant.
- Two moving objects can have the same speed but different velocities. For example, a car travelling north at 30 m/s on a motorway has the same speed as a car travelling south at 30 m/s. But their velocities are not the same because they are moving in opposite directions.

a How far apart are the two cars 10 seconds after they pass each other?

Direction of motion

Figure 1 Speed and velocity

Acceleration

A car maker claims their new car 'accelerates more quickly than any other new car'. A rival car maker is not pleased by this claim and issues a challenge. Each car in turn is tested on a straight track with a velocity recorder fitted.

The results are shown in the table:

Time from a standing start (seconds, s)	0	2	4	6	8	10
Velocity of car X (metre per second, m/s)	0	5	10	15	20	25
Velocity of car Y (metre per second, m/s)	0	6	12	18	18	18

Which car has a greater **acceleration**? The results are plotted on the velocity–time graph in Figure 4. You can see the velocity of Y goes up from zero faster than the velocity of X does. So Y accelerates more in the first 6 seconds.

The acceleration of an object is its change of velocity per second. The unit of acceleration is the metre per second squared, abbreviated to m/s².

Any object with a changing velocity is accelerating. We can work out its acceleration using the equation:

Acceleration

$$\text{Acceleration (metres per second squared, m/s}^2) = \frac{\text{change in velocity in metres per second, m/s}}{\text{time taken for the change in seconds, s}}$$

For an object that accelerates steadily from an initial velocity u to a final velocity v, its change of velocity = final velocity – initial velocity = $v - u$.

Therefore, we can write the equation for acceleration as:

$$\text{acceleration, } a = \frac{v - u}{t}$$

Where:
v = the final velocity in metres per second,
u = the initial velocity in metres per second,
t = time taken in seconds.

 Maths skills

Worked example

In Figure 4, the velocity of Y increases from 0 to 18 m/s in 6 seconds. Calculate its acceleration.

Solution

Change of velocity = $v - u$ = 18 m/s – 0 m/s = 18 m/s

Time taken, $t = 6\,\text{s}$

Acceleration, $a = \dfrac{\text{change in velocity in metres per second, m/s}}{\text{time taken for the change in seconds, s}} = \dfrac{v - u}{t}$

$$= \frac{18\,\text{m/s}}{6\,\text{s}} = 3\,\text{m/s}^2$$

b Calculate the acceleration of X in Figure 4.

Deceleration

A car decelerates when the driver brakes. We use the term **deceleration** or **negative acceleration** for any situation where an object slows down.

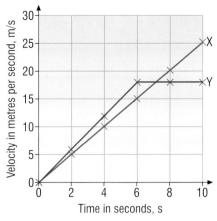

Figure 4 Velocity–time graph

Summary questions

1 Copy and complete **a** to **c** using the words below:

acceleration   speed   velocity

   a An object moving steadily round in a circle has a constant ....
   b If the velocity of an object increases by the same amount every second, its .... is constant.
   c Deceleration is when the .... of an object decreases.

2 The velocity of a car increased from 8 m/s to 28 m/s in 8 s without change of direction. Calculate:
   a its change of velocity
   b its acceleration.

3 The driver of a car increased the speed of the car as it joined the motorway. It then travelled at constant velocity before slowing down as it left the motorway at the next junction.
   a i When did the car decelerate?
     ii When was the acceleration of the car zero?
   b When the car joined the motorway, its velocity increased from 5.0 metres per second to 25 metres per second in 10 seconds. What was its acceleration during this time?

Key points

- Velocity is speed in a given direction.

- Acceleration is change of velocity per second. The unit of acceleration is the metre per second squared (m/s^2).

- Acceleration = change of velocity ÷ time taken.

- Deceleration is the change of velocity per second when an object slows down.

P2 1.3

More about velocity–time graphs

Learning objectives

- What can we say if a velocity–time graph is a horizontal line?

- How can we tell from a velocity–time graph if an object is accelerating or decelerating?

- What does the area under a velocity–time graph represent? **[H]**

Figure 2 Measuring motion using a computer

∞ links

For more information on variables and relationships between them, see H2 Fundamental ideas about how science works.

Investigating acceleration

We can use a motion sensor linked to a computer to record how the velocity of an object changes. Figure 1 shows how we can do this, using a trolley as the moving object. The computer can also be used to display the measurements as a velocity–time graph.

Test A: If we let the trolley accelerate down the runway, its velocity increases with time. Look at the velocity–time graph from a test run in Figure 2.

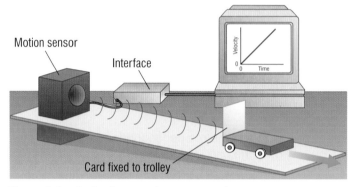

Motion sensor

Interface

Card fixed to trolley

Figure 1 A velocity–time graph on a computer

- The line goes up because the velocity increases with time. So it shows the trolley was accelerating as it ran down the runway.

- The line is straight, which tells us that the increase in velocity was the same every second. In other words, the acceleration of the trolley was constant.

Test B: If we make the runway steeper, the trolley accelerates faster. This would make the line on the graph in Figure 2 steeper than for test A. So the acceleration in test B is greater.

The gradient of a line is a measure of its steepness. The tests show that **the gradient of the line on a velocity–time graph represents acceleration.**

a If you made the runway less steep than in test A, would the line on the graph be steeper or less steep than in A?

Practical

Investigating acceleration

Use a motion sensor and a computer to find out how the gradient of a runway affects a trolley's acceleration.

- Name **i** the independent variable, and **ii** the dependent variable in this investigation.

- What relationship do you find between the variables?

Braking

Braking reduces the velocity of a vehicle. Look at the graph in Figure 3. It is the velocity–time graph for a vehicle that brakes and stops at a set of traffic lights. The velocity is constant until the driver applies the brakes.

Using the gradient of the line:

- The section of the graph for constant velocity is horizontal. The gradient of the line is zero so the acceleration in this section is zero.
- When the brakes are applied, the velocity decreases to zero and the vehicle decelerates. The gradient of the line is negative in this section. So the acceleration is negative.

b How would the gradient of the line differ if the deceleration had taken longer?

Look at the graph in Figure 3 again.

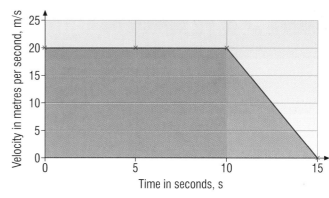

Figure 3 Braking

Using the area under the line

- Before the brakes are applied, the vehicle moves at a velocity of 20 m/s for 10 s. It therefore travels 200 m in this time (= 20 m/s × 10 s). This distance is represented on the graph by the area under the line from 0 s to 10 s. This is the shaded rectangle on the graph.
- When the vehicle decelerates in Figure 3, its velocity drops from 20 m/s to 0 m/s in 5 s. We can work out the distance travelled in this time from the area of the purple triangle in Figure 3. This area is ½ × the height × the base of the triangle. So the vehicle must have travelled a distance of 50 m when it was decelerating.

The area under the line on a velocity–time graph represents distance travelled.

c Would the total distance travelled be greater or smaller if the deceleration had taken longer? **[H]**

> **?? Did you know ...?**
>
> A speed camera flashes when a vehicle travelling over the speed limit has gone past. Some speed cameras flash twice and measure the distance the car travels between flashes.

Higher

Summary questions

1 Match each of the following descriptions to one of the lines, labelled A, B, C and D, on the velocity–time graph.
  1 Accelerated motion throughout
  2 Zero acceleration
  3 Accelerated motion, then decelerated motion
  4 Deceleration

2 Look at the graph in Question 1. Which line represents the object that travelled:
  a the furthest distance? **[H]**
  b the least distance? **[H]**

3 Look again at the graph in Question 1.
  a Show that the object that produced the data for line A (the horizontal line) travelled a distance of 160 m. **[H]**
  b Which one of the other three lines represents the motion of an object that decelerated throughout its journey? **[H]**
  c Calculate the distance travelled by this object. **[H]**

Key points

- If a velocity–time graph is a horizontal line, the acceleration is zero.
- The gradient of the line on a velocity–time graph represents acceleration.
- The area under the line on a velocity–time graph represents distance travelled. **[H]**

P2 1.4 Using graphs ⓚ

Learning objectives

- How can we calculate speed from a distance–time graph? **[H]**
- How can we calculate acceleration from a velocity–time graph? **[H]**
- How can we calculate distance from a velocity–time graph? **[H]**

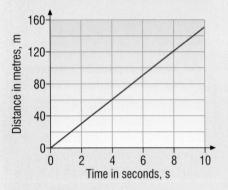

Figure 1 A distance–time graph for constant speed

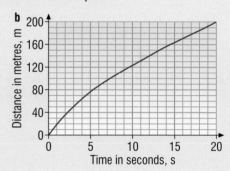

Figure 3 A velocity–time graph for constant acceleration

Using distance–time graphs

For an object moving at constant speed, we saw at the start of this chapter that the distance–time graph is a straight line sloping upwards.

The speed of the object is represented by the gradient of the line. To find the gradient, we need to draw a triangle under the line, as shown in Figure 1. The height of the triangle represents the distance travelled and the base represents the time taken. So

$$\text{the gradient of the line} = \frac{\text{the height of the triangle}}{\text{the base of the triangle}}$$

and this represents the object's speed.

> **a** Find the speed of the object in the graph in Figure 1.

For a moving object with a changing speed, the distance–time graph is not a straight line. The graphs in Figure 2 show two examples.

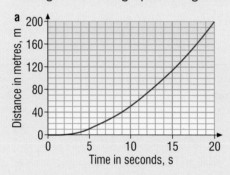

Figure 2 Distance–time graphs for changing speed

In Figure 2a, the gradient of the graph increases gradually, so the object's speed must have increased gradually.

> **b** What can you say about the speed in Figure 2b?

Using velocity–time graphs

Look at the graph in Figure 3. It shows the velocity–time graph of an object X moving with a constant acceleration. Its velocity increases at a steady rate. So the graph shows a straight line that has a constant gradient.

To find the acceleration from the graph, remember the gradient of the line on a velocity–time graph represents the acceleration.

In Figure 3, the gradient is given by the height divided by the base of the triangle under the line.

The height of the triangle represents the change of velocity and the base of the triangle represents the time taken.

Therefore, the gradient represents the acceleration, because:

$$\text{acceleration} = \frac{\text{change of velocity}}{\text{time taken}}$$

Maths skills

Worked example

Use the graph in Figure 3 to find the acceleration of object X.

Solution

The height of the triangle represents an increase of velocity of 8 m/s (= 12 m/s – 4 m/s).

The base of the triangle represents a time of 10 s.

Therefore, the acceleration = $\dfrac{\text{change of velocity}}{\text{time taken}} = \dfrac{8\,\text{m/s}}{10\,\text{s}} = 0.8\,\text{m/s}^2$

To find the distance travelled from the graph, remember the area under a line on a velocity–time graph represents the distance travelled. The shape under the line in Figure 3 is a triangle on top of a rectangle. So the distance travelled is represented by the area of the triangle plus the area of the rectangle under it.

Look at the worked example opposite.

Maths skills

Worked example

Use the graph in Figure 3 to calculate the distance moved by object X.

Solution

The area of the triangle = ½ × height × base.

Therefore, the distance represented by the area of triangle = ½ × 8 m/s × 10 s
= 40 m

The area of the rectangle under the triangle = height × base

Therefore, the distance represented by the area of the rectangle = 4 m/s × 10 s
= 40 m

So the distance travelled by X = 40 m + 40 m = 80 m

Summary questions

1 Copy and complete **a** to **c** using the words below:

acceleration   distance   speed

a The area under the line of a velocity–time graph represents ....
b The gradient of a line on a distance–time graph represents ....
c The gradient of a line on a velocity–time graph represents .... **[H]**

2 The graph shows how the velocity of a cyclist on a straight road changes with time.
a Describe the motion of the cyclist.
b Use the graph to work out
  i the initial acceleration of the cyclist
  ii the distance travelled by the cyclist in the first 40 s. **[H]**

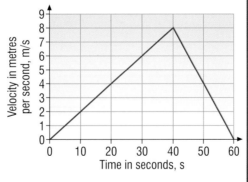

3 In a motorcycle test, the speed from rest was recorded at intervals.

Time (seconds, s)	0	5	10	15	20	25	30
Velocity (metre per second, m/s)	0	10	20	30	40	40	40

a Plot a velocity–time graph of these results.
b What was the initial acceleration?
c How far did it move in:
  i the first 20 s?
  ii the next 10 s? **[H]**

Key points

● The speed of an object is given by the gradient of the line on its distance–time graph. **[H]**

● The acceleration of an object is given by the gradient of the line on its velocity–time graph. **[H]**

● The distance travelled by an object is given by the area under the line of its velocity–time graph. **[H]**

Summary questions

1 A model car travels round a circular track at constant speed.

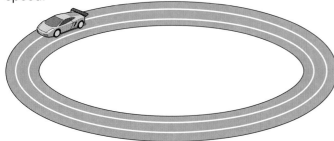

a If you were given a stopwatch, a marker, and a tape measure, how would you measure:

  i the time taken by the car to travel 10 laps
  ii the distance the car travels in 10 laps.

b If the car travels 36 metres in 30 seconds, calculate its speed.

2 A train travels at a constant speed of 35 m/s. Calculate:

a how far it travels in 20 s

b how long it takes to travel a distance of 1400 m.

3 The figure shows the distance–time graph for a car on a motorway.

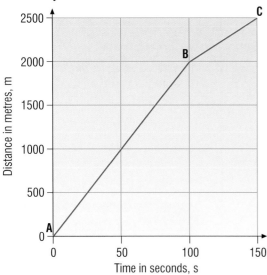

a Which part of the journey was faster, A to B or B to C?

b i How far did the car travel from A to B and how long did it take?

  ii Calculate the speed of the car between A and B.

4 a A car took 8 s to increase its velocity from 8 m/s to 28 m/s. Calculate:

  i its change of velocity
  ii its acceleration.

b A vehicle travelling at a velocity of 24 m/s slowed down and stopped in 20 s. Calculate its deceleration.

5 The figure shows the velocity–time graph of a passenger jet before it took off.

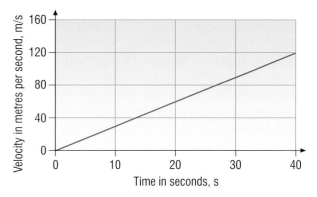

a Calculate the acceleration of the jet.

b Calculate the distance it travelled before it took off. [H]

6 The table below shows how the velocity of a train changes as it travelled from one station to the next.

Time (seconds)	0	20	40	60	80	100	120	140	160
Velocity (m/s)	0	5	10	15	20	20	20	10	0

a Plot a velocity–time graph using this data.

b Calculate the acceleration in each of the three parts of the journey.

c Calculate the total distance travelled by the train.

d Show that the average speed for the train's journey was 12.5 m/s. [H]

7 A motorcyclist started from rest and accelerated steadily to 25 m/s in 5 seconds then slowed down steadily to a halt 30 seconds after she started.

a Draw a velocity–time graph for this journey.

b Show that the acceleration of the motorcyclist in the first 5 seconds was 5.0 m/s².

c Calculate the deceleration of the motorcyclist in the last 25 seconds.

d Use your graph to show that the total distance travelled by the motorcyclist was 375 metres. [H]

AQA Examination-style questions

1 The table gives values of distance and time for a child travelling along a straight track competing in an egg and spoon race.

Time (seconds)	0	5	10	15	20	25
Distance (metres)	0	8	20	20	24	40

a Copy the graph axes below on to graph paper. Plot a graph of distance against time for the child. (3)

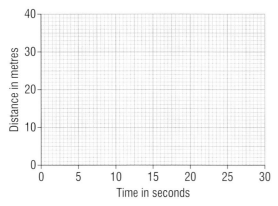

b Name the dependent variable shown on the graph. (1)

c What type of variable is this? (1)

d Use your graph to estimate the distance travelled in 22 seconds. (1)

e Use your graph to estimate the time taken for the child to travel 15 metres. (1)

f Describe the motion of the child between 10 seconds and 15 seconds.
Give a reason for your answer. (2)

2 The graph shows how far a runner travels during a charity running race.

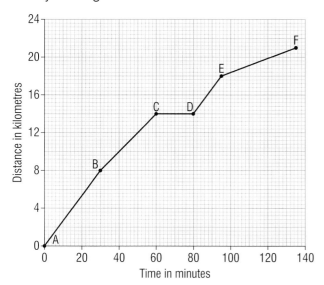

a What was the distance of the race? (1)

b How long did it take the runner to complete the race? (1)

c For how long did the runner rest during the race? (1)

d Between which two points was the runner moving the fastest?
Give a reason for your answer. (2)

e Between which two points did the runner travel at the same speed as they did between A and B? (1)

f Calculate the speed of the runner between B and C in metres per second.
Write down the equation you use. Show clearly how you work out your answer. (3)

3 A cyclist is travelling along a straight road. The graph shows how the velocity changes with time for part of the journey.

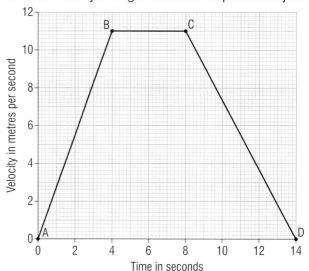

a Explain how the acceleration can be found from a velocity–time graph. (1)

b Copy and complete the following sentences using the list of words and phrases below. Each one can be used once, more than once or not at all.

*is stationary   travels at a constant speed
accelerates   decelerates*

  i Between A and B the cyclist .... (1)
  ii Between B and C the cyclist .... (1)
  iii Between C and D the cyclist .... (1)

c i Use the graph to find the maximum speed of the cyclist. (1)

  ii Use the graph to calculate the distance travelled in metres between 4 and 8 seconds. Show clearly how you work out your answer. (2)

  iii Use the graph to calculate the total distance travelled in metres.
  Show clearly how you work out your answer. [H] (3)

P2 2.1

Forces between objects

Learning objectives

- What can forces do?

- What is the unit of force?

- When two objects interact, what can we say about the forces acting?

??? Did you know ... ?

Quicksand victims sink because they can't get enough support from the sand. The force of gravity on the victim (acting downwards) is greater than the upwards force of the sand on the victim. People caught in quicksand should not struggle but flatten themselves on the surface and crawl to a safe place.

∞ links

For more information on how forces make objects turn, see P3 2.1 Moments.

When you apply a **force** to a tube of toothpaste, be careful not to apply too much force. The force you apply to squeeze the tube changes its shape and pushes toothpaste out of the tube. If you apply too much force, the toothpaste might come out too fast.

A force can change the shape of an object or change its state of rest or its motion.

Equal and opposite forces

Whenever two objects push or pull on each other, they exert equal and opposite forces on one another. The unit of force is the newton (abbreviated as N).

- A boxer who punches an opponent with a force of 100 N experiences a reverse force of 100 N from his opponent.

- Two roller skaters pull on opposite ends of a rope. The skaters move towards each other. This is because they pull on each other with equal and opposite forces. Two newtonmeters could be used to show this.

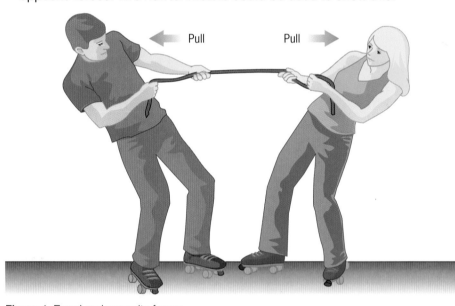

Figure 1 Equal and opposite forces

 Practical

Action and reaction

Test this with a friend if you can, using roller skates and two newtonmeters. Don't forget to wear protective head gear!

- What did you find out?

- Comment on the precision of your readings.

a A hammer hits a nail with a downward force of 50 N. What is the size and direction of the force of the nail on the hammer?

In the mud

A car stuck in mud can be difficult to shift. A tractor can be very useful here. Figure 2 shows the idea. At any stage, the force of the rope on the car is equal and opposite to the force of the car on the rope.

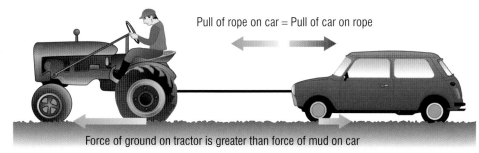

Pull of rope on car = Pull of car on rope

Force of ground on tractor is greater than force of mud on car

Figure 2 In the mud

To pull the car out of the mud, the force of the ground on the tractor needs to be greater than the force of the mud on the car. These two forces aren't necessarily equal to one another because the objects are not the same.

b A lorry tows a broken-down car. When the force of the lorry on the tow rope is 200 N, what is the force of the tow rope on the lorry?

Friction in action

The driving force on a car is the force that makes it move. This is sometimes called the engine force or the **motive force**. This force is due to **friction** between the ground and the tyre of each drive wheel. Friction acts where the tyre is in contact with the ground.

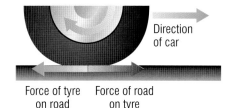

Direction of car

Force of tyre on road   Force of road on tyre

Figure 3 Driving force

When the car moves forwards:

● the force of friction of the ground on the tyre is in the forward direction
● the force of friction of the tyre on the ground is in the reverse direction.

The two forces are equal and opposite to one another.

c What happens if there isn't enough friction between the tyre and the ground?

Summary questions

1 a When the brakes of a moving car are applied, what is the effect of the braking force on the car?
b When you sit on a cushion, what is the effect of your weight on the cushion?
c When you kick a football, what is the effect of the force of your foot on the ball?

2 Copy and complete **a** and **b** using the words below:

downwards   equal   opposite   upwards

a The force on a ladder resting against a wall is .... and .... to the force of the wall on the ladder.
b A book is at rest on a table. The force of the book on the table is .... The force of the table on the book is ....

3 When a student is standing at rest on bathroom scales, the scales read 500 N.
a What is the size and direction of the force of the student on the scales?
b What is the size and direction of the force of the scales on the student?

Key points

● A force can change the shape of an object or change its motion or its state of rest.

● The unit of force is the newton (N).

● When two objects interact, they always exert equal and opposite forces on each other.

Resultant force

Learning objectives

- What is a resultant force?

- What happens if the resultant force on an object is:
  – zero?
  – not zero?

- How do we calculate the resultant force when an object is acted on by two forces acting along the same line?

Wherever you are at this moment, at least two forces are acting on you. These are the force of gravity on you and a force supporting you. Most objects around you are acted on by more than one force. We can work out the effect of the forces on an object by replacing them with a single force, the **resultant force**. This is a single force that has the same effect as all the forces acting on the object.

Zero resultant force

When the resultant force on an object is zero, the object:

- remains stationary if it was at rest, or

- continues to move at the same speed and in the same direction if it was already moving.

If two forces only act on the object, they must be equal to each other and act in opposite directions.

Practical

Investigating forces

Make and test a model hovercraft floating on a cushion of air from a balloon.

And/or:

Use a glider on an air track to investigate the relationship between force and acceleration.

- What relationship do you find between force and acceleration?

links

For more information on using data to draw conclusions, see H8 Using data to draw conclusions.

1 **A glider on a linear air track** floats on a cushion of air. Provided the track is level, the glider moves at constant velocity (i.e. with no change of speed or direction) along the track. That's because friction is absent. The resultant force on the glider is zero.

Figure 1 The linear air track

a What happens to the glider if the air track blower is switched off, and why?

2 **When a heavy crate is pushed across a rough floor at a constant velocity,** the resultant force on the crate is zero. The push force on the crate is equal in size but acts in the opposite direction to the force of friction of the floor on the crate.

b What difference would it make if the floor were smooth?

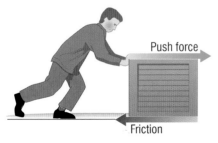

Push force

Friction

Figure 2 Overcoming friction

Non-zero resultant force

When the resultant force on an object is not zero, the movement of the object depends on the size and direction of the resultant force.

1 **When a jet plane is taking off**, the thrust force of its engines is greater than the force of air resistance on it. The resultant force on it is the difference between the thrust force and the force of air resistance on it. The resultant force is therefore non-zero. The greater the resultant force, the quicker the take-off is.

   c What can you say about the thrust force and the force of air resistance when the plane is moving at constant velocity at constant height?

2 **When a car driver applies the brakes**, the braking force is greater than the force from the engine. The resultant force is the difference between the braking force and the engine force. It acts in the opposite direction to the car's direction. So it slows the car down.

   d What can you say about the resultant force if the brakes had been applied harder?

The examples above show that if an object is acted on by two unequal forces acting in opposite directions, the resultant force is:

● equal to the difference between the two forces
● in the direction of the larger force.

Note what happens if the two forces act in the same direction. The resultant force is equal to the sum of the two forces and acts in the same direction as the two forces.

Figure 3 A passenger jet on take-off

Braking force

Figure 4 Braking

Summary questions

1 Copy and complete **a** to **c** using the words below:

   greater than   less than   equal to

   A car starts from rest and accelerates along a straight flat road.
   a The force of air resistance on it is .... the driving force of its engine.
   b The resultant force is .... zero.
   c The downward force of the car on the road is .... the support force of the road on the car.

2 A jet plane lands on a runway and stops.
   a What can you say about the direction of the resultant force on the plane as it lands?
   b What can you say about the resultant force on the plane when it has stopped?

3 A car is stuck in the mud. A tractor tries to pull it out.
   a The tractor pulls the car with a force of 250 N but the car doesn't move. Explain why the car doesn't move.
   b Increasing the tractor force to 300 N pulls the car steadily out of the mud. What is the force of the mud on the car now?

Key points

● The resultant force is a single force that has the same effect as all the forces acting on an object.

● If the resultant force on an object is zero, the object stays at rest or at constant velocity. If the resultant force on an object is not zero, the velocity of the object will change.

● If two forces act on an object along the same line, the resultant force is:
   1 their sum if the forces act in the same direction
   2 their difference if the forces act in opposite directions.

P2 2.3

Force and acceleration ⓚ

Learning objectives

- How does the acceleration of an object depend on the size of the resultant force?

- What effect does the mass of the object have on its acceleration?

- How do we calculate the resultant force on an object from its acceleration and its mass?

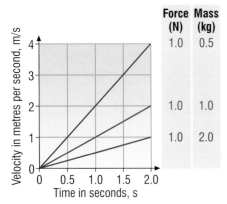

Force (N)	Mass (kg)
1.0	0.5
1.0	1.0
1.0	2.0

Figure 2 Velocity–time graph for different combinations of force and mass

∞ links

For more information on how to work out the acceleration from the gradient of the line, look back at P2 1.4 Using graphs.

 Maths skills

Worked example

Calculate the resultant force on an object of mass 6.0 kg when it has an acceleration of 3.0 m/s².

Solution

Resultant force
= mass × acceleration
= 6.0 kg × 3.0 m/s² = 18.0 N

Practical

Investigating force and acceleration

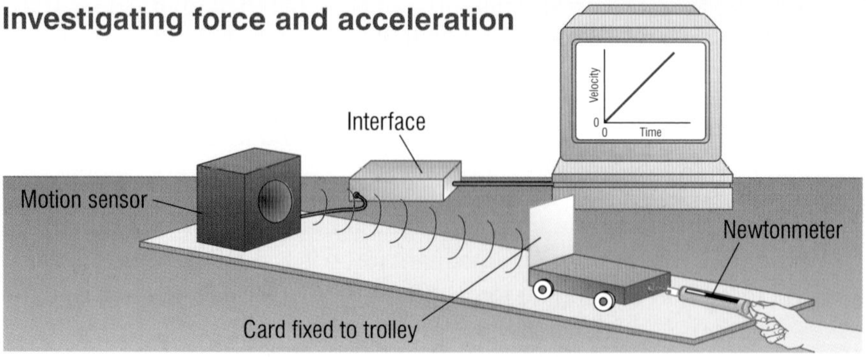

Figure 1 Investigating the link between force and motion

We can use the apparatus above to accelerate a trolley with a constant force.

Use the newtonmeter to pull the trolley along with a constant force.

You can double or treble the total moving mass by using double-deck and triple-deck trolleys.

A motion sensor and a computer record the velocity of the trolley as it accelerates.

- What are the advantages of using a data logger and computer in this investigation?

You can display the results as a velocity–time graph on the computer screen.

Figure 2 shows velocity–time graphs for different masses. You can work out the acceleration from the gradient of the line, as explained in the previous chapter.

Look at some typical results in the table below:

Resultant force (newtons)	0.5	1.0	1.5	2.0	4.0	6.0
Mass (kilograms)	1.0	1.0	1.0	2.0	2.0	2.0
Acceleration (m/s²)	0.5	1.0	1.5	1.0	2.0	3.0
Mass × acceleration (kg m/s²)	0.5	1.0	1.5	2.0	4.0	6.0

The results show that the resultant force, the mass and the acceleration are linked by the equation

resultant force = mass × acceleration
(newtons, N) (kilograms) (metres/second²)

We can write the word equation above using symbols as follows:

resultant force, $F = ma$,

Where F = resultant force in newtons
m = mass in kilograms
a = acceleration in metres/second².

a Calculate the resultant force on a sprinter of mass 80 kg who accelerates at 8 m/s².

Higher

Maths skills

Worked example

Calculate the acceleration of an object of mass 5.0 kg acted on by a resultant force of 40 N.

Solution

Rearranging $F = ma$ gives $a = \dfrac{F}{m} = \dfrac{40\,\text{N}}{5.0\,\text{kg}} = 8.0\,\text{m/s}^2$

b Calculate the acceleration of a car of mass 800 kg acted on by a resultant force of 3200 N.

Speeding up or slowing down

If the velocity of an object changes, it must be acted on by a resultant force. Its acceleration is always in the same direction as the resultant force.

● The velocity of the object increases if the resultant force is in the **same** direction as the velocity. We say its acceleration is positive because it is in the same direction as its velocity.

● The velocity of the object decreases (i.e. it decelerates) if the resultant force is **opposite** in direction to its velocity. We say its acceleration is negative because it is opposite in direction to its velocity.

Summary questions

1 Copy and complete **a** to **c** using the words below. Each word can be used more than once.

acceleration   resultant force   mass   velocity

a A moving object decelerates when a .... acts on it in the opposite direction to its ....

b The greater the .... of an object is, the less its acceleration is when a .... acts on it.

c The .... of a moving object increases when a .... acts on it in the same direction as it is moving in.

2 Copy and complete the following table:

	a	b	c	d	e
Force (newtons, N)		200	840		5000
Mass (kilograms, kg)	20		70	0.40	
Acceleration (metres/ second squared, m/s²)	0.80	5.0		6.0	0.20

3 A car and a trailer have a total mass of 1500 kg.

Tow bar

a Find the force needed to accelerate the car and the trailer at 2.0 m/s².

b The mass of the trailer is 300 kg. Find the force of the tow bar on the trailer.

Maths skills

We can rearrange the equation
$F = ma$ to give

$a = \dfrac{F}{m}$ or $m = \dfrac{F}{a}$

??? Did you know ... ?

If you're in a car that suddenly brakes, your neck pulls on your head and slows it down. The equal and opposite force of your head on your neck can injure your neck.

Figure 3 A 'whiplash' injury

AQA Examiner's tip

● If an object is accelerating, it can be speeding up or changing direction. If it is decelerating, it is slowing down.

● If an object is accelerating or decelerating, there must be a resultant force acting on it.

Key points

● The bigger the resultant force on an object is, the greater its acceleration is.

● The greater the mass of an object is, the smaller its acceleration is for a given force.

● Resultant force (newtons, N) = mass (kilograms) × acceleration (metres/second²)

P2 2.4

On the road

Learning objectives

- What forces oppose the driving force of a car?

- What does the stopping distance of a vehicle depend on?

- What factors can increase the stopping distance of a vehicle?

Did you know ... ?

The mass of a BMW Mini Cooper car is just over 1000 kg.

Did you know ... ?

When the brakes of a car are applied, friction between the brake pads and the car wheels causes kinetic energy to be transferred by heating to the brakes and the brake pads. If the brake pads wear away too much, they need to be replaced.

Practical

Reaction times

Use an electronic stopwatch to test your own reaction time. Ask a friend to start the stopwatch when you are looking at it with your finger on the stop button. The read-out from the watch will give your reaction time.

- How can you make your data as precise as possible?

- What conclusions can you draw?

Forces on the road

For any car travelling at constant velocity, the resultant force on it is zero. This is because the driving force of its engine is balanced by the resistive forces (i.e. friction and air resistance). The resistive forces are mostly due to air resistance. Friction between parts of the car that move against each other also contributes.

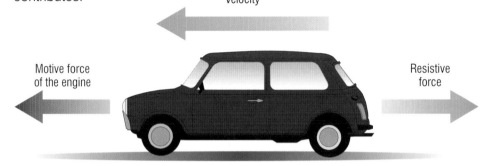

Figure 1 Constant velocity

A car driver uses the accelerator pedal (also called the gas pedal) to vary the driving force of the engine.

a What do you think happens if the driver presses harder on the accelerator?

The **braking force needed to stop a vehicle** in a certain distance depends on:

- the speed of the vehicle when the brakes are first applied
- the mass of the vehicle.

We can see this using the equation 'resultant force = mass × acceleration', in which the braking force is the resultant force.

1 The greater the speed, the greater the deceleration needed to stop the vehicle in a certain distance. So the braking force must be greater than at low speed.

2 The greater the mass, the greater the braking force needed for a given deceleration.

Stopping distances

Driving tests always ask about **stopping distances**. This is the shortest distance a vehicle can safely stop in, and is in two parts:

The thinking distance: the distance travelled by the vehicle in the time it takes the driver to react (i.e. during the driver's reaction time).

The braking distance: the distance travelled by the vehicle during the time the braking force acts.

stopping distance = thinking distance + braking distance

Figure 2 shows the stopping distance for a vehicle on a dry flat road travelling at different speeds. Check for yourself that the stopping distance at 31 m/s (70 miles per hour) is 96 m.

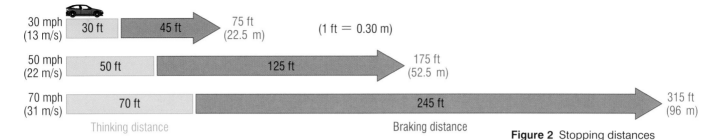

30 mph (13 m/s)	30 ft	45 ft	75 ft (22.5 m)	(1 ft = 0.30 m)
50 mph (22 m/s)	50 ft	125 ft	175 ft (52.5 m)	
70 mph (31 m/s)	70 ft	245 ft	315 ft (96 m)	

Thinking distance   Braking distance

Figure 2 Stopping distances

b What are the thinking distance, the braking distance in metres and the stopping distance at 13 m/s (30 mph)? (1 foot = 0.3 metres).

Maths skills

- The thinking distance is equal to the car's speed multiplied by the driver's reaction time. So it is directly proportional to the car's speed.
- The braking distance is equal to the average speed of the car during braking multiplied by the braking time. Since both of these quantities are directly proportional to the car's speed (before the brakes are applied), the braking distance is directly proportional to the square of the car's speed.

Factors affecting stopping distances

1 **Tiredness, alcohol and drugs** all increase reaction times. Distractions such as using a mobile phone can also affect reaction time. All these factors increase the thinking distance (because thinking distance = speed × reaction time). Therefore, the stopping distance is greater.
2 **The faster a vehicle is travelling**, the further it travels before it stops. This is because the thinking distance and the braking distance both increase with increased speed.
3 **In adverse road conditions**, for example on wet or icy roads, drivers have to brake with less force to avoid skidding. Stopping distances are therefore greater in poor road conditions.
4 **Poorly maintained vehicles**, for example with worn brakes or tyres, take longer to stop because the brakes and tyres are less effective.

c Why are stopping distances greater in poor visibility?

Figure 3 Stopping distances are further than you might think!

Summary questions

1 Each of the following factors affects the thinking distance or the braking distance of a vehicle. Which of these two distances is affected in these?
 a The road surface condition affects the .... distance.
 b The tiredness of a driver increases his or her .... distance.
 c Poorly maintained brakes affects the .... distance.

2 a Use the chart in Figure 2 to work out, in metres, the increase in
  i the thinking distance   ii the braking distance
  iii the stopping distance from 13 m/s (30 mph) to 22 m/s (50 mph).
 b A driver has a reaction time of 0.8 s. Calculate her thinking distance at a speed of   i 15 m/s   ii 30 m/s.

3 When the speed of a car is doubled:
 a Explain why the thinking distance of the driver is doubled, assuming the driver's reaction time is unchanged.
 b Explain why the braking distance is more than doubled.

Key points

- Friction and air resistance oppose the driving force of a car.
- The stopping distance of a car depends on the thinking distance and the braking distance.
- High speed, poor weather conditions and poor maintenance all increase the braking distance. Poor reaction time and high speed both increase the thinking distance.

P2 2.5

Falling objects

Learning objectives

- What is the difference between mass and weight?
- What can we say about the motion of a falling object acted on only by gravity?
- What is terminal velocity?

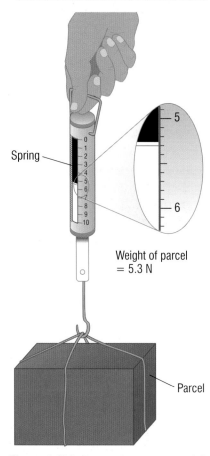

Spring

Weight of parcel = 5.3 N

Parcel

Figure 1 Using a newtonmeter to weigh an object

Maths skills

Worked example

Calculate the weight in newtons of a person of mass 55 kg.

Solution

Weight = mass × gravitational field strength = 55 kg × 10 N/kg
= 550 N

How to reduce your weight

Your weight is due to the gravitational force of attraction between you and the Earth. This force is very slightly weaker at the equator than at the poles. So if you want to reduce your weight, go to the equator. However, your mass will be the same no matter where you are.

- The **weight** of an object is the force of gravity on it. Weight is measured in newtons.
- The **mass** of an object is the quantity of matter in it. Mass is measured in kilograms.

We can measure the weight of an object using a newtonmeter.

The weight of an object:
- of mass 1 kg is 10 N
- of mass 5 kg is 50 N.

The force of gravity on a 1 kg object is the **gravitational field strength** at the place where the object is. The unit of gravitational field strength is the newton per kilogram (N/kg).

The value of the Earth's gravitational field strength at its surface is about 10 N/kg.

If we know the mass of an object, we can calculate the force of gravity on it (i.e. its weight) using the equation:

$$\text{weight} = \text{mass} \times \text{gravitational field strength}$$
$$\text{(newtons, N)} \quad \text{(kilograms, kg)} \quad \text{(newtons/kilogram, N/kg)}$$

We can write the word equation above using symbols as follows:

$$\text{weight, } W = mg,$$

Where:
W = weight in newtons, N
m = mass in kilograms, kg
g = gravitational field strength in newtons per kilogram, N/kg

a Calculate the weight of a steel bar of mass 20 kg.

The forces on falling objects

If we release an object above the ground, it falls because of its weight (i.e. the force of gravity on it).

If the object falls with no other forces acting on it, the resultant force on it is its weight. It accelerates downwards at a constant acceleration of 10 m/s². This is called the acceleration due to gravity. For example, if we release a 1 kg object above the ground:

- the force of gravity on it is 10 N, and
- its acceleration (= force/mass = 10 N/1 kg) = 10 m/s².

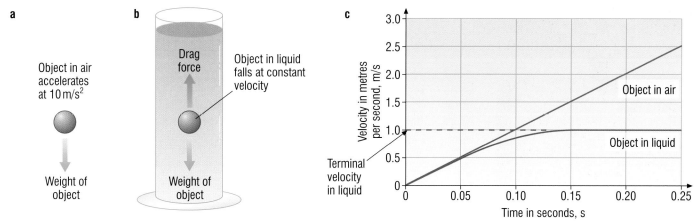

Figure 2 Falling objects. **a** Falling in air, **b** falling in a liquid, **c** velocity–time graph for **a** and **b**.

If the object falls in a fluid, the fluid drags on the object. The **drag force** increases with speed. At any instant, the resultant force on the object is its weight minus the drag force on it.

- The acceleration of the object decreases as it falls. This is because the drag force increases as it speeds up. So the resultant force on it decreases and therefore its acceleration decreases.

- The object reaches a constant velocity when the drag force on it is equal and opposite to its weight. We call this velocity its **terminal velocity**. The resultant force is then zero, so its acceleration is zero.

When an object moves through the air (i.e. the fluid is air) the drag force is called **air resistance**. This is not shown in Figure 2a because air resistance is very small in a short descent.

b Why does an object released in water eventually reach a constant velocity?

Summary questions

1 Copy and complete **a** to **c** using the words below:

equal to   greater than   less than

When an object is released in a fluid:
a The drag force on it is .... its weight before it reaches its terminal velocity.
b Its acceleration is .... zero after it reaches its terminal velocity.
c The resultant force on it is initially .... its weight.

2 The gravitational field strength at the surface of the Earth is 10 N/kg. For the Moon, it is 1.6 N/kg.
a Calculate the weight of a person of mass 50 kg at the surface of the Earth.
b Calculate the weight of the same person if she was on the surface of the Moon.

3 A parachutist of mass 70 kg supported by a parachute of mass 20 kg reaches a constant speed.
a Explain why the parachutist reaches a constant speed.
b Calculate:
i the total weight of the parachutist and the parachute
ii the size and direction of the force of air resistance on the parachute when the parachutist falls at constant speed.

Key points

- The weight of an object is the force of gravity on it. Its mass is the quantity of matter in it.

- An object acted on only by gravity accelerates at about 10 m/s².

- The terminal velocity of a falling object is the velocity it reaches when it is falling in a fluid. The weight is then equal to the drag force on the object.

P2 2.6

Stretching and squashing

Learning objectives

- How do we measure the extension of an object when it is stretched?

- How does the extension of a spring vary with the force applied to it?

- What is the spring constant of a spring?

Figure 2 A flower dipped in nitrogen and then smashed

Table 1 Weight versus length measurements for a rubber strip

Weight (N)	Length (mm)	Extension (mm)
0	120	0
1.0	152	32
2.0	190	70
3.0	250	
4.0		

Squash players know that hitting a squash ball changes the ball's shape briefly. A squash ball is **elastic** because it regains its original shape. A rubber band is also elastic as it regains its original length after it is stretched and then released. Rubber is an example of an elastic material.

An elastic object regains its original shape when the forces deforming it are removed.

Practical

Stretch tests (k)

We can investigate how easily a material stretches by hanging weights from it, as shown in Figure 1.

- The strip of material to be tested is clamped at its upper end. A weight hanger is attached to the material to keep it straight.

- The length of the strip is measured using a metre ruler. This is its original length.

- The weight hung from the material is increased by adding weights one at a time. The strip stretches each time more weight is hung from it.

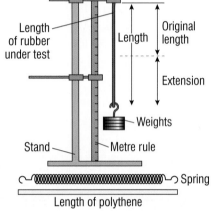

Figure 1 Investigating stretching

- The length of the strip is measured each time a weight is added. The total weight added and the total length of the strip are recorded in a table.

The increase of length from the original is called the **extension**. This is calculated each time a weight is added and recorded, as shown in Table 1.

The extension of the strip of material at any stage = its length at the stage – its original length

The measurements may be plotted on a graph of extension on the vertical axis against weight on the horizontal axis. Figure 3 shows the results for strips of different materials and a steel spring plotted on the same axes.

- The steel spring gives a straight line through the origin. This shows that the extension of the steel spring is **directly proportional** to the weight hung on it. For example, doubling the weight from 2.0 N to 4.0 N doubles the extension of the spring.

- The rubber band does not give a straight line. When the weight on the rubber band is doubled from 2.0 N to 4.0 N, the extension more than doubles.

- The polythene strip does not give a straight line either. As the weight is increased from zero, the polythene strip stretches very little at first then it 'gives' and stretches easily.

a Which part of a plastic shopping bag 'gives' if you overload the bag?

Elastic energy

When an elastic object is stretched, elastic potential energy is stored in the object. This is because work is done on the object by the stretching force.

When the stretching force is removed, the elastic energy stored in the object is released. Some of this energy may be transferred into kinetic energy of the object or may make its atoms vibrate more so it becomes warmer.

Hooke's law 🄚

In the tests above, the extension of a steel spring is directly proportional to the force applied to it. We can use the graph to predict what the extension would be for any given force. But if the force is too large, the spring stretches more than predicted. This is because the spring has been stretched beyond its **limit of proportionality**.

The extension of a spring is directly proportional to the force applied, provided its limit of proportionality is not exceeded.

The above statement is known as **Hooke's law**. If the extension of any stretched object or material is directly proportional to the stretching force, we say it obeys Hooke's law.

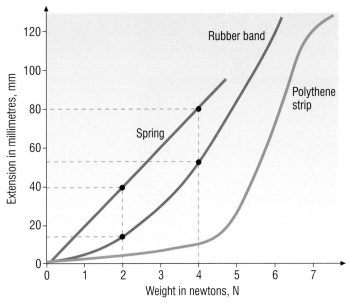

Figure 3 Extension versus weight for different materials

1 The lines on the graph in Figure 3 show that rubber and polythene have a low limit of proportionality. Beyond this limit, they do not obey Hooke's law. A steel spring has a much higher limit of proportionality.

2 Hooke's law may be written as an equation:

Force applied	=	**spring constant**	×	**extension**
(in newtons, N)		(in newtons per metre, N/m)		(in metres, m)

The **spring constant** is equal to the force per unit extension needed to extend the spring, assuming its limit of proportionality is not reached. The stiffer a spring is, the greater its spring constant is.

b A spring has a spring constant of 25 N/m. How much force is needed to make the spring extend by 0.10 m?

> ## Maths skills
>
> We can write the word equation for Hooke's law using symbols as follows:
>
> $$F = k \times e,$$
>
> Where:
> F = force in newtons, N
> k = the spring constant in newtons per metre, N/m
> e = extension in metres, m.

Summary questions

1 Copy and complete **a** to **c** using the words below.

elastic limit   extension   length

  a When a steel spring is stretched, its .... is increased.
  b When a strip of polythene is stretched beyond its .... , its length is permanently increased.
  c When rubber is stretched and unstretched, its .... afterwards is zero.

2 What is meant by:
  a the limit of proportionality of a spring?
  b the spring constant of a spring?

3 **a** In Figure 3, when the weight is 4.0 N, what is the extension of:
    i the spring **ii** the rubber band **iii** the polythene strip?
  b i What is the extension of the spring when the weight is 3.0 N?
    ii Calculate the spring constant of the spring.

> ## Key points
>
> ● The extension is the difference between the length of the spring and its original length.
>
> ● The extension of a spring is directly proportional to the force applied to it, provided the limit of proportionality is not exceeded.
>
> ● The spring constant of a spring is the force per unit extension needed to stretch it.

P2 2.7

Force and speed issues

Learning objectives

- How can the fuel economy of road vehicles be improved?
- What is an average speed camera?

Speed costs

Reducing the speed of a vehicle reduces the fuel it uses. This is because air resistance at high speed is much greater than at low speed. So more fuel is used. Lorry drivers can reduce their fuel usage by fitting a wind deflector over the cab. The deflector reduces the air resistance on the lorry. This means that less engine force and less power are needed to maintain a certain speed. So fuel costs are reduced because less fuel is needed.

Figure 1 A wind deflector on a lorry

a When a vehicle is accelerating, what can you say about the engine force and the air resistance?

Activity

The shape of a wind deflector on a lorry affects air resistance. Investigate the effect of the deflector shape by testing a trolley with a box on (or a toy lorry) without a deflector then fitted with deflectors of different shapes. You could use a hairdryer to blow air at the 'lorry' and use a newtonmeter to measure the force needed to stop it being blown backwards. (See P2 3.1 Figure 2.)

Speed kills!

- At 20 mph, the stopping distance of a car is 12 metres.
- At 40 mph, the stopping distance is 36 metres.
- At 60 mph, the stopping distance is 72 metres.

If someone walks across a road in front of a car, a driver travelling slowly is much more likely to stop safely than a speeding driver. The force on a person struck by a car increases with speed. Even at 20 mph, it can be many times the person's weight. A speed limit of 20 mph is in place outside many schools now.

Speed cameras

Speed cameras are very effective in discouraging motorists from speeding. A speeding motorist caught by a speed camera is fined and can lose his or her driving licence. On some motorways:

- Speed limits can vary according to the amount of traffic on the motorway.
- Speed cameras may be linked. These can catch out motorists who slow down for a speed camera then speed up.

In some areas, residents are supplied with 'mobile' speed cameras to catch speeding motorists. Some motorists think this is going too far and that speed cameras should not be used in this way. Lots of motorists say speed cameras are being used by local councils to increase their income.

Are speed cameras effective?

A report from one police force said that where speed cameras had been introduced:

- average speeds fell by 17%
- deaths and serious injuries fell by 55%.

Another police force reported that, in their area, as a result of installing more speed cameras in 2003:

- There were no child deaths in road accidents for the first time since 1927.
- 420 fewer children were involved in road accidents compared with the previous year.

b Discuss whether or not the statements above prove the argument that speed cameras save lives.

Anti-skid surfaces

Have you noticed that road surfaces near road junctions and traffic lights are often different from normal road surfaces?

- The surface is rougher than normal. This gives increased friction between the surface and a vehicle tyre, so it reduces the chance of skidding when a driver in a car applies the brakes.
- The surface is lighter in colour so it is marked out clearly from a normal road surface.

Skidding happens when the brakes are applied too harshly. The wheels lock and the tyres slide on the road as a result. Increased friction between the tyres and the road allows more force to be applied without skidding happening, so the stopping distance is reduced.

Figure 3 A speed camera

Figure 4 An anti-skid surface

Summary questions

1 The legal limit for a driver with alcohol in the blood is 80 milligrams per litre. Above this level, reaction times become significantly longer. The thinking distance of a normal car driver (i.e. one with no alcohol in the blood) travelling at 30 mph is 9.0 m (30 feet).

  a **i** What would this distance be for a driver whose reaction time is 20% longer than that of a normal driver?

    ii Drivers at the legal limit are 80% more likely to be in a road accident than normal drivers. Researchers think that a reduction of the legal limit to 40 milligrams per litre would cut the risk from 80% to 20%. Discuss whether or not the present legal limit should be reduced.

  b The braking distance for a car at 30 mph is 13.5 m and 6.0 m at 20 mph.

    i Thinking distance is directly proportional to speed. Show that the thinking distance at 20 mph is 6.0 m.

    ii Calculate the reduction in the stopping distance.

  c Many parents want the speed limit outside schools to be reduced to 20 mph. Explain why this would reduce road accidents outside schools significantly.

2 Campaigners in the village of Greystoke want the council to resurface the main road at the traffic lights in the village. A child was killed crossing the road at the traffic lights earlier in the year. The council estimates it would cost £45 000. They say they can't afford it. Campaigners have found some more data to support their case.

- There are about 50 000 road accidents each year in the UK.
- The cost of road accidents is over £8 billion per year.
- Anti-skid surfaces have cut accidents by about 5%.

  a Estimate how much each road accident costs.

  b Imagine you are one of the campaigners. Write a letter to your local newspaper to challenge the council's response that they can't afford to resurface the road.

Key points

- Fuel economy of road vehicles can be improved by reducing the speed or fitting a wind deflector.

- Average speed cameras are linked in pairs and they measure the average speed of a vehicle.

- Anti-skid surfaces increase the friction between a car tyre and the road surface. This reduces skids, or even prevents skids altogether.

Summary questions

1 A student is pushing a box across a rough floor. Friction acts between the box and the floor.

  a Copy and complete sentences **i** and **ii** using the words below:

  in the same direction as   in the opposite direction to

  i The force of friction of the box on the floor is .... the force of friction of the floor on the box.

  ii The force of the student on the box is .... the force of friction of the box on the floor.

  b The student is pushing the box towards a door. Which direction, towards the door or away from the door, is:

  i the force of the box on the student?

  ii the force of friction of the student on the floor?

2 a The weight of an object of mass 100 kg on the Moon is 160 N.

  i Calculate the gravitational field strength on the Moon.

  ii Calculate the weight of the object on the Earth's surface.

  The gravitational field strength near the Earth's surface is 10 N/kg.

  b Calculate the acceleration and the resultant force in each of the following situations.

  i A sprinter of mass 80 kg accelerates from rest to a speed of 9.6 m/s in 1.2 s.

  ii A train of mass 70 000 kg decelerates from a velocity of 16 m/s to a standstill in 40 s without change of direction.

3 The figure shows the velocity–time graphs for a metal object X dropped in air and a similar object Y dropped in a tank of water.

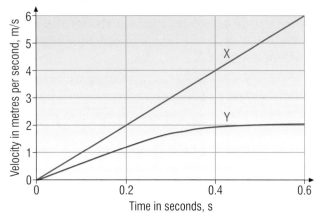

  a What does the graph for X tell you about its acceleration?

  b In terms of the forces acting on Y, explain why it reached a constant velocity.

4 Copy and complete **a** to **c** using the words below:
decreasing   increasing   terminal

  a When the resultant force on an object is not zero and acts in the opposite direction to the object's velocity, its velocity is ....

  b When an object falls in a fluid and the drag force on it is less than its weight, its velocity is ....

  c When the drag force on an object falling in a fluid is equal to its weight, the object moves at its .... velocity.

5 a Explain why the stopping distance of a car is increased if:

  i the road is wet instead of dry

  ii the driver is tired instead of alert.

  b A driver travelling at 18 m/s takes 0.7 s to react when a dog walks into the road 40 m ahead. The braking distance for the car at this speed is 24 m.

  i Calculate the distance travelled by the car in the time it takes the driver to react.

  ii How far in front of the dog does the car stop?

6 In a Hooke's law test on a spring, the following results were obtained.

Weight (N)	Length (mm)	Extension (mm)
0	245	0
1.0	285	40
2.0	324	
3.0	366	
4.0	405	
5.0	446	
6.0	484	

  a Copy and complete the third column of the table.

  b Plot a graph of the extension on the vertical axis against the weight on the horizontal axis.

  c If a weight of 7.0 N is suspended on the spring, what would be the extension of the spring?

  d i Calculate the spring constant of the spring.

  ii An object suspended on the spring gives an extension of 140 mm. Calculate the weight of the object.

7 a A racing cyclist accelerates at 5 m/s² when she starts from rest. The total mass of the cyclist and her bicycle is 45 kg. Calculate:

  i the resultant force that produces this acceleration

  ii the total weight of the cyclist and the bicycle.

  b Explain why she can reach a higher speed by crouching than by staying upright.

AQA Examination-style questions

1 a The tractor is pulling a trailer. The force acting on the trailer is labelled A, and the force acting on the tractor is labelled B.

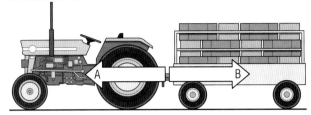

Copy and complete the following sentences using the list of words and phrases below. Each one can be used once, more than once or not at all.

A and B are the same   A is greater than B
B is greater than A

   i If the tractor and trailer are accelerating .... (1)
   ii If the tractor and trailer are moving at a constant speed .... (1)

b The driving force from the tractor is 12000N and the total resistive forces are 10000N.
   i Calculate the resultant force. (1)
   ii Calculate the acceleration of the tractor and trailer. Mass of the tractor and trailer = 2300kg
   Write down the equation you use. Show clearly how you work out your answer and give the unit. (3)

2 A car is travelling at 30m/s when the vehicle in front suddenly stops. The car travels 19m before the driver applies the brake.

a What is the name given to this distance? (1)

b Calculate the reaction time of the driver. Write down the equation you use. Show clearly how you work out your answer. (2)

c The driver applies the brakes and stops 6 seconds later. Calculate the deceleration of the car. Write down the equation you use. Show clearly how you work out your answer. (2)

d The braking distance is 81 m. What is the total stopping distance in metres? (1)

e Give two factors that would increase reaction time. (2)

3 The diagram shows the forces acting on a dragster just before it reaches its top speed. The resistive forces are represented by arrow **X**. The driving force is shown by arrow **Y**.

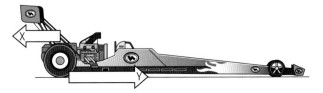

a What is the main type of resistive force acting on the dragster? (1)

b If the driving force remains the same, what will happen to force X?
Give a reason for your answer. (2)

c The dragster slows down by applying its brakes and using a parachute. The velocity–time graph shows the motion of the dragster from a stationary start until it stops.

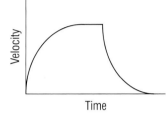

Explain, in terms of energy changes, the shape of the graph when the brakes are applied. (3)

4 A student carries out an experiment to find if extension is proportional to the force applied for an elastic hair bobble. She measures the extension with one and then two 0.1 kg masses. She holds the bobble with one hand and the ruler in the other.

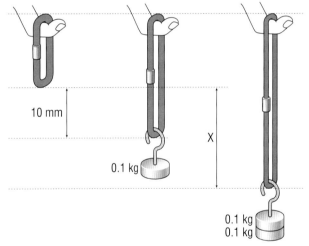

a If the extension is proportional to the force applied, what value should the student expect to obtain for distance X? (1)

b Give the name of the form of energy stored in the stretched hair bobble. (1)

c Calculate the weight of one of the 0.1 kg masses. (g = 10 N/kg) (2)

d *In this question you will be assessed on using good English, organising information clearly and using specialist terms where appropriate.*
The student is unable to draw a valid conclusion because she has not carried out the investigation with sufficient precision. Describe the improvements she could make in order to carry out the investigation more precisely and gain sufficient data to draw a valid conclusion. (6)

P2 3.1

Energy and work ⓚ

Working out

In a fitness centre or a gym, you have to work hard to keep fit. Raising weights and pedalling on an exercise bike are just two ways to keep fit. Whichever way you choose to keep fit, you have to apply a force to move something. So the work you do causes **transfer** of energy.

> **a** When you pedal on an exercise bike, where does the energy transferred go to?

When an object is moved by a force, we say **work** is done on the object by the force. The force therefore transfers energy to the object. The amount of energy transferred to the object is equal to the work done on it. For example, to raise an object, you need to apply a force to it to 'overcome' the force of gravity on it. If the work you do on the object is 20 J, the energy transferred to it must be 20 J. So its gravitational potential energy increases by 20 J.

Figure 1 Working out

Energy transferred = work done

The work done by a force depends on the size of the force and the distance moved. We use the following equation to calculate the work done by a force when it moves an object:

work done = force applied × distance moved in the direction of the force
(joules, J)　　(newtons, N)　　　　　　　　　　(metres, m)

We can write the word equation above using symbols:

$$W = F \times d$$

Where:
W = work done in joules, J
F = force in newtons, N
d = distance moved in metres in the direction of the force, m.

 Maths skills

Worked example

A builder pushed a wheelbarrow a distance of 5.0 m across flat ground with a force of 50 N. How much work was done by the builder?

Solution

Work done = force applied × distance moved = 50 N × 5.0 m = 250 J

> **b** How much work is done when a force of 2000 N pulls a truck through a distance of 40 m in the direction of the force?

Practical

Doing work

Carry out a series of experiments to calculate the work done in performing the tasks below. Use a newtonmeter to measure the force applied and a metre ruler to measure the distance moved.

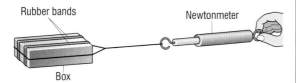

Figure 2 At work

1 Drag a small box a measured distance across a rough surface.
2 Repeat the test above with two rubber bands wrapped around the box as shown in Figure 2.

● What is the resolution of your measuring instruments? Repeat your tests and comment on the precision of your repeat measurements. Can you be confident about the accuracy of your results?

c Why is more work done than the calculated value in the practical with rubber bands?

Friction at work

Work done to overcome friction is mainly transferred into energy by heating.

1 If you rub your hands together vigorously, they become warm. Your muscles do work to overcome the friction between your hands. The work you do is transferred into energy that warms your hands.

2 Brake pads become hot if the brakes are applied for too long a time. Friction between the brake pads and the wheel discs opposes the motion of the wheel. The kinetic energy of the vehicle is transferred into energy that heats the brake pads and the wheel discs, as well as the surrounding air. A small proportion of the energy will be transferred to the surroundings by sound waves if the brakes 'squeal'.

Summary questions

1 Copy and complete **a** and **b** using the words below:

gravitational potential   kinetic   sound   wasted

a When a rower pulls on an oar, the work done by the rower is transferred into .... energy of the boat and .... energy by heating the water.

b When an electric motor is used to raise a car park barrier, the work done by the motor is transferred into .... energy of the barrier and .... energy.

2 A car is brought to a standstill when the driver applies the brakes.
a Explain why the brake pads become warm.
b The car travelled a distance of 20 metres after the brakes were applied. The braking force on the car during this time was 7000 N. Calculate the work done by the braking force.

3 Calculate the work done when:
a a force of 20 N makes an object move 4.8 m in the direction of the force
b an object of weight 80 N is raised through a height of 1.2 m.

Key points

● Work is done on an object when a force makes the object move.

● Energy transferred = work done

● Work done (joules) = force (newtons) × distance moved in the direction of the force (metres).

● Work done to overcome friction is transferred as energy that heats the objects that rub together and the surroundings.

P2 3.2 Gravitational potential energy

Learning objectives

- What does the gravitational potential energy of an object depend on?

- What happens to the gravitational potential energy of an object when it moves up or down?

- How can we calculate the change of gravitational potential energy of an object when it moves up or down?

Maths skills

Worked example

A student of weight 300 N climbs on a platform which is 1.2 m higher than the floor. Calculate the increase of her gravitational potential energy.

Solution

Increase of GPE = 300 N × 1.2 m
               = 360 J

Note: We often use the abbreviation 'GPE' or E_p for gravitational potential energy.

Did you know … ?

You use energy when you hold an object stationary in your outstretched hand. The biceps muscle of your arm is in a state of contraction. Energy must be supplied to keep the muscles contracted. No work is done on the object because it doesn't move. The energy supplied heats the muscles and is transferred by heating to the surroundings.

Gravitational potential energy transfers

Every time you lift an object up, you do some work. Some of your muscles transfer chemical energy from your muscles into **gravitational potential energy** of the object.

Gravitational potential energy is energy stored in an object because of its position in the Earth's gravitational field.

The force you need to lift an object steadily is equal and opposite to the force of gravity on the object. Therefore, the upward force you need to apply to it is equal to its weight. For example, a force of 80 N is needed to lift a box of weight 80 N.

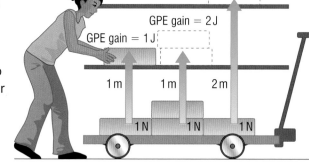

Figure 1 Using joules

- **When an object is moved up**, its gravitational potential energy increases. The increase of its gravitational potential energy is equal to the work done on it by the lifting force.

- **When an object moves down**, its gravitational potential energy decreases. The decrease of its gravitational potential energy is equal to the work done by the force of gravity acting on it.

The work done when an object moves up or down depends on:

1 how far it is moved vertically (its change of height)

2 its weight.

Using the formula $W = F \times d$ (work done = force applied × distance moved in the direction of the force), we can therefore say:

> the change of its gravitational = its weight × its change of height
> potential energy (in joules)   (in newtons)   (in metres)

a Read the 'Did you know?' box. What happens to the energy supplied to the muscles to keep them contracted?

Gravitational potential energy and mass

Astronauts on the Moon can lift objects much more easily than they can on the Earth. This is because, at their surfaces, the gravitational field strength of the Moon is only about a sixth of the Earth's gravitational field strength.

In P2 2.5, 'Falling objects', we saw that the weight of an object in newtons is equal to its mass × the gravitational field strength.

Therefore, when an object is lifted or lowered, because its change of gravitational potential energy is equal to its weight × its change of height:

> change of gravitational = mass × gravitational field × change of height
> potential energy (in J)   (in kg)   strength (in N/kg)   (in metres)

We can write the word equation on the previous page using symbols:

$$E_\text{p} = m \times g \times h$$

Where:

E_p = change of GPE in joules, J

m = mass in kilograms, kg

g = gravitational field strength in newtons per kilogram, N/kg

h = change in height in metres, m.

 Maths skills

Worked example

A 2.0 kg object is raised through a height of 0.4 m. Calculate the gain of gravitational potential energy of the object. The gravitational field strength of the Earth at its surface is 10 N/kg.

Solution

Gain of GPE = mass × gravitational field strength × height gain

= 2.0 kg × 10 N/kg × 0.4 m

= 8.0 J

Power and energy

Power is the rate of transfer of energy. If energy E (in joules) is transferred in time t (in seconds):

$$\text{power, } P \text{ (in watts)} = \frac{E}{t}$$

b A weightlifter raises a 20 kg metal bar through a height of 1.5 m.

i Calculate the gain of gravitational potential energy. The gravitational field strength of the Earth at its surface is 10 N/kg.

ii The bar is raised by the weightlifter in 0.5 seconds. Calculate the power of the weightlifter.

Summary questions

1 Copy and complete **a** to **c** using the words below. Each word can be used more than once.

decreases   increases   stays the same

a When a ball falls, its gravitational potential energy ....

b When a car travels along a level road, the gravitational potential energy of the car ....

c When a child on a swing moves from one extreme to the opposite extreme, her gravitational potential energy .... then ....

2 A student of weight 450 N steps on a box of height 0.20 m.

a Calculate the gain of gravitational potential energy of the student.

b Calculate the work done by the student if she steps on and off the box 50 times.

3 a A weightlifter raises a steel bar of mass 25 kg through a height of 1.2 m. Calculate the change of gravitational potential energy of the bar. The gravitational field strength at the surface of the Earth is 10 N/kg.

b The weightlifter then drops the bar and it falls vertically to the ground. Assume air resistance is negligible. What is the change of its gravitational potential energy in this fall?

P2 3.3 Kinetic energy

Learning objectives

- What does the kinetic energy of an object depend on?
- How can we calculate kinetic energy?
- What is elastic potential energy?

Did you know ... ?

Sports scientists design running shoes:

- to reduce the force of each impact when the runner's foot hits the ground
- to return as much kinetic energy as possible to each foot in each impact.

Figure 2 A sports shoe

a Some of the kinetic energy of the runner's foot is wasted in each impact. What is this energy transferred into?

Practical

Investigating a catapult

Use rubber bands to 'catapult' a trolley along a horizontal runway. Find out how the speed of the trolley depends on how much the catapult is pulled back before the trolley is released. For example, see if the distance needs to be doubled to double the speed. Figure 1 shows how the speed of the trolley can be measured.

Practical

Investigating kinetic energy

- The kinetic energy of an object is the energy it has due to its motion. It depends on its mass and its speed.

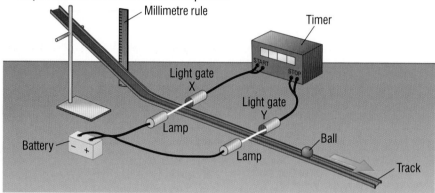

Figure 1 Investigating kinetic energy

Figure 1 shows how we can investigate how the **kinetic energy** of a ball depends on its speed.

1 The ball is released on a slope from a measured height above the foot of the slope. We can calculate the gravitational potential energy it loses from its mass × gravitational field strength × its drop of height. This is equal to its gain of kinetic energy.

2 The ball is timed, using light gates, over a measured distance between X and Y after the slope.

- Why do light gates improve the quality of the data you can collect in this investigation?

Some sample measurements for a ball of mass 0.5 kg are shown in the table:

Height drop to foot of slope (metres, m)	0.05	0.10	0.16	0.20
Initial kinetic energy of ball (joules, J)	0.25	0.50	0.80	1.00
Time to travel 1.0 m from X to Y (seconds, s)	0.98	0.72	0.57	0.50
Speed (metres/second, m/s)	1.02			2.00

Work out the speed in each case. The first and last values have been worked out for you. Can you see a link between speed and the height drop? The results show that the greater the height drop, the faster the speed is. So we can say that the kinetic energy of the ball increases if the speed increases.

The kinetic energy formula

The table shows that when the height drop is increased by four times from 0.05 m to 0.20 m, the speed doubles. The height drop is directly proportional to the (speed)^2. Since the height drop is a measure of the ball's kinetic energy, we can say that the ball's kinetic energy is directly proportional to the square of its speed.

b Check the other measurements in the table to see if they fit this rule.

The exact link between the kinetic energy of an object and its speed is given by the equation:

kinetic energy = **½ × mass** × **speed²**
(joules, J)   (kilograms, kg)   (metres/second, m/s)²

Maths skills

Worked example

Calculate the kinetic energy of a vehicle of mass 500 kg moving at a speed of 12 m/s.

Solution

kinetic energy = ½ × mass × speed² = 0.5 × 500 kg × (12 m/s)² = 36 000 J.

Kinetic energy recovery systems (KERS) in vehicles store energy when the vehicle brakes and use it later. In 2009, some Formula 1 racing cars were fitted with a flywheel. The kinetic energy of the vehicle could be transferred to the flywheel in braking and used later to boost the vehicle's speed when overtaking. Other vehicles, including hybrid cars, use an electric generator to transfer kinetic energy into electrical energy, which is then stored in a battery.

Using elastic potential energy

When you stretch a rubber band or a bowstring, the work you do is stored in it as **elastic potential energy**. Figure 3 shows one way you can **transfer** elastic potential energy into kinetic energy.

An object is **elastic** if it regains its shape after being stretched or squashed. A rubber band is an example of an elastic object.

Elastic potential energy is the energy stored in an elastic object when work is done on it to change its shape.

Summary questions

1 Copy and complete **a** and **b** using the words below.

elastic potential   *kinetic*   *gravitational potential*

A student on a trampoline falls on to the trampoline and rebounds.
   a Before she rebounds, the impact decreases her .... energy to zero.
   b During the rebound, ... energy changes into .... energy and .... energy.

2 **a** A catapult is used to fire an object into the air. Describe the energy transfers when the catapult is:
   i stretched
   ii released.
   b An object of weight 2.0 N fired vertically upwards from a catapult reaches a maximum height of 5.0 m. Calculate:
   i the gain of gravitational potential energy of the object
   ii the kinetic energy of the object when it left the catapult.

3 A car moving at a constant speed has 360 000 J of kinetic energy. When the driver applies the brakes, the car stops in a distance of 100 m.
   a Calculate the force that stops the vehicle.
   b The speed of the car was 30 m/s when its kinetic energy was 360 000 J. Calculate its mass.

 Maths skills

We can write this word equation using symbols:

$$E_K = ½ × m × v^2$$

Where:
E_K = kinetic energy in joules, J
m = mass in kilograms, kg
v = speed in metres/second, m/s.

AQA **Examiner's tip**

Don't forget to square the speed when calculating kinetic energy. This is a common mistake.

Figure 3 Using elastic potential energy

Key points

- The **kinetic energy** of a moving object depends on its mass and its speed.

- Kinetic energy (J) = ½ × mass (kg) × speed² (m/s)²

- **Elastic potential energy** is the energy stored in an elastic object when work is done on the object.

P2 3.4 Momentum

Learning objectives

- How can we calculate momentum?
- What is the unit of momentum?
- What happens to the total momentum of two objects when they collide?

Momentum is important to anyone who plays a contact sport. In a game of rugby, a player with a lot of momentum is very difficult to stop.

The momentum of a moving object = mass × velocity.

So momentum has a size and a direction.

The unit of momentum is the kilogram metre/second (kg m/s).

We can write the word equation above using symbols: $p = m \times v$

Where:
p = momentum in kilograms metres/second, kg m/s
m = mass in kilograms, kg
v = speed in metres/second, m/s

Figure 1 A contact sport

Maths skills

Worked example

Calculate the momentum of a sprinter of mass 50 kg running at a velocity of 10 m/s.

Solution

Momentum = mass × velocity = 50 kg × 10 m/s = 500 kg m/s

a Calculate the momentum of a 40 kg person running at 6 m/s.

Practical

Investigating collisions

When two objects collide, the momentum of each object changes. Figure 2 shows how to use a computer and a motion sensor to investigate a collision between two trolleys.

Trolley A is given a push so it collides with a stationary trolley B. The two trolleys stick together after the collision. The computer gives the velocity of A before the collision and the velocity of both trolleys afterwards.

- What does each section of the velocity–time graph show?

1 **For two trolleys of the same mass**, the velocity of trolley A is halved by the impact. The combined mass after the collision is twice the moving mass before the collision. So the momentum (= mass × velocity) after the collision is the same as before the collision.

2 **For a single trolley pushed into a double trolley**, the velocity of A is reduced to one-third. The combined mass after the collision is three times the initial mass. So once again, the momentum after the collision is the same as the momentum before the collision.

In both tests, the total momentum is unchanged (i.e. is conserved) by the collision. This is an example of the **conservation of momentum**. It applies to any system of objects provided the system is a closed system, which means no resultant force acts on it.

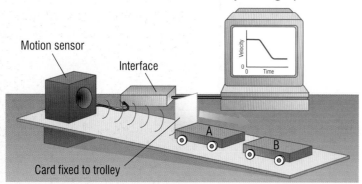

Motion sensor
Interface
Velocity
0
0 Time
A
B
Card fixed to trolley

Figure 2 Investigating collisions

Figure 3 A 'shunt' collision

 Did you know ... ?

If a vehicle crashes into the back of a line of cars, each car in turn is 'shunted' into the one in front. Momentum is transferred along the line of cars to the one at the front.

In general, the **law of conservation of momentum** states that

in a closed system, the total momentum before an event is equal to the total momentum after the event.

We can use this law to predict what happens whenever objects collide or push each other apart in an 'explosion'. Momentum is conserved in any collision or explosion provided no external forces act on the objects.

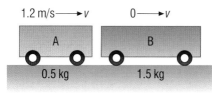

Figure 4 Worked example

 Maths skills

Worked example

A 0.5 kg trolley A is pushed at a velocity of 1.2 m/s into a stationary trolley B of mass 1.5 kg as shown in Figure 4. The two trolleys stick to each other after the impact.

Calculate:

a the momentum of the 0.5 kg trolley before the collision

b the velocity of the two trolleys straight after the impact.

Solution

a Momentum = mass × velocity = 0.5 kg × 1.2 m/s = 0.6 kg m/s.

b The momentum after the impact = the momentum before the impact = 0.6 kg m/s

(1.5 kg + 0.5 kg) × velocity after the impact = 0.6 kg m/s

the velocity after the impact $= \dfrac{0.6 \text{ kg m/s}}{2 \text{ kg}} = 0.3 \text{ m/s}$

b Calculate the speed after the collision if trolley A had a mass of 1.0 kg.

 Maths skills

Worked example

A 3000 kg truck moving at a velocity of 16 m/s crashes into the back of a stationary 1000 kg car. The two vehicles move together immediately after the impact. Calculate their velocity.

Solution

Let *v* represent the velocity of the vehicles after the impact.

momentum of the truck before the impact = 48 000 kg m/s

momentum of car before impact = 0 m/s

momentum of truck after impact = 3000 kg × *v*

momentum of car after impact = 1000 kg × *v*

3000 *v* + 1000 *v* = 48 000 + 0

4000 *v* = 48 000; *v* = 12 m/s

Summary questions

1 Complete **a** and **b** using the words below:

force   mass   momentum   velocity

a The momentum of a moving object is its .... × its ....

b .... is conserved when objects collide, provided no external .... acts.

2 a Calculate the momentum of an 80 kg rugby player running at a velocity of 5 m/s.

b An 800 kg car moves with the same momentum as the rugby player in part **a**. Calculate the velocity of the car.

3 A 1000 kg rail wagon moving at a velocity of 5.0 m/s on a level track collides with a stationary 1500 kg wagon. The two wagons move together after the collision.

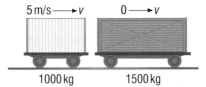

Figure 5

a Calculate the momentum of the 1000 kg wagon before the collision.

b Show that the wagons move at a velocity of 2.0 m/s after the collision.

Key points

● Momentum = mass × velocity

● The unit of momentum is kg m/s.

● Momentum is conserved whenever objects interact, provided the objects are in a closed system so that no external forces act on them.

P2 3.5   Explosions

Learning objectives

- Why does momentum have a direction as well as size?
- When two objects push each other apart
  - do they move away at different speeds?
  - why is their total momentum zero?

If you are a skateboarder, you will know that the skateboard can shoot away from you when you jump off it. Its momentum is in the opposite direction to your own momentum. What can we say about the total momentum of objects when they fly apart from each other?

Practical

Investigating a controlled explosion

Figure 1 shows controlled explosion using trolleys. When the trigger rod is tapped, a bolt springs out and the trolleys recoil (spring back) from each other.

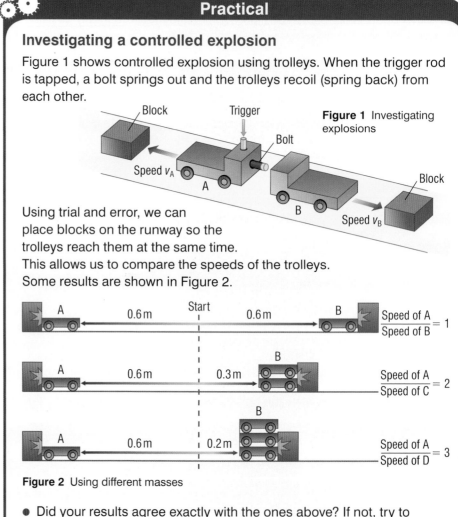

Figure 1 Investigating explosions

Using trial and error, we can place blocks on the runway so the trolleys reach them at the same time.
This allows us to compare the speeds of the trolleys.
Some results are shown in Figure 2.

Figure 2 Using different masses

- Did your results agree exactly with the ones above? If not, try to explain why.

- Two single trolleys travel equal distances in the same time. This shows that they recoil at equal speeds.
- A double trolley only travels half the distance that a single trolley does. Its speed is half that of the single trolley.

In each test:

1 the mass of the trolley × the speed of the trolley is the same, and
2 they recoil in opposite directions.

So momentum has size and direction. The results show that the trolleys recoil with equal and opposite momentum.

  a Why does a stationary rowing boat recoil when someone jumps off it?

Conservation of momentum

In the trolley examples:

- momentum of A after the explosion = (mass of A × velocity of A)
- momentum of B after the explosion = (mass of B × velocity of B)
- total momentum before the explosion = 0 (because both trolleys were at rest).

Using conservation of momentum gives:

(mass of A × velocity of A) + (mass of B × velocity of B) = 0

Therefore

(mass of A × velocity of A) = – (mass of B × velocity of B)

The minus sign after the equal sign tells us that the momentum of B is in the opposite direction to the momentum of A. The equation tells us that A and B move apart with equal and opposite amounts of momentum. So the total momentum after the explosion is the same as before it.

Momentum in action

When a shell is fired from an artillery gun, the gun barrel recoils backwards. The recoil of the gun barrel is slowed down by a spring. This lessens the backwards motion of the gun.

b In the worked example, if the mass of the gun had been much greater than 2000 kg, why would the speed of the shell have been greater?

Summary questions

1 A 60 kg skater and a 80 kg skater standing in the middle of an ice rink push each other away. Copy and complete **a** to **c** using the words below:

force   momentum   velocity

80 kg   60 kg

Figure 4

a They move apart with equal and opposite ....
b The 60 kg skater moves away with a bigger .... than the other skater.
c They push each other with equal and opposite ....

2 In Question **1**, the 60 kg skater moves away at 2.0 m/s. Calculate:
a her momentum
b the velocity of the other skater.

3 A 600 kg cannon recoils at a speed of 0.5 m/s when a 12 kg cannon ball is fired from it.
a Calculate the velocity of the cannon ball when it leaves the cannon.
b Calculate the kinetic energy of:
i the cannon
ii the ball.

🖩 Maths skills

Worked example

An artillery gun of mass 2000 kg fires a shell of mass 20 kg at a velocity of 120 m/s. Calculate the recoil velocity of the gun.

Solution

Applying the conservation of momentum gives:

mass of gun × recoil velocity of gun = – (mass of shell × velocity of shell)

If we let V represent the recoil velocity of the gun,

$2000\,kg \times V = -(20\,kg \times 120\,m/s)$

$V = \dfrac{2400\,kg\,m/s}{2000\,kg} = -1.2\,m/s$

Figure 3 An artillery gun in action

Key points

- Momentum is mass × velocity and velocity is speed in a certain direction.

- When two objects push each other apart, they move apart:
  – with different speeds if they have unequal masses
  – with equal and opposite momentum so their total momentum is zero.

P2 3.6

Impact forces

Learning objectives

- When vehicles collide, what does the force of the impact depend on?

- How does the impact force depend on the impact time?

- What can we say about the impact forces and the total momentum when two vehicles collide?

Crumple zones at the front end and rear end of a car are designed to lessen the force of an impact. The force changes the momentum of the car.

- In a front-end impact, the momentum of the car is reduced.

- In a rear-end impact (where a vehicle is struck from behind by another vehicle), the momentum of the car is increased.

In both cases the effect of a crumple zone is to increase the impact time and so lessen the impact force.

Practical

Investigating impacts

We can test an impact using a trolley and a brick, as shown in Figure 1. When the trolley hits the brick, the plasticine flattens on impact, making the impact time longer. This is the key factor that reduces the impact force.

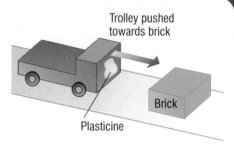

Figure 1 Investigating impacts

Figure 2 A crash test. Car makers test the design of a crumple zone by driving a remote control car into a brick wall.

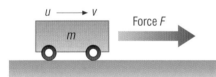

Figure 3 Impact force

a Why is rubber matting under a child's swing a good idea?

Impact time

Let's see why making the impact time longer reduces the impact force.

Suppose a moving trolley hits another object and stops. The impact force on the trolley acts for a certain time (the impact time) and causes it to stop. A soft pad on the front of the trolley would increase the impact time and would allow the trolley to travel further before it stops. The momentum of the trolley would be lost over a longer time and its kinetic energy would be transferred over a greater distance.

1 The kinetic energy of the trolley is transferred to the pad as work done by the impact force in squashing the pad.

2 Since work done = force × distance, the impact force is therefore reduced because the distance is increased.

The longer the impact time is, the more the impact force is reduced.

If we know the impact time, we can calculate the impact force as follows:

- From P2 1.2, since acceleration = change of velocity ÷ time taken, we can work out the deceleration by dividing the change of velocity by the impact time.

- From P2 2.3, since force = mass × acceleration, we can now calculate the impact force by multiplying the mass of the trolley by the deceleration.

The above method shows how much the impact force can be reduced by increasing the impact time. Car safety features such as crumple zones and side bars increase the impact time and so reduce the impact force.

b In a car crash, why does wearing a car seat belt reduce the impact force on the wearer?

??? Did you know ...?

Scientists at Oxford University have developed new lightweight material for bullet-proof vests. The material is so strong and elastic that bullets bounce off it.

Maths skills

Worked example

A bullet of mass 0.004 kg moving at a velocity of 90 m/s is stopped by a bulletproof vest in 0.0003 s.

Calculate **a** the deceleration and **b** the impact force.

Solution

a Initial velocity of bullet = 90 m/s

Final velocity of bullet = 0

Change of velocity = final velocity − initial velocity

$$= 0 - 90\,\text{m/s} = -90\,\text{m/s}$$

(where the minus sign tells us the change of velocity is a decrease)

$$\text{Deceleration} = \frac{\text{change of velocity}}{\text{impact time}} = \frac{-90\,\text{m/s}}{0.0003\,\text{s}} = -300\,000\,\text{m/s}^2$$

b Using 'force = mass × acceleration', impact force = 0.004 kg × $-300\,000\,\text{m/s}^2 = -1200\,\text{N}$

 Did you know ...?

We sometimes express the effect of an impact on an object or person as a force to weight ratio. We call this the **g-force**. For example, a g-force of 2 g means the force on an object is twice its weight. You would experience a g-force of:

- about 3–4 g on a fairground ride that whirls you round
- about 10 g in a low-speed car crash
- more than 50 g in a high-speed car crash. You would be lucky to survive!

c Calculate the impact force if the impact time had been 0.0002 s.

Two-vehicle collisions

When two vehicles collide, they exert equal and opposite impact forces on each other at the same time. The change of momentum of one vehicle is therefore equal and opposite to the change of momentum of the other vehicle. The total momentum of the two vehicles is the same after the impact as it was before the impact, so momentum is conserved – assuming no external forces act.

For example, suppose a fast-moving truck runs into the back of a stationary car. The impact decelerates the truck and accelerates the car. Assuming the truck's mass is greater than the mass of the car, the truck loses momentum and the car gains momentum.

Summary questions

1 Copy and complete **a** to **c** using the words below:

equal   greater   smaller

a The greater the mass of a moving object is the .... the force needed to stop it in a certain time.

b When two objects collide, they exert .... forces on each other.

c When two vehicles collide, the vehicle with the .... mass has a greater change of velocity.

2 **a** An 800 kg car travelling at 30 m/s is stopped safely when the brakes are applied. What deceleration and braking force is required to stop it in **i** 6.0 s?   **ii** 30 s?

b If the vehicle in part **a** had been stopped in a collision lasting less than a second, explain why the force on it would have been much greater.

3 A 2000 kg van moving at a velocity of 12 m/s crashes into the back of a stationary truck of mass 10 000 kg. Immediately after the impact, the two vehicles move together.

a Show that the velocity of the van and the truck immediately after the impact was 2 m/s.

b The impact lasted for 0.3 seconds. Calculate the **i** deceleration of the van   **ii** force of the impact on the van.

Key points

- When vehicles collide, the force of the impact depends on mass, change of velocity, and the duration of the impact.

- The longer the impact time is, the more the impact force is reduced.

- When two vehicles collide,
  - they exert equal and opposite forces on each other
  - their total momentum is unchanged.

P2 3.7

Car safety

Learning objectives

- Why do seat belts and air bags reduce the force on people in car accidents?
- How do side impact bars and crumple zones work?
- How can we work out if a car in a crash was 'speeding'?

When you travel in a car, you want to feel safe if the car is in a crash. In this topic, we look at different car safety features that are designed to keep us safe.

Clunk click!

When seat belts were first introduced, some car users claimed they should not be forced by law to wear them. A very successful campaign was launched to convince car users to 'belt up'. It included the catchy phrase '*Clunk click every trip*'. As a result, deaths and injuries in road accidents fell significantly.

A **seat belt** stops its wearer from continuing forwards when the car suddenly stops. Someone without a seat belt would hit the windscreen in a 'short sharp' impact and suffer major injury.

- The time taken to stop someone in a car is longer if they are wearing a seat belt than if they are not. So the decelerating force is reduced by wearing a seat belt.
- The seat belt acts across the chest so it spreads the force out. Without the seat belt, the force would act on the head when it hits the windscreen.

   a A seat belt 'locks' when in an impact. What would happen to the wearer if it didn't lock?

Figure 1 An air bag in action

Air bags

Most new cars are fitted with front air bags that protect the driver and the front passenger. Some new cars also have side air bags. These bags protect people in the car from an impact on the side of the car. In a car crash, an inflated air bag spreads the force of an impact across the upper part of the body. It also increases the duration of the impact time. So the effect of the force is lessened compared with a seat belt.

Child car seats

Any baby or child in a car must be strapped in a child car seat. This law applies to children up to 12 years old or up to 1.35 metres in height. Different types of child car seat must be used for babies up to 9 months old, infants up to about 4 years old and children over 4.

- Baby seats must face backwards.
- Children under 4 years old should usually be in a child car seat fitted to a back seat.

The law was brought in to reduce deaths and serious injuries of children in cars. Before the law was passed, dozens of children were killed and hundreds were seriously injured each year in car accidents. Many such accidents happened during the school run. The driver is responsible for making sure every child in their car is seated safely in a correct type of seat.

   b Why are ordinary car seat belts unsafe for children?

Figure 2 A child car seat

Safety costs

Car makers need to sell cars. If their cars are too expensive, people won't buy them. Safety features add to the cost of a new car. Some safety features (e.g. seat belts) are required by law and some (e.g. side impact bars) are optional.

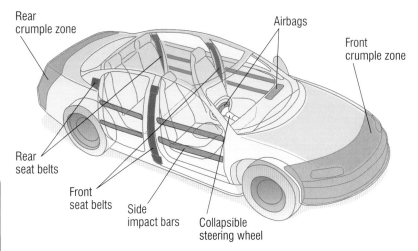

Figure 3 Car safety features

Activity

a With the help of your friends, find out what safety features are in new cars. Find out if they are compulsory or optional. List the price (including tax) of each car.

b Use your information to say if cheaper cars have fewer safety features than more expensive cars.

Activity

Brakes are very important vehicle safety features! Flywheel brakes can transfer large amounts of energy very quickly and very efficiently, unlike ordinary friction brakes which can overheat and wear away. Electric brakes (see P2 3.3) waste energy due to the heating effect of the electric current.

a State and explain the advantages of fitting flywheel brakes in addition to friction brakes in a racing car.

b Explain why flywheel brakes would be better than electric brakes for additional braking on a racing car.

Summary questions

1 Why are rear-facing car seats for babies safer than front-facing seats?

2 Explain why an inflated air bag in front of a car user reduces the force on a user in a 'head-on' crash.

3 A car crashed into a lorry that was crossing a busy road. The speed limit on the road was 60 miles per hour (27 m/s).

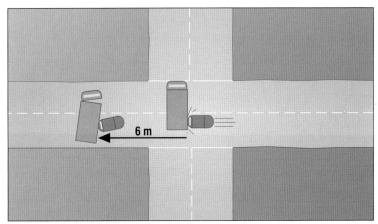

Figure 4 A road accident

The following measurements were made by police officers at the scene of a road crash:

- The car and lorry ended up 6 m from the point of impact.
- The car's mass was 750 kg and the lorry's mass was 2150 kg.
The speed of a vehicle for a braking distance of 6 m is 9 m/s.

a Use this speed to calculate the momentum of the car and the lorry immediately after the impact.

b Use conservation of momentum to calculate the velocity of the car immediately before the collision.

c Was the car travelling over the speed limit before the crash?

Key points

- Seat belts and air bags spread the force across the chest and they also increase the impact time.

- Side impact bars and crumple zones 'give way' in an impact so increasing the impact time.

- We can use the conservation of momentum to find the speed of a car before an impact.

Summary questions 🄚

1 a Copy and complete **i** and **ii** using the words below. Each term can be used once, twice or not at all.

equal to   *greater than*   *less than*

When a braking force acts on a vehicle and slows it down,

  i the work done by the force is .... the energy transferred from the object

  ii the kinetic energy after the brakes have been applied is .... the kinetic energy before they were applied.

b A student pushes a trolley of weight 150 N up a slope of length 20 m. The slope is 1.2 m high.

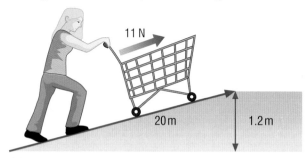

11 N

20 m   1.2 m

  i Calculate the gravitational potential energy gained by the trolley.

  ii The student pushed the trolley up the slope with a force of 11 N. Show that the work done by the student was 220 J.

  iii Give one reason why all the work done by the student was not transferred to the trolley as gravitational potential energy.

2 A 700 kg car moving at 20 m/s is stopped in a distance of 80 m when the brakes are applied.

a Show that the kinetic energy of the car at 20 m/s is 140 000 J.

b Calculate the braking force on the car.

3 A student of mass 40 kg standing at rest on a skateboard of mass 2.0 kg jumps off the skateboard at a speed of 0.30 m/s. Calculate:

a the momentum of the student

b the recoil velocity of the skateboard.

4 A car bumper is designed not to bend in impacts at less than 4 m/s. It was fitted to a car of mass 900 kg and tested by driving the car into a wall at 4 m/s. The time of impact was measured and found to be 1.8 s.

Show that the deceleration of the car was 2.2 m/s².

5 a Copy and complete **i** and **ii** using the words below. Each term can be used once, twice or not at all.

elastic potential energy   *kinetic energy*
gravitational potential energy

An object is catapulted from a catapult.

  i .... is stored in the catapult when it is stretched.

  ii The object has .... when it leaves the catapult.

b A stone of mass 0.015 kg is catapulted into the air and it reaches a height of 20 m before it descends and hits the ground some distance away.

  i Calculate the increase of gravitational potential energy of the stone when it reached its maximum height (g = 10 N/kg).

  ii State two reasons why the catapult stored more energy than that calculated in part **b i**?

6 A 1200 kg rail wagon moving at a velocity of 3.0 m/s on a level track collides with a stationary wagon of mass 800 kg. The 1200 kg truck is slowed down to a velocity of 1.0 m/s as a result of the collision.

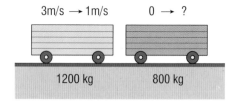

3m/s → 1m/s   0 → ?

1200 kg   800 kg

a Calculate the momentum of the 1200 kg wagon
  i before the collision
  ii after the collision.

b Calculate
  i the momentum, and
  ii the velocity of the 800 kg wagon after the collision.

c Calculate the kinetic energy of:
  i the 1200 kg wagon before the collision
  ii the 1200 kg wagon after the collision
  iii the 800 kg wagon after the collision.

d Give a reason why the total kinetic energy after the collision is not equal to the total kinetic energy before the collision.

AQA Examination-style questions

1 a Copy and complete the following sentences using the list of words and phrases below. Each one can be used once.

kinetic energy   work   power
gravitational potential energy

   i   Energy is transferred when .... is done.   (1)
   ii .... is the energy that an object has by virtue of its position in a gravitational field.   (1)
   iii The .... of an object depends on its mass and speed.   (1)
   iv .... is the energy transferred in a given time.   (1)

b Explain why a meteorite 'burns up' as it enters the Earth's atmosphere. Use ideas about work and energy.   (3)

2 The diagram shows three cars, **A**, **B** and **C**, travelling along a straight, level road.

A   Speed
                                                     40 m/s

650 kg

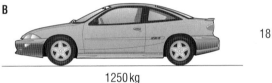

B

                                                     18 m/s

1250 kg

C

                                                     15 m/s

1500 kg

a Calculate the momentum of each of the vehicles and explain which one has the greatest momentum. Write down the equation you use. Show clearly how you work out your answer and give the unit.   (3)

b Car **C,** travelling at 15 m/s, crashes into the back of car **A** when car **A** is stationary. The cars move together after the collision.
   i Calculate the total momentum of the cars just after the collision.   (1)
   ii Calculate the speed of the two cars just after the collision.   (2)

c Explain, using ideas about momentum changes, how the crumple zone at the front of car **C** may reduce the chance of injury to the occupants during the collision.   (3)

3 When ploughing a field a horse and plough move 170 m and the horse pulls with a force of 800 N.

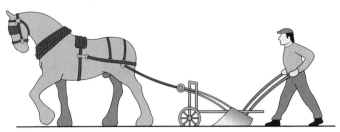

a Calculate the work done by the horse.

   Write down the equation you use. Show clearly how you work out your answer and give the unit.   (3)

b i The horse takes 3 minutes to plough 170 m. Calculate the power of the horse. Write down the equation you use. Show clearly how you work out your answer and give the unit.   (3)
   ii Calculate the kinetic energy of the horse. Write down the equation you use. Show clearly how you work out your answer and give the unit. Mass of horse = 950 kg   (3)

c Explain why the horse has to do more work if the field slopes uphill than it would do on level ground.   (2)

4 The picture shows a catapult.

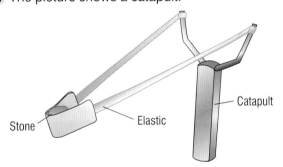

When a force is applied to the stone, work is done in stretching the elastic and the stone moves backwards.

a Calculate the work done if the average force applied to the stone is 20 N. The force moves it backwards 0.15 m. Write down the equation you use. Show clearly how you work out your answer and give the unit.   (3)

b Calculate the maximum speed of the stone after the catapult is released. The mass of the stone is 0.049 kg. Assume all the work done is transferred to the stone as kinetic energy when the catapult is released. Write down the equation you use. Show clearly how you work out your answer and give the unit.   (3)

P2 4.1

Electrical charges

Learning objectives

- What happens when insulating materials are rubbed together?
- What is transferred when objects become charged?
- What happens when charges are brought together?

??? Did you know ... ?

Take off a woolly jumper and listen out! You can hear it crackle as tiny sparks from static electricity are created. If the room is dark, you can even see the sparks.

You can get charged up just by sitting in a plastic chair. If this happens, you may feel a slight shock from static electricity when you stand up.

Have you ever stuck a balloon on a ceiling? All you need to do is to rub the balloon on your clothing before you touch it on the ceiling. The rubbing action charges the balloon with **static electricity**. In other words, the balloon becomes electrically charged. The charge on the balloon attracts it to the ceiling.

a Why does a TV screen crackle when you switch it on?

Demonstration

The Van de Graaff generator

A Van de Graaff generator can make your hair stand on end. The dome charges up when the generator is switched on. Massive sparks are produced if the charge on the dome builds up too much.

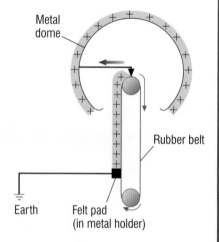

Figure 1 The Van de Graaff generator

The Van de Graaff generator charges up because:
- the belt rubs against a felt pad and becomes charged
- the belt carries the charge onto an insulated metal dome
- sparks are produced when the dome can no longer hold any more charge.

- Why should you keep away from a Van de Graaff generator?

Inside the atom

The **protons** and **neutrons** make up the nucleus of the atom. Electrons move about in the space round the nucleus.
- A proton has a positive charge.
- An electron has an equal negative charge.
- A neutron is uncharged.

An uncharged atom has equal numbers of electrons and protons. Only electrons can be transferred to or from an atom. A charged atom is referred to as an **ion**.

1 Adding electrons to an uncharged atom makes it negative (because the atom then has more electrons than protons).
2 Removing electrons from an uncharged atom makes it positive (because the atom has fewer electrons than protons).

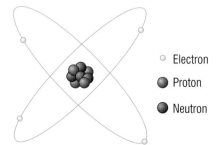

Figure 2 Inside an atom

Charging by friction

Some insulators become charged by rubbing them with a dry cloth.

- Rubbing a polythene rod with a dry cloth transfers electrons to the surface atoms of the rod from the cloth. So the polythene rod becomes negatively charged.
- Rubbing a perspex rod with a dry cloth transfers electrons from the surface atoms of the rod on to the cloth. So the perspex rod becomes positively charged.

b Glass is charged positively when it is rubbed with a cloth.
Does glass gain or lose electrons when it is charged?

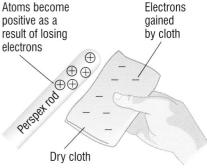

Figure 3 Charging by friction

Practical

The force between two charged objects

Two charged objects exert a force on each other. Figure 4 shows how you can investigate this force.

- What happens?

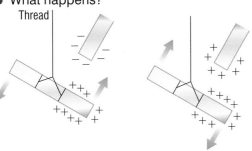

Figure 4 The law of force for charges

Your results in the experiment above should show that:

- two objects with the same type of charge (i.e. like charges) repel each other
- two objects with different types of charge (i.e. unlike charges) attract each other.

Like charges repel. Unlike charges attract.

c What force keeps the electrons inside an atom?

Summary questions

1 Copy and complete **a** and **b** using the words below. Each word can be used more than once.

to   from   loses   gains

a When a polythene rod is charged using a dry cloth, it becomes negative because it .... electrons that transfer .... it .... the cloth.

b When a perspex rod is charged using a dry cloth, it becomes positive because it .... electrons that transfer .... it .... the cloth.

2 When rubbed with a dry cloth, perspex becomes positively charged. Polythene and ebonite become negatively charged. State whether or not attraction or repulsion takes place when:

a a perspex rod is held near a polythene rod

b a perspex rod is held near an ebonite rod

c a polythene rod is held near an ebonite rod.

Key points

- Certain insulating materials become charged when rubbed together.

- Electrons are transferred when objects become charged:
  - Insulating materials that become positively charged when rubbed lose electrons.
  - Insulating materials that become negatively charged when rubbed gain electrons.

- Like charges repel; unlike charges attract.

P2 4.2

Electric circuits

Learning objectives

- Why are electric circuits represented by circuit diagrams?
- What is the difference between a battery and a cell?
- What determines the size of an electric current?
- How can we calculate the size of an electric current from the charge flow and the time taken?

An electric torch can be very useful in a power cut at night. But it needs to be checked to make sure it works. Figure 1 shows what is inside a torch. The circuit shows how the torch bulb is connected to the switch and the two cells.

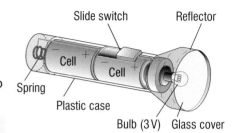

Figure 1 An electric torch

a Why does the switch have to be closed to turn the torch bulb on?

A circuit diagram shows us how the components in a circuit are connected together. Each component has its own symbol. Figure 2 shows the symbols for some of the components you will meet in this course. The function of each component is also described. You need to recognise these symbols and remember what each component is used for – otherwise you'll get mixed up in your exams. More importantly, you could get a big shock if you mix them up!

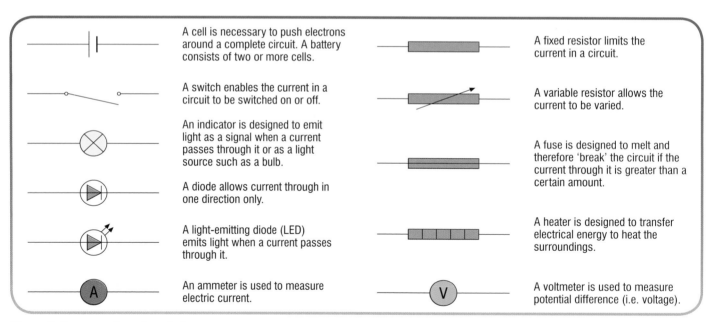

A cell is necessary to push electrons around a complete circuit. A battery consists of two or more cells.

A switch enables the current in a circuit to be switched on or off.

An indicator is designed to emit light as a signal when a current passes through it or as a light source such as a bulb.

A diode allows current through in one direction only.

A light-emitting diode (LED) emits light when a current passes through it.

An ammeter is used to measure electric current.

A fixed resistor limits the current in a circuit.

A variable resistor allows the current to be varied.

A fuse is designed to melt and therefore 'break' the circuit if the current through it is greater than a certain amount.

A heater is designed to transfer electrical energy to heat the surroundings.

A voltmeter is used to measure potential difference (i.e. voltage).

Figure 2 Components and symbols

⬭ links

See P2 4.4 for two other symbols you need to know.

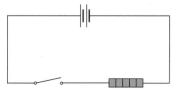

Figure 3

b What components are in the circuit diagram in Figure 3?

Electric current

An electric current is a flow of charge. When an electric torch is on, millions of **electrons** pass through the torch bulb and through the cell every second. Each electron carries a negative charge. Metals contain lots of electrons that move about freely between the positively charged metal ions. These electrons stop the ions moving away from each other. The electrons pass through the bulb because its filament is made of a metal. The electrons transfer energy from the cell to the torch bulb.

The size of an electric current is the rate of flow of electric charge. This is the flow of charge per second. The greater the number of electrons that pass through a component, the bigger the current passing through it.

Electric charge is measured in **coulombs (C)**. Electric current is measured in **amperes** (A) sometimes abbreviated as 'amps'.

An electric current of 1 ampere is a rate of flow of charge of 1 coulomb per second. If a certain amount of charge flows steadily through a wire or a component in a certain time,

$$\text{the current in amperes} = \frac{\text{charge flow in coulombs}}{\text{time taken in seconds}}$$

We can write the equation above using symbols as follows:

$$I = \frac{Q}{t}$$

Where:
I = current in amperes, A
Q = charge in coulombs, C
t = time taken in seconds, s.

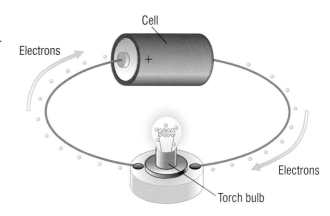

Cell

Electrons

Torch bulb

Electrons

Figure 4 Electrons on the move

 Maths skills

Worked example

A charge of 8.0 C passes through a bulb in 4.0 seconds. Calculate the current through the bulb.

Solution

$$I = \frac{Q}{t} = \frac{8.0\,\text{C}}{4.0\,\text{s}} = 2.0\,\text{A}$$

Practical

Circuit tests

Connect a variable resistor in series with the torch bulb and a cell, as shown in Figure 6.

Adjust the slider of the variable resistor. This alters the amount of current flowing through the bulb and therefore affects its brightness.

Figure 6 Using a variable resistor

- In Figure 6, the torch bulb goes dim when the slider is moved one way. What happens if the slider is moved back again?
- What happens if you include a diode in the circuit?

Summary questions

1 Name the numbered components in the circuit diagram in Figure 7.

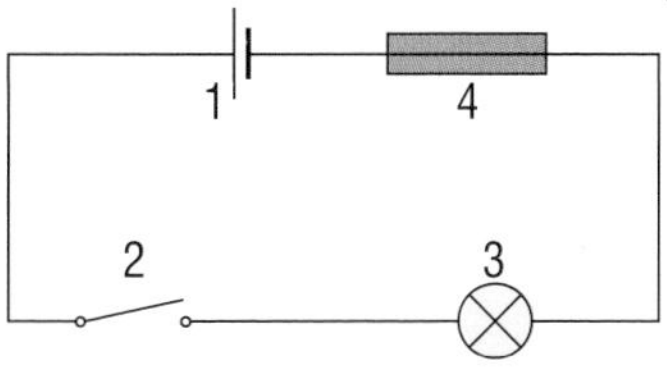

2 a Redraw the circuit diagram in Question 1 with a diode in place of the switch so it allows current through.
   b What further component would you need in this circuit to alter the current in it?

3 a What is a light-emitting diode?
   b What is a variable resistor used for?

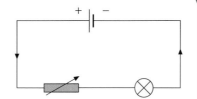

Figure 7

You would damage a portable radio if you put the batteries in the wrong way round, unless a diode is in series with the battery. The diode only allows current through when it is connected as shown in Figure 5.

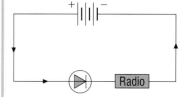

Radio

Figure 5 Using a diode

c Would the radio in Figure 5 work if the diode was turned round in the circuit?

Key points

- Every component has its own agreed symbol. A circuit diagram shows how components are connected together.

- A battery consists of two or more cells connected together.

- The size of an electric current is the rate of flow of charge.

- Electric current = charge flow/time taken

P2 4.3 Resistance ⓚ

Learning objectives

- What do we mean by potential difference?
- What is resistance and what is its unit?
- What is Ohm's law?
- What happens if you reverse the current in a resistor?

Ammeters and voltmeters

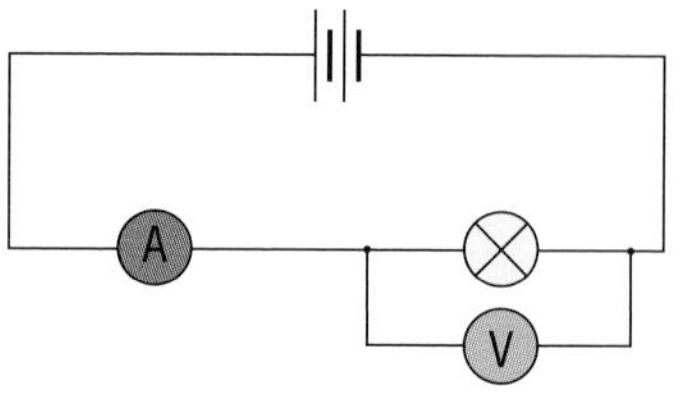

Look at the circuit in Figure 1. The battery forces electrons to pass through the ammeter and the bulb.

- The ammeter measures the current through the torch bulb. It is connected in **series** with the bulb so the current through them is the same. The ammeter reading gives the current in amperes (or milliamperes (mA) for small currents, where 1 mA = 0.001 A).

- The voltmeter measures the **potential difference** (pd) across the torch bulb. This is the amount of work done or energy transferred to the bulb by each coulomb of charge that passes through it. The unit of potential difference is the **volt (V)**. We sometimes use the word 'voltage' for potential difference.

- The voltmeter is connected in **parallel** with the torch bulb so it measures the pd across it. The voltmeter reading gives the pd in volts (V).

Figure 1 Using an ammeter and a voltmeter

When charge flows steadily through a component,

$$\text{the potential difference across the component in volts} = \frac{\text{work done in joules}}{\text{charge in coulombs}}$$

We can write the equation above using symbols as:

$$V = \frac{W}{Q}$$

Where:
V = the potential difference in volts, V
W = work done or energy transferred in joules, J
Q = charge in coulombs, C.

Electrons passing through a torch bulb have to push their way through lots of vibrating ions in the metal filament. The ions resist the passage of electrons through the torch bulb.

Higher

We define the **resistance** of an electrical component as:

$$\text{Resistance (ohms)} = \frac{\text{potential difference (volts)}}{\text{current (amperes)}}$$

The unit of resistance is the **ohm**. The symbol for the ohm is the Greek letter Ω (omega). Note that a resistor in a circuit limits the current. For a given pd, the larger the resistance of a resistor, the smaller the current is.

We can write the definition above as:

$$R = \frac{V}{I}$$

Where:
R = resistance (ohms)
V = potential difference (volts)
I = current (amperes).

a The current through a wire is 0.5 A when the potential difference across it is 4.0 V. Calculate the resistance of the wire.

Maths skills

Worked example

The energy transferred to a bulb is 24 J when 8.0 C of charge passes through it. Calculate the potential difference across the bulb.

Solution

$$V = \frac{W}{Q} = \frac{24\,\text{J}}{8.0\,\text{C}} = 3.0\,\text{V}$$

AQA *Examiner's tip*

Ammeters are always connected in series and voltmeters are always connected in parallel.

Maths skills

Rearranging the equation

$R = \dfrac{V}{I}$ gives $V = IR$ or $I = \dfrac{V}{R}$

Practical

Investigating the resistance of a wire

Does the resistance of a wire change when the current through it is changed? Figure 2 shows how we can use a variable resistor to change the current through a wire. Make your own measurements and use them to plot a current–potential difference graph like the one in Figure 2.

- Discuss how your measurements compare with the ones from the table used to plot the graph in Figure 2.
- Calculate the resistance of the wire you tested.

Current (A)	0	0.05	0.10	0.15	0.20	0.25
Potential difference (V)	0	0.50	1.00	1.50	2.00	2.50

a

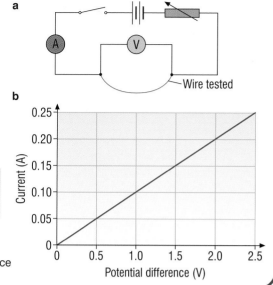

b

Figure 2 Investigating the resistance of a wire.
a Circuit diagram **b** A current–potential difference graph for a wire

b Calculate the resistance of the wire that gave the results in the practical table.

Current–potential difference graphs

The graph in Figure 2 is a straight line through the origin. This means that the current is directly proportional to the potential difference. In other words, the resistance (= pd ÷ current) is constant. This was first discovered for a wire at constant temperature by Georg Ohm and is known as **Ohm's law**:

The current through a resistor at constant temperature is directly proportional to the potential difference across the resistor.

We say a wire is an **ohmic conductor** because its resistance is constant. As shown in Figure 3, reversing the pd makes no difference to the shape of the line. The resistance is the same whichever direction the current is in.

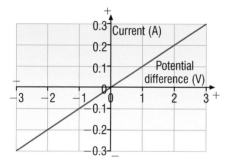

Figure 3 A current–potential difference graph for a resistor

Summary questions

1 Copy and complete **a** and **b** using the words below. Each word can be used once, twice or not at all.

decreases   increases   reverses   stays the same

a If the current through a resistor is decreased, the pd across the resistor ....

b If the current through a resistor is reversed, the pd across the resistor .... and the resistance of the resistor ....

2 Calculate the missing value in each line of the table, using the equation $V = IR$ or a rearrangement of it.

Resistor	Current (A)	Potential difference (V)	Resistance (Ω)
W	2.0	12.0	
X	4.0		20
Y		6.0	3.0

Key points

- The potential difference across a component (in volts) =

$$\frac{\text{work done or energy transferred (in joules)}}{\text{charge (in coulombs)}}$$

- Resistance (in ohms) =

$$\frac{\text{potential difference (volts)}}{\text{current (amperes)}}$$

- Ohm's law states that the current through a resistor at constant temperature is directly proportional to the potential difference across the resistor.

- Reversing the current through a component reverses the pd across it.

P2 4.4

More current–potential difference graphs

Learning objectives

- What happens to the resistance of a filament bulb as its temperature increases?

- How does the current through a diode depend on the potential difference across it?

- What happens to the resistance of a thermistor as its temperature increases and of an LDR as the light level increases?

Have you ever switched a light bulb on only to hear it 'pop' and fail? Electrical appliances can fail at very inconvenient times. Most electrical failures are because too much current passes through a component in the appliance.

Practical

Investigating different components

We can use the circuit in Figure 2 on the previous page to find out if the resistance of a component depends on the current. We can also see if reversing the component in the circuit has any effect.

Make your own measurements using a resistor, a filament bulb and a diode.

Plot your measurements on a current–potential difference graph. Plot the 'reverse' measurements on the negative section of each axis.

- Why can you use a line graph to display your data? (See H3 Using data.)

Using current–potential difference graphs

A filament bulb

Figure 1 shows the graph for a torch bulb (i.e. a low-voltage filament bulb).

- The line **curves** away from the current axis. So the current is *not* directly proportional to the potential difference. The filament bulb is a non-ohmic conductor.

- The resistance (= potential difference/current) increases as the current increases. So the resistance of a filament bulb increases as the filament temperature increases.

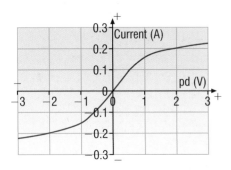

Figure 1 A current–potential difference graph for a filament bulb

The resistance of the metal filament increases as its temperature increases. This is because the ions in the metal filament vibrate more as the temperature increases. So they resist the passage of the electrons through the filament more.

Higher

- Reversing the potential difference makes no difference to the shape of the curve. The resistance is the same for the same current, regardless of its direction.

  a Calculate the resistance of the filament bulb at **i** 0.1 A **ii** 0.2 A.

The diode

Look at Figure 2, a graph for a diode.

- In the 'forward' direction, the line curves towards the current axis. So the current is not directly proportional to the potential difference. A diode is not an ohmic conductor.

- In the reverse direction, the current is negligible. So its resistance in the reverse direction is much higher than in the forward direction.

Note that a light-emitting diode (LED) emits light when a current passes through it in the forward direction.

  b What can we say about the forward resistance of a diode as the current increases?

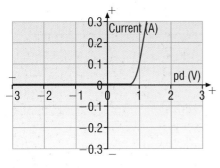

Figure 2 A current–potential difference graph for a diode

Practical

Thermistors and light-dependent resistors (LDRs)

We use thermistors and LDRs in sensor circuits. A thermistor is a temperature-dependent resistor. The resistance of an LDR depends on how much light is on it.

Test a thermistor and then an LDR in series with a battery and an ammeter.

● What did you find out about each component tested?

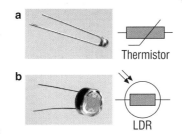

Figure 3 a A thermistor and its circuit symbol **b** An LDR and its circuit symbol

Current–potential difference graphs for a thermistor and an LDR

For a thermistor, Figure 4 shows the current–potential difference graph at two different temperatures.

● At constant temperature, the line is straight so its resistance is constant.
● If the temperature is increased, its resistance decreases.

For a light-dependent resistor, Figure 5 shows the current–potential difference graph in bright light and in dim light.

c What does the graph tell us about an LDR's resistance if the light intensity is constant?
d If the light intensity is increased, what happens to the resistance of the LDR?

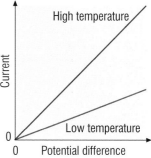

Figure 4 Thermistor graph

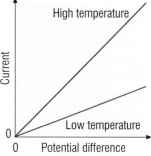

Figure 5 LDR graph

Did you know ... ?

When a light bulb fails, it usually happens when you switch it on. Because resistance is low when the bulb is off, a large current passes through it when you switch it on. If the current is too large, it burns the filament out.

Summary questions

1 Copy and complete sentences **a** to **d** using the words below:

diode filament bulb resistor thermistor

a The resistance of a .... decreases as its temperature increases.
b The resistance of a .... depends on which way round it is connected in a circuit.
c The resistance of a .... increases as the current through it increases.
d The resistance of a .... does not depend on the current through it.

2 A thermistor is connected in series with an ammeter and a 3.0 V battery, as shown.

Figure 6

a At 15 °C, the current through the thermistor is 0.2 A and the potential difference across it is 3.0 V. Calculate its resistance at this temperature.
b State and explain what happens to the ammeter reading if the thermistor's temperature is increased.

3 The thermistor in Figure 6 is replaced by a light-dependent resistor (LDR). State and explain what happens to the ammeter reading when the LDR is covered.

Key points

● *Filament bulb:* resistance increases with increase of the filament temperature.
● *Diode:* 'forward' resistance low; 'reverse' resistance high.
● *Thermistor:* resistance decreases if its temperature increases.
● *LDR:* resistance decreases if the light intensity on it increases.

Series circuits

- What can we say about the current and potential difference for components in a series circuit?

- How can we find the total resistance of resistors in series?

- What can we say about the potential difference of several cells in series?

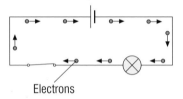

Figure 1 A torch bulb circuit

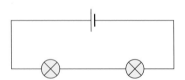

Figure 2 Bulbs in series

Circuit rules

In the torch circuit in Figure 1, the bulb, the cell and the switch are connected in series with each other. The same number of electrons passes through each component every second. So the same current passes through each component.

The same current passes through components in series with each other.

a If the current through the bulb is 0.12 A, what is the current through the cell?

In Figure 2, each electron from the cell passes through two bulbs. The electrons are pushed through each bulb by the cell. The potential difference (or **voltage**) of the cell is a measure of the energy transferred from the cell by each electron that passes through it. Since each electron in the circuit in Figure 2 goes through both bulbs, the potential difference of the cell is shared between the bulbs. This rule applies to any series circuit.

The total potential difference of the voltage supply in a series circuit is shared between the components.

b In Figure 2, if the potential difference of the cell is 1.2 V and the potential difference across one bulb is 0.8 V, what is the potential difference across the other bulb?

Cells in series

What happens if we use two or more cells in series in a circuit? Provided we connect the cells so they act in the same direction, each electron gets a push from each cell. So an electron would get the same push from a battery of three 1.5 V cells in series as it would from a single 4.5 V cell.

In other words, provided the cells act in the same direction:

The total potential difference of cells in series is the sum of the potential difference of each cell.

Table 1

Filament bulb	Voltmeter V_1 (volts)	Voltmeter V_2 (volts)
normal	1.5	0.0
dim	0.9	0.6
very dim	0.5	1.0

Practical

Investigating potential differences in a series circuit

Figure 3 shows how to test the potential difference rule for a series circuit. The circuit consists of a filament bulb in series with a variable resistor and a cell. We can use the variable resistor to see how the voltmeter readings change when we alter the current. Make your own measurements.

- How do they compare with the data in Table 1?

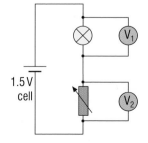

Figure 3 Voltage tests

The measurements in the table show that the voltmeter readings for each setting add up to 1.5 V. This is the potential difference of the cell. The share of the cell's potential difference across each component depends on the setting of the variable resistor.

c What would voltmeter V_2 read if voltmeter V_1 showed 0.4 V?

The resistance rule for components in series

In Figure 3, suppose the current through the bulb is 0.1 A when the bulb is dim.

Using data from Table 1:

● the resistance of the bulb would then be 9 Ω (= 0.9 V ÷ 0.1 A),
● the resistance of the variable resistor at this setting would be 6 Ω (= 0.6 V ÷ 0.1 A).

If we replaced these two components by a single resistor, what should its resistance be for the same current of 0.1 A? We can calculate this because we know the potential difference across it would be 1.5 V (from the cell). So the resistance would need to be 15 Ω (= 1.5 V ÷ 0.1 A). This is the sum of the resistance of the two components. The rule applies to any series circuit.

The total resistance of components in series is equal to the sum of the resistance of each component.

d What is the total resistance of a 2 Ω resistor in series with a 3 Ω resistor?

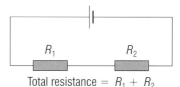

Total resistance = $R_1 + R_2$

Figure 4 Resistors in series

AQA **Examiner's tip**

Remember that in a series circuit the same current passes through all the components.

Summary questions

1 Copy and complete **a** and **b** using the words below. Each word can be used once, twice or not at all.

greater than   less than   the same as

For the circuit in Figure 5:

a The current through the battery is .... the current through resistor P.

b The potential difference across resistor Q is .... the potential difference across the battery.

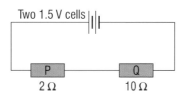

Two 1.5 V cells

P   2 Ω
Q   10 Ω

Figure 5

2 A 1.5 V cell is connected to a 3.0 Ω resistor and 2.0 Ω resistor in series with each other.
   a Draw the circuit diagram for this arrangement.
   b Calculate:
      i the total resistance of the two resistors
      ii the current through the resistors.

3 For the circuit in Question 1, each cell has a potential difference of 1.5 V.
   a Calculate:
      i the total resistance of the two resistors
      ii the total potential difference of the two cells.
   b Show that the current through the battery is 0.25 A.
   c Calculate the potential difference across each resistor.

Key points

● For components in series:
   – the current is the same in each component
   – adding the potential differences gives the total potential difference

● Adding the resistances gives the total resistance of resistors in series.

● For cells in series, acting in the same direction, the total potential difference is the sum of their individual potential differences.

P2 4.6 Parallel circuits

Did you know ... ?

A bypass is a parallel route. A heart bypass is another route for the flow of blood. A road bypass is a road that passes a town centre instead of going through it. For components in parallel, charge flows separately through each component. The total flow of charge is the sum of the flow through each component.

Practical

Investigating parallel circuits

Figure 1 shows how you can investigate the current through two bulbs in parallel with each other. You can use ammeters in series with the bulbs and the cell to measure the current through each component.

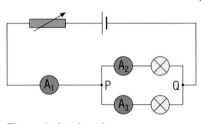

Figure 1 At a junction

Set up your own circuit and collect your data.

- How do your measurements compare with the ones for different settings of the variable resistor shown in Table 1 below?

- Discuss if your own measurements show the same pattern.

Look at the sample data below.

Table 1

Ammeter A_1 (A)	Ammeter A_2 (A)	Ammeter A_3 (A)
0.50	0.30	0.20
0.30	0.20	0.10
0.18	0.12	0.06

In each case, the reading of ammeter A_1 is equal to the sum of the readings of ammeters A_2 and A_3.

This shows that the current through the cell is equal to sum of the currents through the two bulbs. This rule applies wherever components are in parallel.

a If ammeter A_1 reads 0.40 A and A_2 reads 0.1 A, what would A_3 read?

The total current through the whole circuit is the sum of the currents through the separate components.

Potential difference in a parallel circuit

Figure 2 shows two resistors X and Y in parallel with each other. A voltmeter is connected across each resistor. The voltmeter across resistor X shows the same reading as the voltmeter across resistor Y. This is because each electron from the cell either passes through X or through Y. So it delivers the same amount of energy from the cell, whichever resistor it goes through. In other words:

For components in parallel, the potential difference across each component is the same.

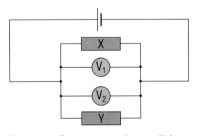

Figure 2 Components in parallel

Calculations on parallel circuits

Components in parallel have the same potential difference across them. The current through each component depends on the resistance of the component.

- The bigger the resistance of the component, the smaller the current through it. The resistor which has the largest resistance passes the smallest current.
- We can calculate the current using the equation:

$$\text{current (amperes)} = \frac{\text{potential difference (volts)}}{\text{resistance (ohms)}}$$

b A $3\,\Omega$ resistor and a $6\,\Omega$ resistor are connected in parallel in a circuit. Which resistor passes the most current?

 Maths skills

Worked example

The circuit diagram shows three resistors $R_1 = 1\,\Omega$, $R_2 = 2\,\Omega$ and $R_3 = 6\,\Omega$ connected in parallel to a 6V battery.

Calculate:

a the current through each resistor

b the current through the battery.

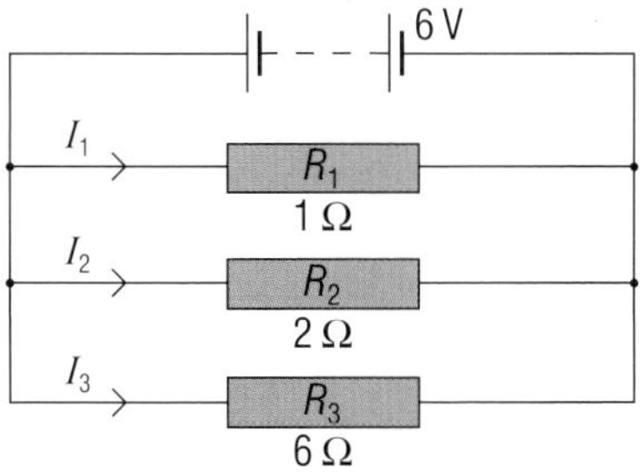

Solution

a $I_1 = \dfrac{V_1}{R_1} = \dfrac{6}{1} = 6\,\text{A}$

$I_2 = \dfrac{V_2}{R_2} = \dfrac{6}{2} = 3\,\text{A}$

$I_3 = \dfrac{V_3}{R_3} = \dfrac{6}{6} = 1\,\text{A}$

b The total current from the battery $= I_1 + I_2 + I_3 = 6\,\text{A} + 3\,\text{A} + 1\,\text{A} = 10\,\text{A}$

Figure 3

Summary questions

1 Copy and complete **a** and **b** using the words below:

current   potential difference

   a Components in parallel with each other have the same ....

   b For components in parallel, each component has a different ....

2 A 1.5V cell is connected across a $3\,\Omega$ resistor in parallel with a $6\,\Omega$ resistor.

   a Draw the circuit diagram for this circuit.

   b Show that the current through:

     i the $3\,\Omega$ resistor is 0.50 A   **ii** the $6\,\Omega$ resistor is 0.25 A.

   c Calculate the current passing through the cell.

3 The circuit diagram shows three resistors $R_1 = 2\,\Omega$, $R_2 = 3\,\Omega$ and $R_3 = 6\,\Omega$ connected to each other in parallel and to a 6V battery.

Calculate:

   a the current through each resistor

   b the current through the battery.

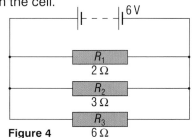

Figure 4

Key points

- For components in parallel:
  - the total current is the sum of the currents through the separate components
  - the bigger the resistance of a component, the smaller its current is.

- In a parallel circuit the potential difference is the same across each component.

- To calculate the current through a resistor in a parallel circuit, use this equation:

  current (amperes)

  $= \dfrac{\text{potential difference (volts)}}{\text{resistance (ohms)}}$

Summary questions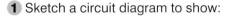

1 Sketch a circuit diagram to show:

a a torch bulb, a cell and a diode connected in series so that the torch bulb is on

b a variable resistor, two cells in series and a torch bulb whose brightness can be varied by adjusting the variable resistor.

2 Match each component in the list to each statement **a** to **d** that describes it.

diode   filament bulb   resistor   thermistor

a Its resistance increases if the current through it increases.

b The current through it is proportional to the potential difference across it.

c Its resistance decreases if its temperature is increased.

d Its resistance depends on which way round it is connected in a circuit.

3 a Sketch a circuit diagram to show two resistors P and Q connected in series to a battery of two cells in series with each other.

b In the circuit in part **a**, resistor P has a resistance of 4 Ω, resistor Q has a resistance of 2 Ω and each cell has a potential difference of 1.5 V. Calculate:
i the total potential difference of the two cells
ii the total resistance of the two resistors
iii the current in the circuit
iv the potential difference across each resistor.

4 a Sketch a circuit diagram to show two resistors R and S in parallel with each other connected to a single cell.

b In the circuit in part **a**, resistor R has a resistance of 2 Ω, resistor S has a resistance of 4 Ω and the cell has a potential difference of 2 V. Calculate:
i the current through resistor R
ii the current through resistor S
iii the current through the cell in the circuit.

5 Copy and complete **a** and **b** using the phrases below. Each option can be used once, twice or not at all.

different from   equal to

a For two components X and Y in series, the potential difference across X is usually .... the potential difference across Y.

b For two components X and Y in parallel, the potential difference across X is .... the potential difference across Y.

6 Figure 1 shows a light-dependent resistor is series with a 200 Ω resistor, a 3.0 V battery and an ammeter.

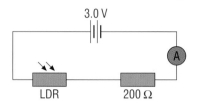

3.0 V

LDR   200 Ω

Figure 1

a With the LDR in daylight, the ammeter reads 0.010 A.
i Calculate the potential difference across the 200 Ω resistor when the current through it is 0.010 A.
ii Show that the potential difference across the LDR is 1.0 V when the ammeter reads 0.010 A.

b If the LDR is then covered, explain whether the ammeter reading increases or decreases or stays the same.

7 In Figure 1 in Question 6, the LDR is replaced by a 100 Ω resistor and a voltmeter connected in parallel with this resistor.

a Draw the circuit diagram for this circuit.

b Calculate:
i the total resistance of the two resistors in the circuit
ii the current through the ammeter
iii the voltmeter reading
iv the potential difference across the 200 Ω resistor.

8 Figure 2 shows a light-emitting diode (LED) in series with a resistor and a 3.0 V battery.

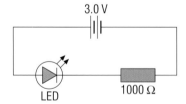

3.0 V

LED   1000 Ω

Figure 2

a The LED in the circuit emits light. The potential difference across it when it emits light is 0.6 V.
i Explain why the potential difference across the 1000 Ω resistor is 2.4 V.
ii Calculate the current in the circuit.

b If the LED in the circuit is reversed, what would be the current in the circuit? Give a reason for your answer.

9 State and explain how the resistance of a filament bulb changes when the current through the filament is increased.
[H]

AQA Examination-style questions

1 A plastic rod is rubbed with a dry cloth.

a Explain how the rod becomes negatively charged. (3)

b What charge is left on the cloth? (1)

c What happens if the negatively charged rod is brought close to another negatively charged rod? (1)

2 **a** Copy and complete the table of circuit symbols and their names. (5)

Circuit symbol	Name
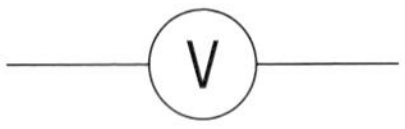	i
ii	ammeter
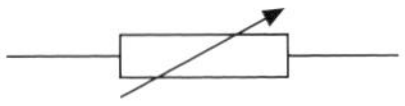	iii
iv	LDR
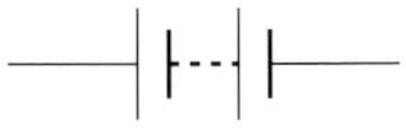	v

b Copy and complete the following sentences using the list of words and phrases below. Each word can be used once, more than once or not at all.

energy transferred   charge   resistance   voltage

Electric current is a flow of ....

The potential difference between two points in a circuit is the .... per unit of .... that passes between the points.

The greater the .... the lower the current for a given potential difference. (4)

3 Complete the following calculations. Write down the equation you use. Show clearly how you work out your answer and give the unit.

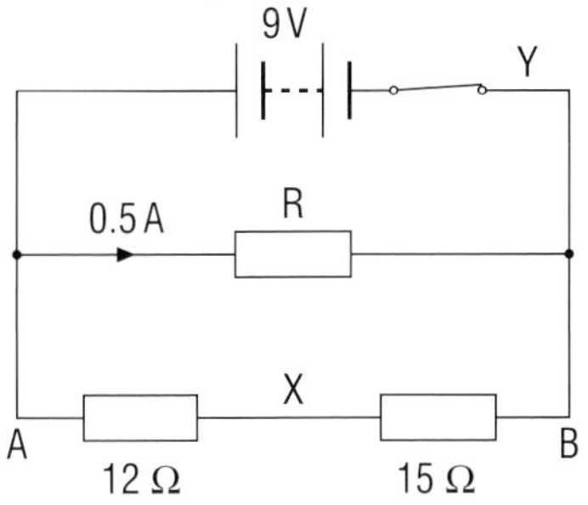

a **i** Calculate the potential difference between A and B. (1)

ii The potential difference across the 15 Ω resistor is 5 V.
Calculate the potential difference across the 12 Ω resistor. (1)

b **i** Calculate the combined resistance of the 12 Ω and the 15 Ω resistors in series. (1)

ii Calculate the current that flows through the circuit at X. (2)

iii Calculate the current flowing through the circuit at Y. (1)

c Calculate the resistance of the resistor labelled R. (2)

d Calculate the charge that flows through resistor R in 2 minutes. (3)

e Calculate the work done (energy transferred) by the cell if the total charge that has flowed through it is 3000 C. (2)

4 **a** Sketch and label a graph of current against potential difference for a diode. (3)

b The graph of current against potential difference for a filament bulb is shown.

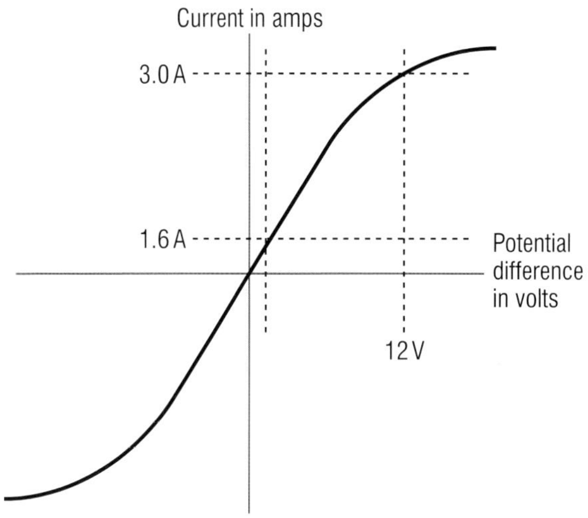

i Calculate the potential difference when the resistance of the filament bulb is 2 Ω when the current is 1.6 A. Write down the equation you use. Show clearly how you work out your answer and give the unit. (2)

ii Calculate the resistance at a potential difference of 12 V. Write down the equation you use. Show clearly how you work out your answer and give the unit. (3)

c *In this question you will be assessed on using good English, organising information clearly and using specialist terms where appropriate.*

Explain the change in resistance of the filament bulb in terms of ions and electrons. [H] (6)

P2 5.1 Alternating current ⓚ

Learning objectives

- What is meant by direct current and alternating current?
- What do we mean by the peak voltage of an alternating potential difference?
- What do we mean by the live wire and the neutral wire of a mains circuit?
- How do we use an oscilloscope to measure the frequency of an alternating current?   [H]

The battery in a torch makes the current to go round the circuit in one direction only. We say the current in the circuit is a **direct current** (dc) because it is in one direction only.

When you switch a light on at home, you use **alternating current** (ac) because mains electricity is an ac supply. An alternating current repeatedly reverses its direction. It flows one way then the opposite way in successive cycles. Its **frequency** is the number of cycles it passes through each second.

In the UK, the mains frequency is 50 cycles per second (or 50 Hz). A light bulb works just as well at this frequency as it would with a direct current.

a Why would a much lower frequency than 50 Hz be unsuitable for a light bulb?

Mains circuits

Every mains circuit has a **live wire** and a **neutral wire**. The current through a mains appliance alternates. That's because the mains supply provides an alternating potential difference between the two wires.

The neutral wire is **earthed** at the local substation. The potential difference between the live wire and 'earth' is usually referred to as the 'potential' or voltage of the live wire. The live wire is dangerous because its voltage repeatedly changes from + to − and back every cycle. It reaches over 300 V in each direction, as shown in Figure 1.

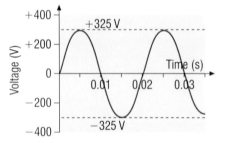

Figure 1 Mains voltage v. time

Practical

The oscilloscope

We use an oscilloscope to show how an alternating potential difference (pd) changes with time.

1 Connect a signal generator to an oscilloscope, as shown in Figure 2.

- The trace on the oscilloscope screen shows electrical waves. They are caused by the potential difference increasing and decreasing continuously.

- The highest potential difference is reached at each peak. The peak potential difference or

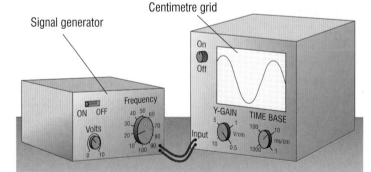

Figure 2 Using an oscilloscope

the **peak voltage** is the difference in volts between the peak and the middle level of the waves. Increasing the pd of the ac supply makes the waves on the screen taller.

- Increasing the frequency of the ac supply increases the number of cycles you see on the screen. So the waves on the screen get squashed together.

- How would the trace change if the pd of the ac supply were reduced?

2 Connect a battery to the oscilloscope. You should see a flat line at a constant potential difference.

- What difference is made by reversing the battery?

Measuring an alternating potential difference

We can use an **oscilloscope** to measure the peak potential difference and the frequency of a low voltage ac supply. For example, in Figure 2:

- the peak voltage is 2.1 V if the peaks are 8.4 cm above the troughs. Each peak is 4.2 cm above the middle which is at zero pd. The **Y-gain control** at 0.5 V/cm tells us each centimetre of height is due to a potential difference of 0.5 V. So the peak potential difference is 2.1 V (= 0.5 V/cm × 4.2 cm).

Higher

- the frequency is 12.5 Hz if each cycle on the screen is 8 cm across. The **time base control** at 10 milliseconds per centimetre (ms/cm) tells us each centimetre across the screen is a time interval of 10 ms. So the time taken for one cycle is 80 ms (= 10 ms/cm × 8 cm). The frequency is therefore 12.5 Hz (= 1/80 ms or 1/0.08 s).

Note: the frequency of ac supply = $\dfrac{1}{\text{the time for one cycle}}$

More about mains circuits

Look at Figure 1 again. It shows how the potential of the live wire varies with time.

- The live wire alternates between +325 V and −325 V. In terms of electrical power, this is equivalent to a direct voltage of 230 V. So we say the voltage of the mains is 230 V.

Higher

Each cycle in Figure 1 takes 0.02 second. The frequency of the mains supply (the number of cycles per second) is therefore 50 Hz (= $\dfrac{1}{0.02 \text{ seconds}}$)

b What is the maximum potential difference between the live wire and the neutral wire in Figure 1?

Summary questions

1 Choose the correct potential difference from the list for each appliance **a** to **d**.

   1.5 V   12 V   230 V   325 V

   a a car battery   **c** a torch cell
   b the mains voltage   **d** the maximum potential of the live wire.

2 In Figure 2, how would the trace on the screen change if the frequency of the ac supply was:
   a increased   **b** reduced?

3 In Figure 2, what is the frequency if one cycle measures 4 cm across the screen for the same time base setting?   [H]

4 **a** How does an alternating current differ from a direct current?
   b Figure 4 shows a diode and a resistor in series with each other connected to an ac supply. Explain why the current in the circuit is a direct current not an alternating current.

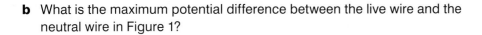

ac supply

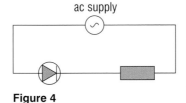

Figure 4

Key points

- Direct current is in one direction only. Alternating current repeatedly reverses its direction.

- The peak voltage of an alternating potential difference is the maximum voltage measured from zero volts.

- A mains circuit has a live wire that is alternately positive and negative every cycle and a neutral wire at zero volts.

- To measure the frequency of an a.c. supply, we measure the time period of the waves then use the formula:
  frequency = $\dfrac{1}{\text{time taken for 1 cycle}}$.   [H]

P2 5.2 — Cables and plugs

Learning objectives

- What is the casing of a mains plug or socket made from and why?
- What is in a mains cable?
- What colour are the live, neutral and earth wires?
- Why does a 3-pin plug include an earth pin?

Did you know … ?

Mains electricity is dangerous. By law, mains wiring must be done by properly qualified electricians.

When you plug in a heater with a metal case into a wall **socket**, you 'earth' the metal case automatically. This stops the metal case becoming 'live' if the live wire breaks and touches the case. If the case did become live and you touched it, you would be electrocuted.

Plastic materials are very good insulators. An appliance with a plastic case is doubly-insulated and carries the double insulation ⊡ symbol.

Plugs, sockets and cables

The outer casings of **plugs**, sockets and **cables** of all mains circuits and appliances are made of hard-wearing electrical insulators. That's because plugs, sockets and cables contain live wires. Most mains appliances are connected via a wall socket to the mains using a cable and a **three-pin plug**.

Sockets are made of stiff plastic materials with the wires inside. Figure 1 shows part of a wall socket circuit. It has an earth wire as well as a live wire and a neutral wire.

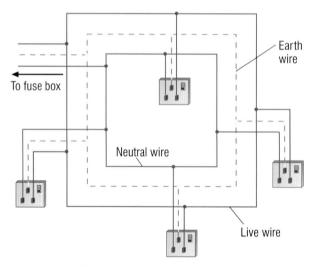

Figure 1 A wall socket circuit

- The earth wire of this circuit is connected to the ground at your home.
- The longest pin of a three-pin plug is designed to make contact with the earth wire of a wall socket circuit. So when you plug an appliance with a metal case to a wall socket, the case is automatically earthed.

  a Why are sockets wired in parallel with each other?

Plugs have cases made of stiff plastic materials. The live pin, the neutral pin and the earth pin, stick out through the plug case. Figure 2 shows inside a three-pin plug.

- The pins are made of brass because brass is a good conductor and does not rust or oxidise. Copper isn't as hard as brass even though it conducts better.
- The case material is an electrical insulator. The inside of the case is shaped so the wires and the pins cannot touch each other when the plug is sealed.
- The plug contains a fuse between the live pin and the live wire. If too much current passes through the wire in the fuse, it melts and cuts the live wire off.

b Why is brass, an alloy of copper and zinc, better than copper for the pins of a three-pin plug?

- The brown wire is connected to the live pin.
- The blue wire is connected to the neutral pin.
- The green and yellow wire (of a three-core cable) is connected to the earth pin. A two-core cable does not have an earth wire.

Cables used for mains appliances (and for mains circuits) consist of two or three insulated copper wires surrounded by an outer layer of rubber or flexible plastic material.

- Copper is used for the wires because it is a good electrical conductor and it bends easily.
- Plastic is a good electrical insulator and therefore prevents anyone touching the cable from receiving an electric shock.
- Two-core cables are used for appliances which have plastic cases (e.g. hairdryers, radios).
- Cables of different thicknesses are used for different purposes. For example, the cables joining the wall sockets in a house must be much thicker than the cables joining the light fittings. This is because more current passes along wall socket cables than along lighting circuits. So the wires in them must be much thicker. This stops the heating effect of the current making the wires too hot.

c Why are cables that are worn away or damaged dangerous?
d In Figure 3, which wire in each cable is the earth wire?

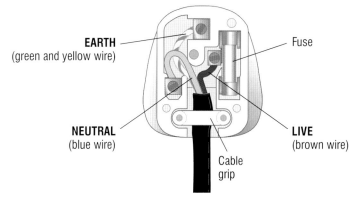

Figure 2 Inside a three-pin plug

Figure 3 Mains cables

Summary questions

1 Copy and complete **a** to **e** using the words below:

earth   live   neutral   series   parallel

a The .... wire in a mains plug is blue.
b If too much current passes through the fuse, it blows and cuts the .... wire off.
c Appliances plugged into the same mains circuit are in .... with each other.
d The metal frame of an appliance is connected to the .... wire of a mains circuit when it is plugged in.
e The fuse in a plug is in .... with the live wire.

2 **a** Match the list of parts 1–4 in a three-pin plug with the list of materials A–D.
  1 cable insulation   **A** brass
  2 case   **B** copper
  3 pin   **C** rubber
  4 wire   **D** stiff plastic
  b Explain your choice of material for each part in **a**.

3 **a** Why is each of the three wires in a three-core mains cable insulated?
  b How is the metal case of an electrical appliance connected to earth?

Key points

- **Sockets** and **plug cases** are made of stiff plastic materials that enclose the electrical connections. Plastic is used because it is a good electrical insulator.

- **Mains cable** consists of two or three insulated copper wires surrounded by an outer layer of flexible plastic material.

- In a **three-pin plug** or a three-core cable, the live wire is brown, the neutral wire is blue, and the earth wire is green and yellow.

- The earth wire is connected to the longest pin and is used to earth the metal case of a mains appliance.

P2 5.3

Fuses

Learning objectives

- What do we use a fuse for?
- Why is a fuse always on the 'live' side of an appliance?
- What is a circuit breaker?
- Why are appliances with plastic cases not earthed?

Did you know ... ?

If a live wire inside the appliance touches a neutral wire, a very large current passes between the two wires at the point of contact. We call this a short circuit. If the fuse blows, it cuts the current off.

If you need to buy a **fuse** for a mains appliance, make sure you know the fuse rating. Otherwise, the new fuse might 'blow' as soon as it is used. Worse still, it might let too much current through and cause a fire.

- A fuse contains a thin wire that heats up and melts if too much current passes through it. If this happens, we say the fuse 'blows'.
- The rating of a fuse is the maximum current that can pass through it without melting the fuse wire.
- The fuse should always be in series with the live wire between the live wire and the appliance. If the fuse blows, the appliance is then cut off from the live wire.

A fuse in a mains plug must always have the correct current rating for the appliance. If the current rating is too large, the fuse will not blow when it should. The heating effect of the current could set the appliance or its connecting cable on fire. Provided the correct fuse is fitted, the connecting cable must be thick enough (so its resistance is small enough) to make the heating effect of the current in the cable insignificant.

a What would happen if the current rating of the fuse was too small?

a   b

Figure 1   **a** A cartridge fuse   **b** A rewireable fuse

The importance of earthing

Figure 2 shows why an electric heater is made safer by earthing its metal frame.

In Figure 2a, the heater works normally and its frame is earthed. The frame is safe to touch.

In Figure 2b, the earth wire is broken. The frame would become live if the live wire touched it.

AQA Examiner's tip

The earth wire protects the user and the fuse protects the appliance and the wiring of the circuit.

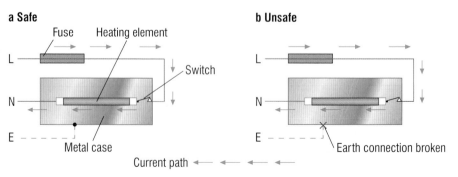

Figure 2 a and b Earthing an electric heater

In Figure 2c, the heater element has touched the unearthed frame so the frame is live. Anyone touching it would be electrocuted. The fuse provides no protection to the user as a current of no more than 20 mA can be lethal.

In Figure 2d, the earth wire has been repaired but the heater element still touches the frame. The current is greater than in **a** or **b** because it only passes through part of the heater element. Because the frame is earthed, anyone touching it would not be electrocuted. But Figure 2d is still dangerous. This is because although the current might not be enough to blow the fuse, it might cause the wires of the appliance to overheat.

b Why is the current in Figure 2d greater than normal?

Circuit breakers

A **circuit breaker** is an electromagnet switch that opens (switches off or 'trips') when there is a fault. This stops the current in the live wire flowing. The electromagnet is in series with the live wire. If the current in the live wire is too large, the magnetic field of the electromagnet is strong enough to pull the switch contacts apart. Once the switch is open, it stays open. It can then be reset once the fault that made it trip has been put right.

Circuit breakers are used instead of fuses. They work faster than fuses and can be reset more quickly.

The **Residual Current Circuit Breaker (RCCB)** works even faster than the ordinary circuit breaker described above. An RCCB cuts off the current in the live wire when it is different from the current in the neutral wire. The RCCB can be used where there is no earth connection. The RCCB is also more sensitive than either a fuse or an ordinary circuit breaker.

c What should you do if a circuit breaker trips again after being reset?

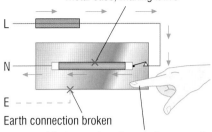

c Deadly   Heating element touches the metal case, making it live

Earth connection broken

Victim touches the metal case, and because the earth wire is broken, conducts current to earth

d Still dangerous as it may overheat

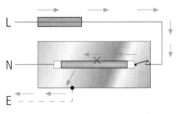

Figure 2 c and d

Figure 3 A circuit breaker

Summary questions

1 a What is the purpose of a fuse in a mains circuit?
  b Why is the fuse of an appliance always on the live side?
  c What advantages does a circuit breaker have compared with a fuse?

2 Figure 4 shows the circuit of an electric heater that has been wired incorrectly.
  a Does the heater work when the switch is closed?
  b When the switch is open, why is it dangerous to touch the element?
  c Redraw the circuit correctly wired.

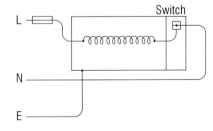

Switch

Figure 4

3 a What is the difference between an ordinary circuit breaker and a Residual Current Circuit Breaker (RCCB)?
  b What are the advantages of an RCCB mains socket compared with an ordinary mains socket with a fuse in it?

Key points

- A fuse contains a thin wire that heats up, melts, and cuts off the current if the current is too large.

- A fuse is always fitted in series with the live wire. This cuts the appliance off from the live wire if the fuse blows.

- A circuit breaker is an electromagnetic switch that opens (i.e. 'trips') and cuts off the current if too much current passes through the circuit breaker.

- A mains appliance with a plastic case does not need to be earthed because plastic is an insulator and cannot become live.

P2 5.4

Electrical power and potential difference

Learning objectives

- What is the relationship between power and energy?

- How can we calculate electrical power and what is its unit?

- How can we calculate the correct current for a fuse?

Did you know ...?

A surgeon fitting an artificial heart in a patient needs to make sure the battery will last a long time. Even so, the battery may have to be replaced every few years.

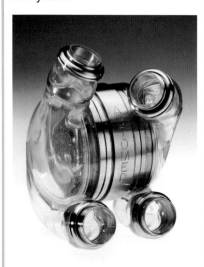

Figure 1 An artificial heart

When you use an electrical appliance, it transfers electrical energy into other forms of energy. The **power** of the appliance, in watts, is the energy it transfers, in joules per second. We can show this as the following equation:

$$\text{Power (watts, W)} = \frac{\text{energy transferred (joules, J)}}{\text{time (seconds, s)}}$$

We can write the equation for the power of an appliance as:

$$P = \frac{E}{t}$$

Where:
P = power in watts, W
E = energy transferred in joules, J
t = time taken in seconds, s.

Maths skills

Worked example

A light bulb transfers 30 000 J of electrical energy when it is on for 300 s. Calculate its power.

Solution

$$\text{Power} = \frac{\text{energy transferred}}{\text{time}} = \frac{30\,000\,\text{J}}{300\,\text{s}} = 100\,\text{W}$$

a The human heart transfers about 30 000 J of energy in about 8 hours. Calculate an estimate of the power of the human heart.

Calculating power

Millions of electrons pass through the circuit of an artificial heart every second. Each electron transfers a small amount of energy to it from the battery. So the total energy transferred to it each second is large enough to enable the appliance to work.

For any electrical appliance:

- the current through it is the charge that flows through it each second

- the potential difference across it is the energy transferred to the appliance by each coulomb of charge that passes through it

- the power supplied to it is the energy transferred to it each second. This is the electrical energy it transfers every second.

Therefore:

the energy transfer to the appliance each second = the charge flow per second × the energy transfer per unit charge.

In other words:

power supplied = current × potential difference
(watts, W)   (amperes, A)   (volts, V)

The equation can be written as:

$$P = I \times V$$

Where:
P = electrical power in watts, W
I = current in amperes, A
V = potential difference in volts, V.

For example, the power supplied to:

- a 4A, 12V electric motor is 48W (= 4A × 12V)
- a 0.1A, 3V torch lamp is 0.3W (= 0.1A × 3.0V).

b Calculate the power supplied to a 5A, 230V electric heater.

Rearranging the equation $P = I \times V$ gives:

potential difference, $V = \dfrac{P}{I}$ or

current, $I = \dfrac{P}{V}$

Figure 2 Changing a fuse

Choosing a fuse

Domestic appliances are often fitted with a 3A, 5A or 13A fuse. If you don't know which one to use for an appliance, you can work it out. You use the power rating of the appliance and its potential difference (voltage). The next time you change a fuse, do a quick calculation to make sure its rating is correct for the appliance (see the worked example opposite).

c Why would a 13A fuse be unsuitable for a 230V, 100W table lamp?

 Maths skills

Worked example

a Calculate the normal current through a 500W, 230V heater.

b Which fuse, 3A, 5A or 13A, would you use for the appliance?

Solution

a Current = $\dfrac{500\,W}{230\,V}$ = 2.2A

b You would use a 3A fuse.

Summary questions

1 Copy and complete **a** and **b** using the words below. Each word can be used more than once.

current   potential difference   power

a When an electrical appliance is on, ... is supplied to it as a result of .... passing through it.

b When an electrical appliance is on, a .... is applied to it which causes .... to pass through it.

2 **a** Calculate the power supplied to each of the following devices in normal use:

i a 12V, 3A light bulb
ii a 230V, 2A heater.

b Which type of fuse, 3A, 5A or 13A, would you select for:

i a 24W, 12V heater?
ii a 230V, 800W microwave oven?

3 **a** Why would a 3A fuse be unsuitable for a 230V, 800W microwave oven?

b The heating element of a 12V heater has a resistance of 4.0Ω. When the heating element is connected to a 12V power supply, calculate:

i the current through it
ii the electrical power supplied to it.

Key points

- The power supplied to a device is the energy transferred to it each second.

- Electrical power supplied (watts)

  = current (amperes) × potential difference (volts)

- Correct rating (in amperes) for a fuse:

  = $\dfrac{\text{electrical power (watts)}}{\text{potential difference (volts)}}$

P2 5.5

Electrical energy and charge

Learning objectives

- What is an electric current?

- How do we calculate the flow of electric charge from the current?

- What energy transfers take place when charge flows through a resistor?

- How is the energy transferred by a flow of charge related to potential difference?   [H]

- What can we say about the electrical energy supplied by the battery in a circuit and the electrical energy transferred to the components?   [H]

Charge flow = current × time

Figure 1 Charge and current

 Maths skills

Worked example

Calculate the charge flow when the current is 8 A for 80 s.

Solution

Charge flow = current × time
$= 8\,\text{A} \times 80\,\text{s}$
$= 640\,\text{C}$

Calculating charge

When an electrical appliance is on, electrons are forced through the appliance by the potential difference of the power supply unit. The potential difference causes a flow of charge through the appliance carried by electrons.

As explained in P2 4.2, the electric current is the rate of flow of charge through the appliance. The unit of charge, the **coulomb (C)**, is the amount of charge flowing through a wire or a component in 1 s when the current is 1 A.

The charge passing along a wire or through a component in a certain time depends on the current and the time.

We can calculate the charge using the equation:

$$\underset{\text{(coulombs)}}{\text{charge}} = \underset{\text{(amperes)}}{\text{current}} \times \underset{\text{(seconds)}}{\text{time}}$$

The equation can be written as: $Q = I \times t$

Where:
Q = charge in coulombs, C
I = current in amperes, A
t = time in seconds, s.

a Calculate the charge flowing in 50 s when the current is 3 A.

Energy and potential difference

When a resistor is connected to a battery, electrons are made to pass through the resistor by the battery. Each electron repeatedly collides with the vibrating metal ions of the resistor, transferring energy to them. The ions of the resistor therefore gain kinetic energy and vibrate even more. The resistor becomes hotter.

When charge flows through a resistor, energy is transferred to the resistor so the resistor becomes hotter.

The energy transferred in a certain time in a resistor depends on:

- the amount of charge that passes through it

- the potential difference across the resistor.

Because energy = power × time = potential difference × current × time, we can calculate the energy transferred using the equation:

$$\underset{\text{(joules, J)}}{\text{energy transferred}} = \underset{\text{(volts, V)}}{\text{potential difference}} \times \underset{\text{(coulombs, C)}}{\text{charge}}$$

The equation can be written as:

$$E = V \times Q$$

Where: E = energy transferred in joules, J
V = potential difference in volts, V
Q = charge in coulombs, C.

b Calculate the energy transferred when the charge flow is 30 C and the pd is 4 V.

Higher

Energy transfer in a circuit

The circuit in Figure 2 shows a 12 V battery in series with a torch bulb and a variable resistor. When the voltmeter reads 10 V, the potential difference across the variable resistor is 2 V.

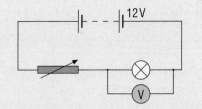

Figure 2 Energy transfer in a circuit

Each coulomb of charge:

- leaves the battery with 12 J of energy (because energy from the battery = charge × battery potential difference)
- transfers 10 J of energy to the torch bulb (because energy transfer to bulb = charge × potential difference across bulb)
- transfers 2 J of energy to the variable resistor.

The energy transferred to the bulb makes the bulb hot and emit light. The energy transferred to the variable resistor makes the resistor warm, so energy is therefore transferred to the surroundings by both bulb and resistor.

So the energy from the battery is equal to the sum of the energy transferred to the bulb and to the variable resistor.

Maths skills

Worked example

Calculate the energy transferred in a component when the charge passing through it is 30 C and the potential difference is 20 V.

Solution

Energy transferred = 20 V × 30 C
= 600 J

AQA Examiner's tip

Make sure you know and understand the relationship between charge, current and time.

Summary questions

1 Copy and complete **a** to **d** using the words below:

charge   current   energy   potential difference

  a The coulomb is the unit of ....
  b Charge flowing through a resistor transfers .... to the resistor.
  c A .... is the rate of flow of charge.
  d Energy transferred = ... × charge.   **[H]**

2 **a** Calculate the charge flow for:
    i a current of 4 A for 20 s
    ii a current of 0.2 A for 60 minutes.
  b Calculate the energy transfer:
    i for a charge flow of 20 C when the potential difference is 6.0 V
    ii for a current of 3 A that passes through a resistor for 20 s, when the potential difference is 5 V.   **[H]**

3 In Figure 3, a 4.0 Ω resistor and an 8.0 Ω resistor in series with each other are connected to a 6.0 V battery.

  Calculate:

Figure 3

  a the resistance of the two resistors in series
  b the current through the resistors
  c the charge flow through each resistor in 60 seconds
  d the potential difference across each resistor
  e the energy transferred to each resistor in 60 seconds
  f the energy supplied by the battery in 60 seconds.   **[H]**

Key points

- An electric current is the rate of flow of charge.

- Charge (coulombs) = current (amperes) × time (seconds). **[H]**

- When an electrical charge flows through a resistor, energy transferred to the resistor makes it hot.

- Energy transferred (joules) = potential difference (volts) × charge flow (coulombs). **[H]**

- When charge flows round a circuit for a certain time, the electrical energy supplied by the battery is equal to the electrical energy transferred to all the components in the circuit. **[H]**

P2 5.6

Electrical issues

An electrical fault is dangerous. It could give someone a nasty shock or even electrocute them, resulting in death. Also, a fault can cause a fire. This happens when too much current passes through a wire or an appliance and heats it up.

Fault prevention

Electrical faults can happen if sockets, plugs, cables or appliances are damaged. Users need to check for loose fittings, cracked plugs and sockets and worn cables. Any such damaged items need to be repaired or replaced by a qualified electrician.

- If a fuse blows or a circuit breaker trips when a mains appliance is in use, switch the appliance off. Then don't use it until it has been checked by a qualified electrician.
- If an appliance (or its cable or plug or socket) overheats and/or you get a distinctive burning smell from it, switch it off. Again, don't use it until it has been checked.

Too many appliances connected to a socket may cause the socket to overheat. If this happens, switch the appliances and the socket off and disconnect the appliances from the socket.

Smoke alarms and infrared sensors connected to an alarm system are activated if a fire breaks out. An electrical fault could cause an appliance or a cable to become hot and could set fire to curtains or other material in a room. Smoke alarms and sensors should be checked regularly to make sure they work properly.

An electrician selecting a cable for an appliance needs to use:

- a two-core cable if the appliance is 'double-insulated' and no earth wire is needed
- a three-core cable if an earth wire is needed because the appliance has a metal case
- a cable with conductors of suitable thickness so the heating effect of the current in the cable is insignificant.

> **a i** If a mains appliance suddenly stops working, why is it a mistake to replace the fuse straightaway?
>
> **ii** Should the cable of an electric iron be a two-core or a three-core cable?

New bulbs for old

When choosing an electrical appliance, most people compare several different appliances. The cost of the appliance is just one factor that may need to be considered. Other factors might include the power of the appliance and its efficiency.

If you want to replace a bulb, a visit to an electrical shop can present you with a bewildering range of bulbs.

A filament bulb is very inefficient. The energy from the hot bulb gradually makes the plastic parts of the bulb socket brittle and they crack.

Learning objectives

- Why are electrical faults dangerous?
- How can we prevent electrical faults?
- When choosing an electrical appliance, what factors in addition to cost should we consider?
- How do different forms of lighting compare in terms of cost and energy efficiency?

Activity

Spot the hazards!

How many electrical faults and hazards can you find in Shockem Hall? See how many you can spot in the main hall.

Figure 1 Shockem Hall

??? Did you know ...?

What kills you – current or voltage? Mains electricity is dangerous. A current of no more than about 0.03 A through your body would give you a severe shock and might even kill you. Your body has a resistance of about 1000 Ω including contact resistance at the skin. If your hands get wet, your resistance is even lower.

Low energy bulbs are much more efficient so they don't become hot like filament bulbs do. Different types of low energy bulb are now available:

- **Low-energy compact fluorescent bulbs (CFLs)** are now used for room lighting instead of filament bulbs.
- **Low-energy light-emitting diodes (LEDs)** used for spotlights are usually referred to as high-power LEDs. They operate at low voltage and low power. They are much more efficient than filament bulbs or halogen bulbs and they last much longer.

This table gives more information about these different bulbs.

Type	Power	Efficiency	Lifetime in hours	Cost of bulb	Typical use
Filament bulb	100 W	20%	1000	50p	room lighting
Halogen bulb	100 W	25%	2500	£2.00	spotlight
Low-energy compact fluorescent bulb (CFL)	25 W	80%	15 000	£2.50	room lighting
Low-energy light-emitting diode (LED)	2 W	90%	30 000	£7.00	spotlight

b A householder wants to replace a 100 W room light with a row of low-energy LEDs with the same light output. Use the information in the table above to answer the following questions.

   i How many times would the filament bulb need to be replaced in the lifetime of an LED?

   ii How many LEDs would be needed to give the same light output as a 100 W filament bulb?

   iii The householder reckons the cost of the electricity for each LED at 10p per kWh over its lifetime of 30 000 hours would be £6. Show that the cost of the electricity for a 100 W bulb over this time would be £300.

   iv Use your answers above to calculate how much the householder would save by replacing the filament bulb with LEDs.

Summary questions

1 An 'RCCB' socket should be used for mains appliances such as lawnmowers where there is a possible hazard when the appliance is used. Such a socket contains a residual current circuit breaker instead of a fuse. This type of circuit breaker switches the current off if the live current and the neutral current differ by more than 30 mA. This can happen, for example, if the blades of a lawnmower cut into the cable.

Create a table to show a possible 'electrical' hazard for each of these appliances: lawnmower, electric drill, electric saw, hairdryer, vacuum cleaner. The first entry has been done for you.

Appliance	Hazard
Lawnmower	The blades might cut the cable.

Did you know ... ?

All new appliances like washing machines and freezers sold in the EU are labelled clearly with an efficiency rating. The rating is from A (very efficient) to G (lowest efficiency). Light bulbs are also labelled in this way on the packaging.

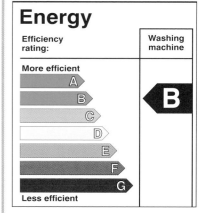

Figure 2 Efficiency measures

Key points

- Electrical faults are dangerous because they can cause electric shocks and fires.

- Never touch a mains appliance (or plug or socket) with wet hands. Never touch a bare wire or a terminal at a potential of more than 30 V.

- Check cables, plugs and sockets for damage regularly. Check smoke alarms and infrared sensors regularly.

- When choosing an electrical appliance, the power and efficiency rating of the appliance need to be considered.

- Filament bulbs and halogen bulbs are much less efficient than low energy bulbs.

Summary questions

1 **a** In a mains circuit, which wire:
   i is earthed at the local sub-station
   ii alternates in potential?

   b An oscilloscope is used to display the potential difference of an alternating voltage supply unit. How would the trace change if:
   i the pd is increased
   ii the frequency is increased?

2 Copy and complete **a** and **b** using the words below. Each word can be used more than once.

   earth   live   neutral

   a When a mains appliance is switched on, current passes through it via the .... wire and the .... wire.

   b In a mains circuit:
   i the .... wire is blue
   ii the .... wire is brown
   iii the .... wire is green and yellow.

3 **a** Copy and complete the following sentences:
   i Wall sockets are connected in .... with each other.
   ii A fuse in a mains plug is in .... with the appliance and cuts off the .... wire if too much current passes through the appliance.

   b **i** What is the main difference between a fuse and a circuit breaker?
   ii Give two reasons why a circuit breaker is safer than a fuse.

4 **a** **i** Calculate the current in a 230 V, 2.5 kW electric kettle.
   ii Which fuse, 3 A, 5 A or 13 A, would you fit in the kettle plug?

   b Calculate the power supplied to a 230 V electric toaster when the current through it is 4.0 A.

5 A 5 Ω resistor is in series with a bulb, a switch and a 12 V battery.

   a Draw the circuit diagram.

   b When the switch is closed for 60 seconds, a direct current of 0.6 A passes through the resistor. Calculate:
   i the energy supplied by the battery
   ii the energy transferred to the resistor
   iii the energy transferred to the bulb. **[H]**

   c The bulb is replaced by a 25 Ω resistor.
   i Calculate the total resistance of the two resistors.
   ii Show that a current of 0.4 A passes through the battery.

iii Calculate the power supplied by the battery and the power delivered to each resistor.

6 When a 6 V bulb operates normally, the electrical power supplied to it is 15 W.
   a Calculate:
   i the current through the bulb when it operates normally
   ii the resistance of the bulb when it operates normally.

   b If the bulb is connected to a 3 V battery, state and explain why its resistance is less than at 6 V.

7 A 12 V 36 W bulb is connected to a 12 V supply.

   a Calculate:
   i the current through the bulb.
   ii the charge flow through the bulb in 200 s. **[H]**

   b **i** Show that 7200 J of electrical energy is delivered to the bulb in 200 s.

   ii Calculate the energy delivered to the bulb by each coulomb of charge that passes through it. **[H]**

8 An electrician has the job of connecting a 6.6 kW electric oven to the 230 V mains supply in a house.

   a Calculate the current needed to supply 6.6 kW of electrical power at 230 V.

   b The table below shows the maximum current that can pass safely through five different mains cables. For each cable the cross-sectional area (csa) of each conductor is given in square millimetres (mm²).

	Cross-sectional area of conductor (mm²)	Maximum safe current (A)
A	1.0	14
B	1.5	18
C	2.5	28
D	4.0	36
E	6.0	46

   i To connect the oven to the mains supply, which cable should the electrician choose? Give a reason for your answer.
   ii State and explain what would happen if she chose a cable with thinner conductors?

AQA Examination-style questions

1 The pictures show situations in which electricity is not being used safely.

For each picture **a**, **b** and **c**, explain how electricity is not being used safely.

a

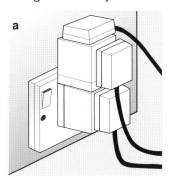

(2)

b

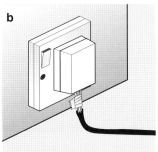

(2)

c

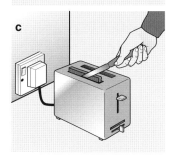

(2)

d The colour of the earth wire in a plug is .... (1)

e The pins of the plug are made of brass because it is a good .... (1)

f The voltage on the neutral wire is about .... V. (1)

g RCCB stands for .... (1)

2 Most domestic appliances are connected to the 230 V mains supply with a 3-pin plug containing a fuse. 3 A, 5 A and 13 A fuses are available.

a A bulb for a desk lamp has a normal current of 0.26 A.
  i Which of the three fuses should be used? (1)
  ii Calculate the power of the lamp. (2)
  iii Calculate how many coulombs of charge pass through the lamp if it is left on for 1 hour. **[H]** (3)

b i Calculate the current passing through a 1.15 kW electric fan heater. (2)
  ii Which fuse should be used in the plug for this heater? (1)

c Calculate how much electrical energy is transferred when the fan heater is left on for 30 minutes. Write down the equation you use. Show clearly how you work out your answer and give the unit. (3)

d *In this question you will be assessed on using good English, organising information clearly and using specialist terms where appropriate.*

The heater is made of metal and has an earth wire connected to it. Explain how the fuse and earth wire together protect the wiring of the circuit. (6)

3 A kettle is connected to the UK mains supply and boiled. An energy monitoring device measures that 420 000 J has been transferred to the kettle in the time it takes to boil.

a Calculate how much charge has flowed through the kettle. Write down the equation you use. Show clearly how you work out your answer and give the unit. **[H]** (3)

b The power of the kettle is 2.2 kW. How long did the kettle take to boil? (3)

4 An oscilloscope is connected to a power supply. The trace is shown on a centimetre grid.

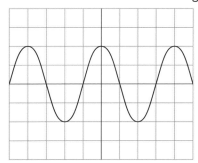

a Explain how you know that it is an ac supply being measured. (1)

b Give the peak voltage if each division on the y-axis is 2 V/cm. (1)

c Each x-axis division is 0.01 s/cm.
  i Calculate the time period of the supply. (1)
  ii Calculate the frequency of the supply. **[H]** (2)

d Describe the position and appearance of the trace on the screen if the supply was switched to 6 V dc. (2)

P2 6.1

Observing nuclear radiation

Learning objectives

- What is a radioactive substance?

- What types of radiation are given out from a radioactive substance?

- When does a radioactive source give out radiation (radioactive decay)?

- Where does background radiation come from?

A key discovery

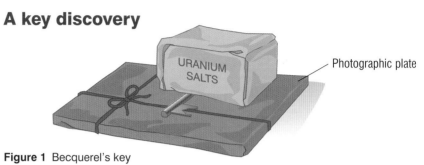

URANIUM SALTS

Photographic plate

Figure 1 Becquerel's key

If your photos showed a mysterious image, what would you think? In 1896, the French physicist, **Henri Becquerel**, discovered the image of a key on a film he developed. He remembered the film had been in a drawer under a key. On top of that there had been a packet of uranium salts. The uranium salts must have sent out some form of radiation that passed through paper (the film wrapper) but not through metal (the key).

Marie Curie

Becquerel asked a young research worker, **Marie Curie**, to investigate. She found that the salts gave out radiation all the time. It happened no matter what was done to them. She used the word **radioactivity** to describe this strange new property of uranium.

She and her husband, Pierre, did more research into this new branch of science. They discovered new radioactive elements. They named one of the elements **polonium**, after Marie's native country, Poland.

a You can stop a lamp giving out light by switching it off. Is it possible to stop uranium giving out radiation?

Becquerel and the Curies were awarded the Nobel Prize for the discovery of radioactivity. When Pierre died in a road accident, Marie went on with their work. She was awarded a second Nobel Prize in 1911 for the discovery of polonium and radium. She died in 1934 from leukaemia, a disease of the blood cells. It was probably caused by the radiation from the radioactive materials she worked with.

Figure 2 Marie Curie 1867–1934

Practical

Investigating radioactivity

We can use a **Geiger counter** to detect radioactivity. Look at Figure 3. The counter clicks each time a particle of radiation from a radioactive substance enters the Geiger tube.

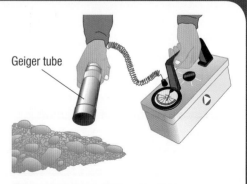

Geiger tube

Figure 3 Using a Geiger counter

What stops the radiation? Ernest Rutherford carried out tests to answer this question about a century ago. He put different materials between the radioactive substance and a detector.

He discovered two types of radiation:

- One type (**alpha radiation**, symbol α) was stopped by paper.
- The other type (**beta radiation**, symbol β) went through the paper.

Scientists later discovered a third type, **gamma radiation** (symbol γ), even more penetrating than beta radiation.

b Can gamma radiation go through paper?

A radioactive puzzle

Why are some substances radioactive? Every atom has a nucleus made up of protons and neutrons. Electrons move about in energy levels (or shells) surrounding the nucleus.

Most atoms each have a stable nucleus that doesn't change. But the atoms of a radioactive substance each have a nucleus that is unstable. An unstable nucleus becomes stable by emitting alpha, beta or gamma radiation. We say an unstable nucleus **decays** when it emits radiation.

We can't tell when an unstable nucleus will decay. It is a **random** event that happens without anything being done to the nucleus.

c Why is the radiation from a radioactive substance sometimes called 'nuclear radiation'?

The origins of background radiation

A Geiger counter clicks even when it is not near a radioactive source. This effect is due to **background radiation**. This is radiation from radioactive substances:

- in the environment (e.g. in the air or the ground or in building materials), or
- from space (cosmic rays), or
- from devices such as X-ray tubes.

Some of these radioactive substances are present because of nuclear weapons testing and nuclear power stations. But most of it is from naturally occurring substances in the Earth. For example, radon gas is radioactive and is a product of the decay of uranium found in the rocks in certain areas.

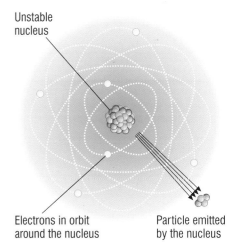

Figure 4 Radioactive decay

Unstable nucleus

Electrons in orbit around the nucleus

Particle emitted by the nucleus

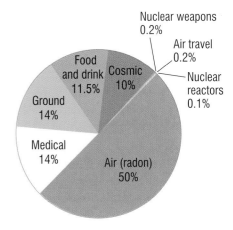

Figure 5 The origins of background radiation

Nuclear weapons 0.2%
Air travel 0.2%
Nuclear reactors 0.1%
Food and drink 11.5%
Cosmic 10%
Ground 14%
Medical 14%
Air (radon) 50%

Key points

- A radioactive substance contains unstable nuclei that become stable by emitting radiation.

- There are three main types of radiation from radioactive substances – alpha, beta and gamma radiation.

- Radioactive decay is a random event – we cannot predict or influence when it will happen.

- Background radiation is from radioactive substances in the environment or from space or from devices such as X-ray machines.

Summary questions

1 Copy and complete **a** and **b** using the words below. Each word can be used more than once.

protons   neutrons   nucleus   radiation

  a The .... of an atom is made up of .... and ....
  b When an unstable .... decays, it emits ....

2 **a** The radiation from a radioactive source is stopped by paper. What type of radiation does the source emit?
  b The radiation from a different source goes through paper. What can you say about this radiation?

3 **a** Explain why some substances are radioactive.
  b State two sources of background radioactivity.

P2 6.2

The discovery of the nucleus

Learning objectives

- How was the nuclear model of the atom established?
- Why was the plum pudding model of the atom rejected?
- Why was the nuclear model accepted?

??? Did you know ...?

Ernest Rutherford was awarded the Nobel Prize in 1908 for his discoveries on radioactivity. His famous discovery of the nucleus was made in 1913. He was knighted in 1914 and made a member of the House of Lords in 1931. He hoped that no one would discover how to release energy from the nucleus until people learned to live at peace with their neighbours. He died in 1937 before the discovery of nuclear fission.

Figure 2 Ernest Rutherford

Practical

Lucky strike!

Fix a small metal disc about 2 cm thick at the centre of a table. Hide the disc under a cardboard disc about 20 cm in diameter. See if you can hit the metal disc with a rolling marble.

Ernest Rutherford made many important discoveries about radioactivity. He discovered that alpha and beta radiation consists of different types of particles. He realised alpha (α) particles could be used to probe the atom. He asked two of his research workers, Hans Geiger and Ernest Marsden, to investigate. They used a thin metal foil to scatter a beam of alpha particles. Figure 1 shows the arrangement they used.

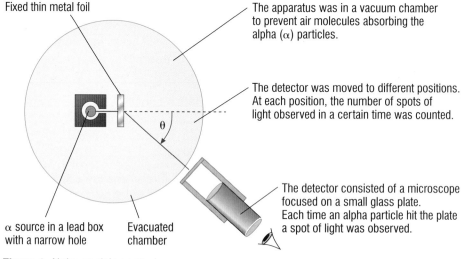

Fixed thin metal foil

The apparatus was in a vacuum chamber to prevent air molecules absorbing the alpha (α) particles.

The detector was moved to different positions. At each position, the number of spots of light observed in a certain time was counted.

The detector consisted of a microscope focused on a small glass plate. Each time an alpha particle hit the plate a spot of light was observed.

α source in a lead box with a narrow hole

Evacuated chamber

Figure 1 Alpha particle scattering

They measured the number of alpha particles deflected per second through different angles. The results showed that:

- most of the alpha particles passed straight through the metal foil
- the number of alpha particles deflected per minute decreased as the angle of deflection increased
- about 1 in 10000 alpha particles were deflected by more than 90°.

> **a** If you kicked a football at an empty goal and the ball bounced back at you, what would you conclude?

Rutherford was astonished by the results. He said it was like firing 'naval shells' at tissue paper and discovering the occasional shell rebounds. He knew that α particles are positively charged. He deduced from the results that there is a nucleus at the centre of every atom that is:

- positively charged because it repels α particles (remember that like charges repel and unlike charges attract)
- much smaller than the atom because most α particles pass through without deflection
- where most of the mass of the atom is located.

Using this model, Rutherford worked out the proportion of α particles that would be deflected for a given angle. He found an exact agreement with Geiger and Marsden's measurements. He used his theory to estimate the diameter of the nucleus. He found it was about 100000 times smaller than the atom.

Rutherford's nuclear model of the atom was quickly accepted because:

- It agreed exactly with the measurements Geiger and Marsden made in their experiments.
- It explains radioactivity in terms of changes that happen to an unstable nucleus when it emits radiation.
- It predicted the existence of the neutron, which was later discovered.

b What difference would it have made if Geiger and Marsden's measurements had not fitted Rutherford's nuclear model?

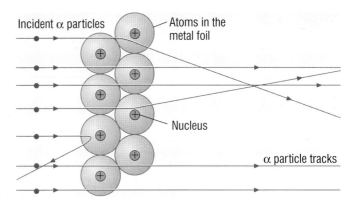

Figure 3 Alpha (α) particle paths

Goodbye to the plum pudding model!

Before the nucleus was discovered in 1914, scientists didn't know what the structure of the atom was. They did know atoms contained electrons and they knew these are tiny negatively charged particles. But they didn't know how the positive charge was arranged in an atom, although there were different models in circulation. Some scientists thought the atom was like a 'plum pudding' with:

- the positively charged matter in the atom evenly spread about (as in a pudding), and
- electrons buried inside (like plums in the pudding).

Rutherford's discovery meant farewell to the 'plum pudding' atom.

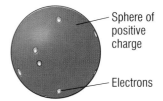

Figure 4 The plum pudding atom

Did you know ... ?

Almost all the mass of an atom is in its nucleus. The density of the nucleus is about a thousand million million times the density of water. A matchbox of nuclear matter would weigh about a million million tonnes!

Summary questions

1 Copy and complete **a** to **c** using the words below:

charge   diameter   mass

a A nucleus has the same type of .... as an alpha particle.

b A nucleus has a much smaller .... than the atom.

c Most of the .... of the atom is in the nucleus.

2 **a** Figure 5 shows four possible paths, labelled A, B, C and D, of an alpha particle deflected by a nucleus. Which path would the alpha particle travel along?

b Explain why each of the other paths in part **a** is not possible.

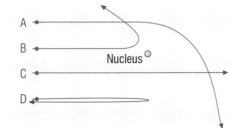

Figure 5

3 **a** Describe two differences between the nuclear model of the atom and the plum pudding model.

b Explain why the alpha-scattering experiment led to the acceptance of the nuclear model of the atom and the rejection of the plum pudding model.

Key points

- Rutherford used the measurements from alpha-scattering experiments to prove that an atom has a small positively charged central nucleus where most of the mass of the atom is located.

- The plum pudding model could not explain why some alpha particles were scattered through large angles.

- The nuclear model of the atom correctly explained why the alpha particles are scattered and why some are scattered through large angles.

P2 6.3

Nuclear reactions

Learning objectives

- What is an isotope?

- How does the nucleus of an atom change when it emits an alpha particle or a beta particle?

- How can we represent the emission of an alpha or a beta particle from a nucleus?

[H]

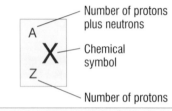

Example: the symbol for the uranium isotope with 92 protons and 146 neutrons is

$$^{238}_{92}U \quad \text{(or sometimes U-238)}$$

Figure 1 Representing an isotope

Table 1

	Relative mass	Relative charge
proton	1	+1
neutron	1	0
electron	0.0005	−1

The nucleus emits an α particle and forms a new nucleus

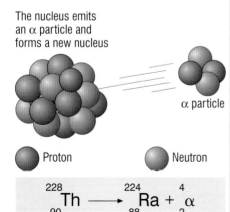

α particle

● Proton   ● Neutron

$$^{228}_{90}Th \longrightarrow {}^{224}_{88}Ra + {}^{4}_{2}\alpha$$

Figure 2 α emission

In α (alpha) or β (beta) decay, the number of protons in a nucleus changes. In α decay, the number of neutrons also changes. We will now look at the changes that happen in α and β decay and how we can represent these changes.

Table 1 gives the relative masses and the relative electric charges of a proton, a neutron and an electron.

Atoms are uncharged. They have equal numbers of protons (+) and electrons (−). A charged particle, called an ion, is formed when an atom gains or loses one or more electrons. Then there are unequal numbers of protons and electrons in the ion.

The atoms of the same element each have the same number of protons. The number of protons in a nucleus is given the symbol Z. It is called the **atomic number** (or **proton number**).

Isotopes are atoms of the same element with different numbers of neutrons. The isotopes of an element have nuclei with the same number of protons but a different number of neutrons.

The number of protons and neutrons in a nucleus is called its **mass number**. We give it the symbol A.

An isotope of an element X, which has Z protons and A protons plus neutrons, is represented by the symbol $^{A}_{Z}X$. For example, the uranium isotope $^{238}_{92}U$ contains 92 protons and 146 neutrons (= 238 − 92) in each nucleus. So its relative mass is 238 and the relative charge of the nucleus is +92.

a How many protons and how many neutrons are in the nucleus of the uranium isotope $^{235}_{92}U$?

Radioactive decay

An unstable nucleus becomes more stable by emitting an α (alpha) or a β (beta) particle or by emitting γ (gamma) radiation.

α emission

An α particle consists of two protons and two neutrons. Its relative mass is 4 and its relative charge is +2. So we can represent it by the symbol $^{4}_{2}\alpha$.

When an unstable nucleus emits an α particle, its atomic number goes down by 2 and its mass number goes down by 4.

For example, the thorium isotope $^{228}_{90}Th$ decays by emitting an α particle. So it forms the radium isotope $^{224}_{88}Ra$.

Figure 2 shows an equation to represent this decay.

- The numbers along the top represent the mass number which is the number of protons and neutrons in each nucleus and in the α particle.
- The equation shows that the total number of protons and neutrons after the change (= 224 + 4) is equal to the total number of neutrons and protons before the change (= 228).
- The numbers along the bottom represent the atomic number which is the number of protons in each nucleus and in the α particle.
- The equation shows that the total number of protons after the change (= 88 + 2) is equal to the total number of protons before the change (= 90).

Higher

b How many protons and how many neutrons are in $^{228}_{90}$Th and $^{224}_{88}$Ra?

β emission

- A β particle is an electron created and emitted by a nucleus which has too many neutrons compared with its protons. A neutron in the nucleus changes into a proton and a β particle. This is instantly emitted at high speed by the nucleus.
- The relative mass of a β particle is effectively zero and its relative charge is −1. So we can represent a β particle by the symbol $^{0}_{-1}$β.

- When an unstable nucleus emits a β particle, the atomic number of the nucleus goes up by 1 but its mass number stays the same (because a neutron changes into a proton).

For example, the potassium isotope $^{40}_{19}$K decays by emitting a β particle. So it forms a nucleus of the calcium isotope $^{40}_{20}$Ca.

- The numbers along the top represent the mass number which is the number of protons and neutrons for each nucleus and −1 for the β particle, as explained below.
- The equation shows that the total number of protons and neutrons after the change (= 40 + 0) is equal to the total number of neutrons and protons before the change (= 40).
- The numbers along the bottom represent the atomic number. This is the number of protons for each nucleus and −1 for the β particle, as explained below.
- The equation shows that the total charge (in relative units) after the change (= 20 −1) is equal to the total charge before the change (= 19).

(Note the relative charge of the β particle is −1 so we represent its atomic number as −1 in these nuclear equations, even though it has no protons at all.)

c How many protons and how many neutrons are in $^{40}_{19}$K and $^{40}_{20}$Ca?

γ emission

γ radiation is emitted by some unstable nuclei after an α particle or a β particle has been emitted. γ radiation is uncharged and has no mass. So it does not change the number of protons or the number of neutrons in a nucleus.

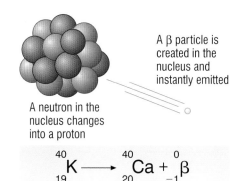

A β particle is created in the nucleus and instantly emitted

A neutron in the nucleus changes into a proton

$$^{40}_{19}K \longrightarrow ^{40}_{20}Ca + ^{0}_{-1}\beta$$

Figure 3 β emission

Key points

- Isotopes of an element are atoms with the same number of protons but different numbers of neutrons. Therefore they have the same atomic number but different mass numbers.

α decay	β decay
Change in the nucleus	
Nucleus loses 2 protons and 2 neutrons	A neutron in the nucleus changes into a proton
Particle emitted	
2 protons and 2 neutrons emitted as an α particle	An electron is created in the nucleus and instantly emitted
Equation	**[H]**
$^{A}_{Z}X \rightarrow ^{A-4}_{Z-2}Y + ^{4}_{2}\alpha$	$^{A}_{Z}X \rightarrow ^{A}_{Z+1}Y + ^{0}_{-1}\beta$

Summary questions

1. How many protons and how many neutrons are there in the nucleus of each of the following isotopes?

   a $^{12}_{6}$C   **b** $^{60}_{27}$Co   **c** $^{235}_{92}$U

2. A substance contains the radioactive isotope $^{238}_{92}$U, which emits alpha radiation. The product nucleus X emits beta radiation and forms a nucleus Y. How many protons and how many neutrons are present in:

   a a nucleus of $^{238}_{92}$U   **b** a nucleus of X   **c** a nucleus of Y?

3. Copy and complete the following equations for α and β decay.

   a $^{235}_{92}U \rightarrow ^{?}_{?}Th + ^{4}_{?}\alpha$ **b** $^{64}_{29}Cu \rightarrow ^{?}_{?}Zn + ^{?}_{-1}\beta$ **[H]**

Higher

More about alpha, beta and gamma radiation

Learning objectives

- How far can each type of radiation travel in air and what stops it?

- What is alpha, beta and gamma radiation?

- How can we separate a beam of alpha, beta and gamma radiation?

- Why is alpha, beta and gamma radiation dangerous?

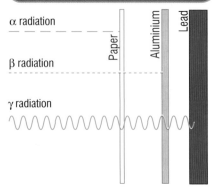

Figure 2 The penetrating powers of α, β and γ radiation

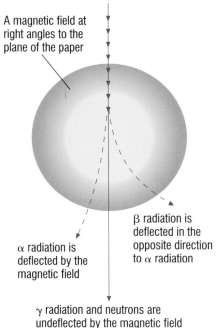

α radiation is deflected by the magnetic field

β radiation is deflected in the opposite direction to α radiation

γ radiation and neutrons are undeflected by the magnetic field

Figure 3 Radiation in a magnetic field

Penetrating power

Alpha radiation can't penetrate paper. But what stops beta and gamma radiation? And how far can each type of radiation travel through air? We can use a Geiger counter to find out, but we must take account of background radiation. To do this we should:

1 Measure the count rate (which is the number of counts per second) without the radioactive source present. This is the background count rate, the count rate due to background radiation.

2 Measure the count rate with the source in place. Subtracting the background count rate from this gives the count rate due to the source alone.

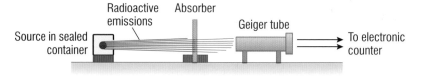

Figure 1 Absorption tests

We can then test absorber materials and the range in air.

- To test different materials, we need to place each material between the tube and the radioactive source. Then we measure the count rate. We can add more layers of material until the count rate due to the source is zero. The radiation from the source has then been stopped by the absorber material.

- To test the range in air, we need to move the tube away from the source. When the tube is beyond the range of the radiation, the count rate due to the source is zero.

The table below shows the results of the two tests.

Type of radiation	Absorber materials	Range in air
alpha (α)	Thin sheet of paper	about 5 cm
beta (β)	Aluminium sheet (about 5 mm thick) Lead sheet (2–3 mm thick)	about 1 m
gamma (γ)	Thick lead sheet (several cm thick) Concrete (more than 1 m thick)	unlimited

Gamma radiation spreads out in air without being absorbed. It does get weaker as it spreads out.

a Why is a radioactive source stored in a lead-lined box?

The nature of alpha, beta and gamma radiation

We can separate these radiations using a magnetic field or an electric field.

Deflection by a magnetic field

- β radiation is easily deflected, in the same way as electrons. So the radiation consists of negatively charged particles. In fact, a β particle is a fast-moving electron. It is emitted by an unstable nucleus that contains too many neutrons.

- α radiation is deflected in the opposite direction to β radiation. So α radiation consists of positively charged particles. α particles are harder to deflect than β radiation. This is because an α particle has a much greater mass than a β particle has. An alpha particle is two protons and two neutrons stuck together, the same as a helium nucleus.
- γ radiation is not deflected by a magnetic field or an electric field. This is because gamma radiation is electromagnetic radiation so is uncharged.

Figure 4 Radiation passing through an electric field

Deflection by an electric field

α and β particles passing through an electric field are deflected in opposite directions, as shown in Figure 4.

- The α particles are attracted towards the negative plate because they are positively charged.
- The β particles are attracted towards the positive plate because they are negatively charged,

In Figures 3 and 4, an alpha particle is deflected much less than the beta particle. The charge of an alpha particle is twice that of a beta particle, so the force is twice as great. But the mass of an alpha particle is about 8000 times that of a beta particle, so the deflection of the alpha particle is much less.

b How do we know that gamma radiation is not made up of charged particles?

Figure 5 Radioactive warnings

Radioactivity dangers

The radiation from a radioactive substance can knock electrons out of atoms. The atoms become charged because they lose electrons. The process is called **ionisation**. (Remember that a charged particle is called an ion.)

X-rays also cause ionisation. Ionisation in a living cell can damage or kill the cell. Damage to the genes in a cell can be passed on if the cell generates more cells. Strict safety rules must always be followed when radioactive substances are used.

Alpha radiation is more dangerous in the body than beta or gamma radiation. This is because it has a greater ionising effect than beta or gamma radiation.

c Why should long-handled tongs be used to move a radioactive source?

Summary questions

1 Copy and complete **a** and **b** using the words below. Each word can be used more than once.

alpha   beta   gamma

   a Electromagnetic radiation from a radioactive substance is called .... radiation.

   b A thick metal plate will stop .... and .... radiation but not .... radiation.

2 Which type of radiation is:

   a uncharged   **b** positively charged   **c** negatively charged?

3 **a** Explain why ionising radiation is dangerous.

   b Explain how you would use a Geiger counter to find the range of the radiation from a source of α radiation.

Key points

- **α radiation** is stopped by paper, has a range of a few centimetres in air and consists of particles, each composed of two protons and two neutrons.

- **β radiation** is stopped by thin metal, has a range of about a metre in air and consists of fast-moving electrons emitted from the nucleus.

- **γ radiation** is stopped by thick lead, has an unlimited range in air and consists of electromagnetic radiation.

- A magnetic or an electric field can be used to separate a beam of alpha, beta and gamma radiation.

- Alpha, beta and gamma radiation ionise substances they pass through. Ionisation in a living cell can damage or kill the cell.

P2 6.5 Half-life

Learning objectives

- What do we mean by the 'half-life' of a radioactive source?

- What do we mean by the activity of a radioactive source?

- What happens to the activity of a radioactive isotope as it decays?

Every atom of an element always has the same number of protons in its nucleus. However, the number of neutrons in the nucleus can differ. Each type of atom is called an isotope. (So isotopes of an element contain the same number of protons but different numbers of neutrons.)

The **activity** of a radioactive isotope is the number of atoms that decay per second. As the nucleus of each unstable atom (the 'parent' atom) decays, the number of parent atoms goes down. Therefore the activity of the sample decreases.

We can use a Geiger counter to monitor the activity of a radioactive sample. We need to measure the **count rate** due to the sample. This is the number of counts per second (or per minute). The graph below shows how the count rate of a sample decreases.

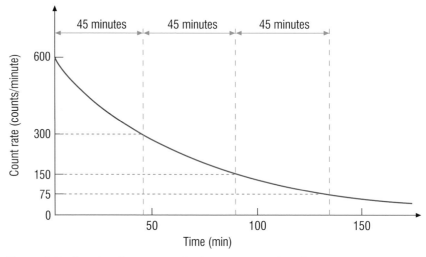

Figure 1 Radioactive decay: a graph of count rate against time

The graph shows that the count rate decreases with time. The count rate falls from:

- 600 counts per minute (c.p.m.) to 300 c.p.m. in the first 45 minutes
- 300 counts per minute (c.p.m.) to 150 c.p.m. in the next 45 minutes.

The average time taken for the count rate (and therefore the number of parent atoms) to fall by half is always the same. This time is called the **half-life**. The half-life shown on the graph is 45 minutes.

a What will the count rate be after 135 minutes from the start?

The half-life of a radioactive isotope is the average time it takes:

- **for the number of nuclei of the isotope in a sample (and therefore the mass of parent atoms) to halve**
- **for the count rate of the isotope in a sample to fall to half its initial value.**

 Did you know … ?

Some radioactive isotopes have half-lives of a fraction of a second, whereas others have half-lives of more than a billion years. The nitrogen isotope N-12 has a half-life of 0.0125 seconds. The uranium isotope U-238 has a half-life of 4.5 billion years.

The random nature of radioactive decay

Radioactive decay is a random process. We can't predict *when* an individual atom will suddenly decay. But we *can* predict how many atoms will decay in a certain time – because there are so many of them. This is a bit like throwing dice. You can't predict what number you will get with a single throw. But if you threw 1000 dice, you would expect one-sixth to come up with a particular number.

Suppose we start with 1000 unstable atoms. Look at the graph on the right:

If 10% decay every hour:

- 100 atoms will decay in the first hour, leaving 900
- 90 atoms (= 10% of 900) will decay in the second hour, leaving 810.

The table below shows what you get if you continue the calculations. The results are plotted as a graph in Figure 2.

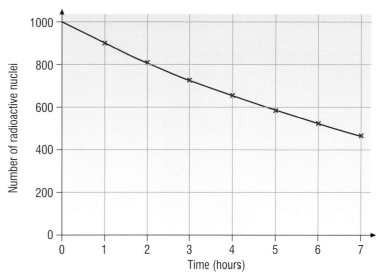

Figure 2 Half-life

Time from start (hours)	0	1	2	3	4	5	6	7
No. of unstable atoms present	1000	900	810	729	656	590	530	477
No. of unstable atoms that decay in the next hour	100	90	81	73	66	59	53	48

b Use the graph in Figure 2 to work out the half-life of this radioactive isotope.

Did you know … ?

Next time you help someone choose numbers for the lottery, think about whether this is something you can predict. The balls come out of the machine at random; is there any way of predicting what they will be?

Summary questions

1 Copy and complete **a** and **b** using the words below. Each word can be used more than once.

half-life   stable   unstable

   a In a radioactive substance, ... atoms decay and become ....
   b The .... of a radioactive isotope is the time taken for the number of .... atoms to decrease to half.

2 A radioactive isotope has a half-life of 15 hours. A sealed tube contains 8 milligrams of this isotope.
   What mass of the isotope is in the tube:
   a 15 hours later?
   b 45 hours later?

3 A sample of a radioactive isotope contains 320 million atoms of the isotope. How many atoms of the isotope are present after:
   a one half-life
   b five half-lives?

Key points

- The **half-life** of a radioactive isotope is the average time it takes for the number of nuclei of the isotope in a sample to halve.

- The activity of a radioactive source is the number of nuclei that decay per second.

- The number of atoms of a radioactive isotope and the activity both decrease by half every half-life.

P2 6.6

Radioactivity at work

Learning objectives

- How do we choose a radioactive isotope for a particular job?
- How can we use radioactivity for monitoring?
- What are radioactive tracers?
- What is radioactive dating?

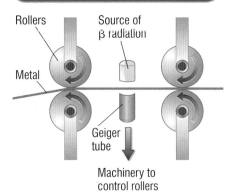

Figure 1 Thickness monitoring using a radioactive source

Radioactivity has many uses. For each use, we need a radioactive isotope that emits a certain type of radiation and has a suitable half-life.

Automatic thickness monitoring

This is used when making metal foil.

Look at Figure 1. The radioactive source emits β radiation. The amount of radiation passing through the foil depends on the thickness of the foil. A detector on the other side of the metal foil measures the amount of radiation passing through it.

- If the thickness of the foil increases too much, the detector reading drops.
- The detector sends a signal to the rollers to increase the pressure on the metal sheet.

This makes the foil thinner again.

  a What happens if the thickness of the foil decreases too much?
  b Why is alpha radiation not used here?

Radioactive tracers

These are used to trace the flow of a substance through a system. For example, doctors use radioactive iodine to find out if a patient's kidney is blocked.

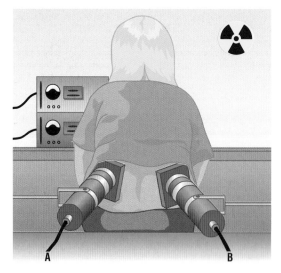

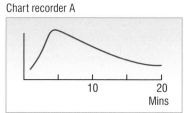

Figure 2 Using a tracer to monitor a patient's kidneys

Before the test, the patient drinks water containing a tiny amount of the radioactive substance. A detector is then placed against each kidney. Each detector is connected to a chart recorder.

- The radioactive substance flows in and out of a normal kidney. So the detector reading goes up then down.
- For a blocked kidney, the reading goes up and stays up. This is because the radioactive substance goes into the kidney but doesn't flow out again.

Radioactive iodine is used for this test because:

- Its half-life is 8 days, so it lasts long enough for the test to be done but decays almost completely after a few weeks.
- It emits gamma radiation, so it can be detected outside the body.
- It decays into a stable product.

c In Figure 2, which kidney is blocked, A or B?

Radioactive dating

This is used to find the age of ancient material. We can use:

- **Carbon dating** – this is used to find the age of ancient wood and other organic material. Living wood contains a tiny proportion of radioactive carbon. This has a half-life of 5600 years. When a tree dies, it no longer absorbs any carbon. So the amount of radioactive carbon in it decreases. To find the age of a sample, we need to measure the count rate from the wood. This is compared with the count rate from the same mass of living wood. For example, suppose the count rate in a sample of wood is half the count rate of an equal mass of living wood. Then the sample must be 5600 years old.

- **Uranium dating** – this is used to find the age of igneous rocks. These rocks contain radioactive uranium, which has a half-life of 4500 million years. Each uranium atom decays into an atom of lead. We can work out the age of a sample by measuring the number of atoms of uranium and lead. For example, if a sample contains 1 atom of lead for every atom of the uranium, the age of the sample must be 4500 million years. This is because there must have **originally** been 2 atoms of uranium for each atom of uranium now present.

d What could you say about an igneous rock with uranium but no lead in it?

Figure 3 A smoke alarm

Summary questions

1 Copy and complete **a** to **c** using the words below. Each word can be used more than once.

alpha   beta   gamma

a In the continuous production of thin metal sheets, a source of .... radiation should be used to monitor the thickness of the sheets.

b A radioactive tracer given to a hospital patient needs to emit .... or .... radiation.

c The radioactive source used to trace a leak in an underground pipeline should be a source of .... radiation.

2 **a** Explain why γ radiation is not suitable for monitoring the thickness of metal foil.

b When a radioactive tracer is used, why is it best to use a radioactive isotope that decays into a stable isotope?

3 **a** What are the ideal properties of a radioactive isotope used as a medical tracer?

b A sample of old wood was carbon dated and found to have 25% of the count rate measured in an equal mass of living wood. The half-life of the radioactive carbon is 5600 years. How old is the sample of wood?

Key points

- The use we can make of a radioactive isotope depends on:

  a its half-life

  b the type of radiation it gives out.

- For monitoring, the isotope should have a long half-life.

- Radioactive tracers should be β or γ emitters that last long enough to monitor but not too long.

- For radioactive dating of a sample, we need a radioactive isotope that is present in the sample which has a half-life about the same as the age of the sample.

Summary questions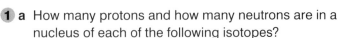

1 a How many protons and how many neutrons are in a nucleus of each of the following isotopes?

   i $^{14}_{6}\text{C}$

   ii $^{228}_{90}\text{Th}$

  b $^{14}_{6}\text{C}$ emits a β particle and becomes an isotope of nitrogen (N).

   i How many protons and how many neutrons are in this nitrogen isotope?

   ii Write down the symbol for this isotope.

  c $^{228}_{90}\text{Th}$ emits an α particle and becomes an isotope of radium (Ra).

   i How many protons and how many neutrons are in this isotope of radium?

   ii Write down the symbol for this isotope.

2 Which type of radiation, alpha, beta or gamma:

  a can pass through lead?

  b travels no further than about 10 cm in air?

  c is stopped by an aluminium metal plate but not by paper?

  d consists of electrons?

  e consists of helium nuclei?

  f is uncharged?

3 The table below gives information about four radioactive isotopes **A**, **B**, **C** and **D**.

Isotope	Type of radiation emitted	Half-life
A californium-241	alpha	4 minutes
B cobalt-60	gamma	5 years
C hydrogen-3	beta	12 years
D strontium-90	beta	28 years

Match each statement 1 to 4 with **A**, **B**, **C** or **D**.

  1 the isotope that gives off radiation with an unlimited range

  2 the isotope that has the longest half-life

  3 the isotope that decays the fastest

  4 the isotope with the smallest mass of each atom.

4 The following measurements were made of the count rate due to a radioactive source.

Time (hours)	0	0.5	1.0	1.5	2.0	2.5
Count rate due to the source (counts per minute)	510	414	337	276	227	188

  a Plot a graph of the count rate (on the vertical axis) against time.

  b Use your graph to find the half-life of the source.

5 In a carbon dating experiment of ancient wood, a sample of the wood gave a count rate of 0.4 counts per minute. The same mass of living wood gave a count rate of 1.6 counts per minute.

  a How many half-lives did the count rate take to decrease from 1.6 to 0.4 counts per minute?

  b The half-life of the radioactive carbon in the wood is 5600 years. What is the age of the sample?

6 In an investigation to find out what type of radiation was emitted from a given source, the following measurements were made with a Geiger counter.

Source	Average count rate (in counts per minute)
No source present	29
Source at 20 mm from tube with no absorber between	385
Source at 20 mm from tube with a sheet of metal foil between	384
Source at 20 mm from tube with a 10 mm thick aluminium plate between	32

  a What caused the count rate when no source was present?

  b What was the count rate due to the source with no absorbers present?

  c What type of radiation was emitted by the source? Explain how you arrive at your answer.

7 Figure 1 shows the path of two α particles labelled A and B that are deflected by the nucleus of an atom.

  a Why are they deflected by the nucleus?

  b Why is B deflected less than A?

  c Why do most α particles directed at a thin metal foil pass straight through it?

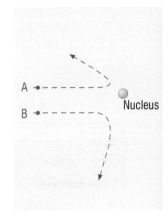

Figure 1

AQA Examination-style questions

1 Diagrams **A** and **B** show two atoms of carbon.

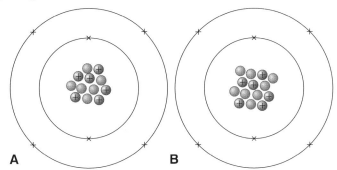

A   **B**

a Copy and complete the following sentences using the list of words and phrases below. Each one can be used once, more than once or not at all.

electrons positive isotopes nuclear
plum pudding negative nucleus ions neutrons

Particles shown by the symbol **x** in the diagram are called .... They orbit the .... of an atom. This is made up of protons and .... Protons have a .... charge. This diagram shows the .... model of the atom which replaced the .... model. (6)

b Explain how a carbon **ion** would be different from atom **A**. (1)

c Give the mass number of atom **A**. (1)

d Give the atomic number of atom **A**. (1)

e Compare atom **B** with atom **A**. (3)

2 a A geologist wishes to know what types of radiation are emitted by three radioactive rock samples. Different absorbers are placed between each sample and a detector. The counts per second are shown in the table.

	Counts per second		
Absorber	**Sample 1**	**Sample 2**	**Sample 3**
1 cm of air	140	80	120
paper	90	50	70
3 mm of aluminium	30	49	0
1 cm of lead	0	1	0

For each sample state which of the three types of radiation (alpha, beta, gamma) are emitted. A rock may emit more than one type. (3)

b Describe the nature of an alpha particle. (2)

c List the three types of nuclear radiation in order of their relative ionising power from the least ionising to the most ionising. (3)

d The source of radiation shown below emits alpha, beta and gamma. When the radiation travels through air in an electric field between two plates, the three types of radiation behave differently.

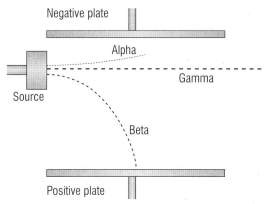

i An alpha particle has more charge than a beta particle. Explain why the beta particle is deflected more by the electric field and in the opposite direction. [H] (4)

ii Explain why the gamma radiation is not affected by the electric field. (1)

iii Explain why the alpha particle does not reach the plate. (1)

3 Technetium-99 is a gamma-emitting radioisotope used as a tracer inside the body in order to diagnose problems with various organs. Cobalt-60 is a gamma emitter used for radiotherapy where the source is used outside the body to kill cancer cells on the inside.

	Half-life	**Radiation**	**Relative ionising power**
technetium-99	6.0 hours	gamma	1
cobalt-60	5.3 years	gamma	10

a Technetium–99 emits a gamma ray and then decays to an isotope of ruthenium (Ru) by beta decay. Balance the nuclear equation by giving the appropriate atomic numbers and mass numbers.

$$^{99}_{43}\text{Tc} \rightarrow \,^{(i)}_{(ii)}\text{Ru} + \,^{(iii)}_{(iv)}\beta$$ [H] (4)

b *In this question you will be assessed on using good English, organising information clearly and using specialist terms where appropriate.*

Explain why cobalt-60 is not used as a medical tracer in humans and why technetium-99 is used for this purpose. (6)

P2 7.1 Nuclear fission ⓚ

Learning objectives

- What is nuclear fission?
- Which radioactive isotopes undergo fission?
- What is a chain reaction?
- How is a chain reaction in a nuclear reactor controlled?

Chain reactions

Energy is released in a nuclear reactor as a result of **nuclear fission**. In this process, the nucleus of an atom of a fissionable substance splits into two smaller 'fragment' nuclei. This event can cause other fissionable nuclei to split. This then produces a **chain reaction** of fission events.

Fission neutrons

When a nucleus undergoes fission, it releases:

- two or three neutrons (referred to as 'fission' neutrons) at high speeds
- energy, in the form of radiation, plus kinetic energy of the fission neutrons and the fragment nuclei.

The fission neutrons may cause further fission resulting in a chain reaction. In a **nuclear fission reactor**, on average, exactly one fission neutron from each fission event goes on to produce further fission. This ensures energy is released at a steady rate in the reactor.

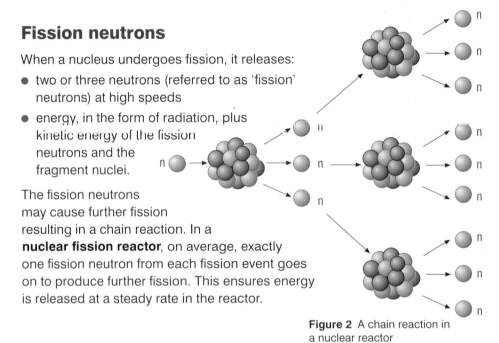

Figure 2 A chain reaction in a nuclear reactor

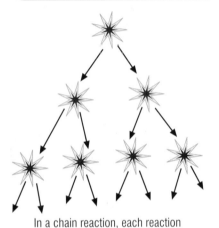

In a chain reaction, each reaction causes more reactions which cause more reactions, etc.

Figure 1 A chain reaction

a What would happen if, on average, more than one fission neutron per event went on to produce further fission?

Fissionable isotopes

The fuel in a nuclear reactor must contain fissionable isotopes.

- Most reactors at the present time are designed to use 'enriched uranium' as the fuel. This consists mostly of the non-fissionable uranium isotope $^{238}_{92}U$ (U-238) and about 2–3% of the uranium isotope $^{235}_{92}U$ (U-235) which is fissionable. In comparison, natural uranium is more than 99% U-238.
- The U-238 nuclei in a nuclear reactor do not undergo fission but they change into other heavy nuclei, including plutonium-239 (the isotope $^{239}_{94}Pu$). This isotope is fissionable. It can be used in a different type of reactor but not in a uranium reactor.

Inside a nuclear reactor

A nuclear reactor consists of uranium fuel rods spaced evenly in the reactor core. Figure 3 shows a cross-section of a pressurised water reactor (PWR).

- The reactor core contains the fuel rods, control rods and water at high pressure. The fission neutrons are slowed down by collisions with the atoms in the water molecules. This is necessary as fast neutrons do not cause further fission of U-235. We say the water acts as a **moderator** because it slows down the fission neutrons.

- **Control rods** in the core absorb surplus neutrons. This keeps the chain reaction under control. The depth of the rods in the core is adjusted to maintain a steady chain reaction.
- The water acts as a **coolant**. Its molecules gain kinetic energy from the neutrons and the fuel rods. The water is pumped through the core. Then it goes through sealed pipes to and from a heat exchanger outside the core. The water transfers energy for heating to the heat exchanger from the core.
- The reactor core is made of thick steel to withstand the very high temperature and pressure in the core. The core is enclosed by thick concrete walls. These absorb radiation that escapes through the walls of the steel vessel.

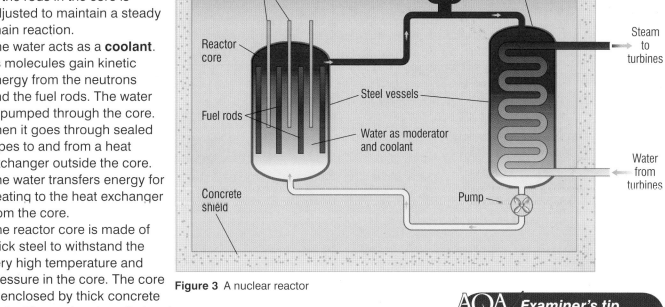

Figure 3 A nuclear reactor

b What would happen if the control rods were removed from the core?

AQA Examiner's tip

During nuclear fission a large nucleus breaks up into two smaller nuclei. Make sure you know how to spell 'fission' – with two s's.

Summary questions

1 Copy and complete **a** and **b** using the words below. Each word can be used more than once.

nucleus uranium-235 uranium-238 plutonium-239

a Nuclear fission happens when a .... of .... or .... splits.

b A nucleus of .... in a nuclear reactor changes without fission into a nucleus of ....

2 Put the statements A to D in the list below into the correct sequence to describe a steady chain reaction in a nuclear reactor.

A a U-235 nucleus splits
B a neutron hits a U-235 nucleus
C neutrons are released
D energy is released

3 Look at the chain reaction shown in Figure 4.

a i Which of the nuclei A to F have been hit by a neutron?

ii What has happened to these nuclei?

iii Which two of the other nuclei A to F could undergo fission from a fission neutron shown?

b State one process that could happen to a fission neutron that does not produce further fission.

Figure 4

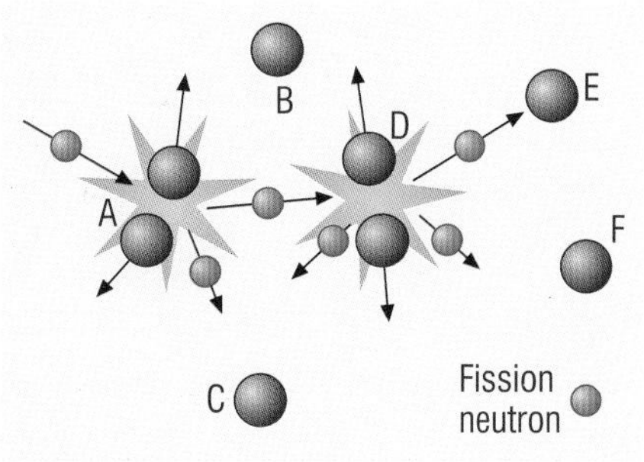

Key points

- Nuclear fission is the splitting of a nucleus into two approximately equal fragments and the release of two or three neutrons.
- Nuclear fission occurs when a neutron hits a uranium-235 nucleus or a plutonium-239 nucleus and the nucleus splits.
- A chain reaction occurs in a nuclear reactor when each fission event causes further fission events.
- In a nuclear reactor, control rods absorb fission neutrons to ensure that, on average, only one neutron per fission goes on to produce further fission.

P2 7.2

Nuclear fusion ⓚ

Energy from the nucleus

Learning objectives

- What is nuclear fusion?
- How can nuclei be made to fuse together?
- Where does the Sun's energy come from?
- Why is it difficult to make a nuclear fusion reactor?

Imagine if we could get energy from water. Stars release energy as a result of fusing small nuclei such as hydrogen to form larger nuclei. Water contains lots of hydrogen atoms. A glass of water could provide the same amount of energy as a tanker full of petrol. But only if we could make a fusion reactor here on Earth.

Fusion reactions

Two small nuclei release energy when they are fused together to form a single larger nucleus. This process is called **nuclear fusion**. It releases energy only if the relative mass of the nucleus formed is no more than about 55 (about the same as an iron nucleus). Energy must be supplied to create bigger nuclei.

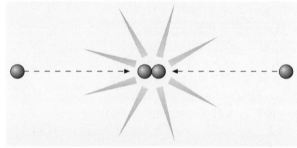

Figure 1 A nuclear fusion reaction

The Sun is about 75 per cent hydrogen and 25 per cent helium. The core is so hot that it consists of a 'plasma' of bare nuclei with no electrons. These nuclei move about and fuse together when they collide. When they fuse, they release energy. Figure 2 shows how protons fuse together to form a $^4_2$He nucleus. Energy is released at each stage.

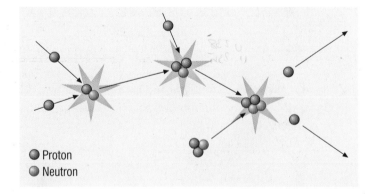

● Proton
● Neutron

Figure 2 Fusion reactions in the Sun

- When two protons (i.e. hydrogen nuclei) fuse, they form a 'heavy hydrogen' nucleus, $^2_1$H. Other particles are created and emitted at the same time.
- Two more protons collide separately with two $^2_1$H nuclei and turn them into heavier nuclei.
- The two heavier nuclei collide to form the helium nucleus $^4_2$He.
- The energy released at each stage is carried away as kinetic energy of the product nucleus and other particles emitted.

a Look at Figure 2 and work out what is formed when a proton collides with a $^2_1$H nucleus.

Fusion reactors

There are enormous technical difficulties with making fusion a useful source of energy. The plasma of light nuclei must be heated to very high temperatures before the nuclei will fuse. This is because two nuclei approaching each other will repel each other due to their positive charges. If the nuclei are moving fast enough, they can overcome the force of repulsion and fuse together.

In a fusion reactor:

● the plasma is heated by passing a very large electric current through it
● the plasma is contained by a magnetic field so it doesn't touch the reactor walls. If it did, it would go cold and fusion would stop.

Scientists have been working on these problems since the 1950s. A successful fusion reactor would release more energy than it uses to heat the plasma. At the present time, scientists working on experimental fusion reactors are able to do this by fusing heavy hydrogen nuclei to form helium nuclei – but only for a few minutes!

Figure 3 An experimental fusion reactor

b Why is a fusion reactor unlikely to explode?

A promising future

Practical fusion reactors could meet all our energy needs.

● The fuel for fusion reactors is readily available as heavy hydrogen and is naturally present in sea water.
● The reaction product, helium, is a non-radioactive inert gas, so is harmless.
● The energy released could be used to generate electricity.

In comparison, fission reactors mostly use uranium, which is only found in certain parts of the world. Also, they produce nuclear waste that has to be stored securely for many years. However, fission reactors have been in operation for over 50 years, unlike fusion reactors, which are still under development.

Summary questions

1 Copy and complete **a** and **b** using the words below:

large   small   stable

  a When two .... nuclei moving at high speed collide, they form a .... nucleus.

  b Energy is released in nuclear fusion if the product nucleus is not as .... as an iron nucleus.

2 **a** Why does the plasma of light nuclei in a fusion reactor need to be very hot?

  b Why would a fusion reactor that needs more energy than it produces not be much use?

3 **a** How many protons and how many neutrons are present in a $^2_1$H nucleus?

  b Copy and complete the equation below to show the reaction that takes place when two $^2_1$H nuclei fuse together to form a helium nucleus.

$$^2_1H + {}^2_1H \rightarrow {}^?_?He$$

    [H]

Key points

● Nuclear fusion is the process of forcing two nuclei close enough together so they form a single larger nucleus.

● Nuclear fusion can be brought about by making two light nuclei collide at very high speed.

● Energy is released when two light nuclei are fused together. Nuclear fusion in the Sun's core releases energy.

● A fusion reactor needs to be at a very high temperature before nuclear fusion can take place. The nuclei to be fused are difficult to contain.

P2 7.3   Nuclear issues

∞ **links**

For more information on ionising radiation, look back at P2 6.4 More about alpha, beta and gamma radiation.

⁇? Did you know ...?

Nuclear waste

Used fuel rods are very hot and very radioactive.

● After removal from a reactor, they are stored in large tanks of water for up to a year. The water cools the rods down.

● Remote-control machines are then used to open the fuel rods. The unused uranium and plutonium are removed chemically from the used fuel. These are stored in sealed containers so they can be used again.

● The remaining material contains many radioactive isotopes with long half-lives. This radioactive waste must be stored in secure conditions for many years.

Figure 2 Storage of nuclear waste

Radioactivity all around us

When we use a Geiger counter, it clicks even without a radioactive source near it. This is due to background **radiation.** Radioactive substances are found naturally all around us.

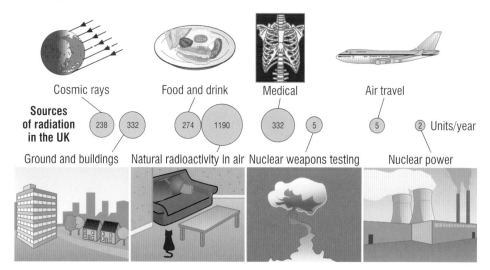

Figure 1 Radioactivity

Figure 1 shows the sources of background radiation. As explained in P2 6.4, the radiation from radioactive substances is hazardous, as it ionises substances it passes through. The numbers in Figure 1 tell you the **radiation dose** or how much radiation on average each person gets in a year from each source.

● Medical sources include X-rays as well as radioactive substances, as X-rays have an ionising effect. People who work in jobs that involve the use of ionising radiation have to wear personal radiation monitors to ensure they are not exposed to too much ionising radiation.

● Background radiation in the air is due mostly to radon gas that seeps through the ground from radioactive substances in rocks deep underground. Radon gas emits alpha particles so it is a health hazard if it is breathed in. It can seep into homes and other buildings in certain locations. In homes and buildings where people are present for long periods, methods need to be taken to reduce exposure to radon gas. For example, pipes under the building can be installed and fitted to a suction pump to draw the gas out of the ground before it seeps into the building.

a What is the biggest source of background radioactivity?
b Which source in the chart contributes least to background radioactivity?

Chernobyl

In 1986, a nuclear reactor in Ukraine exploded. Emergency workers and scientists struggled for days to contain the fire. A cloud of radioactive material from the fire drifted over many parts of Europe, including Britain. More than 100 000 people were evacuated from Chernobyl and the surrounding area. Over 30 people died in the accident. Many more have developed leukaemia or cancer since then. It was and remains (up to now) the world's worst nuclear accident.

Could it happen again?

- Most nuclear reactors are of a different design.
- The Chernobyl accident did not have a high-speed shutdown system like most reactors have.
- The operators at Chernobyl ignored safety instructions.
- There are thousands of nuclear reactors in the world. They have been working safely for many years.

Radioactive risks

The effect on living cells of radiation from radioactive substances depends on:

- the type and the amount of radiation received (the dose)
- whether the source of the radiation is inside or outside the body
- how long the living cells are exposed to the radiation.

	Alpha radiation	Beta radiation	Gamma radiation
source inside the body	**very dangerous** – affects all the surrounding tissue	**dangerous** – reaches cells throughout the body	
source outside the body	**some danger** – absorbed by skin; damages skin cells		

- The larger the dose of radiation someone gets, the greater the risk of cancer. High doses kill living cells.
- The smaller the dose, the less the risk – but it is never zero. So there is a very low level of risk to each and every one of us because of background radioactivity.

Workers who are at risk from ionising radiations cut down their exposure to the radiation by:

- keeping as far as possible from the source of radiation, using special handling tools with long handles
- spending as little time as possible in 'at-risk' areas
- shielding themselves from the radiation by staying behind thick concrete barriers and/or using thick lead plates.

o Why does radioactive waste need to be stored **i** securely **ii** for many years?

d Why is a source of alpha radiation very dangerous inside the body but not outside it?

Figure 3 Chernobyl

Key points

- Radon gas is an α-emitting isotope that seeps into houses in certain areas through the ground.
- There are thousands of fission reactors safely in use in the world. None of them are of the same type as the Chernobyl reactors that exploded.
- Nuclear waste is stored in safe and secure conditions for many years after unused uranium and plutonium (to be used in the future) is removed from it.

Summary questions

1 In some locations, the biggest radiation hazard comes from radon gas which seeps up through the ground and into buildings. The dangers of radon gas can be minimised by building new houses that are slightly raised on brick pillars and modifying existing houses. Radon gas is an α-emitting isotope.
   a Why is radon gas dangerous in a house?
   b Describe one way of making an existing house safe from radon gas.

2 Should the UK government replace our existing nuclear reactors with new reactors, either fission or fusion or both? Answer this question by discussing the benefits and drawbacks of new fission and fusion reactors.

P2 7.4

The early universe

The Big Bang that created the universe was about 13 thousand million (13 billion) years ago. Space, time and radiation were created in the Big Bang. At first, the universe was a hot glowing ball of radiation and matter. As it expanded, its temperature fell. Now the universe is cold and dark, except for hot spots we call stars.

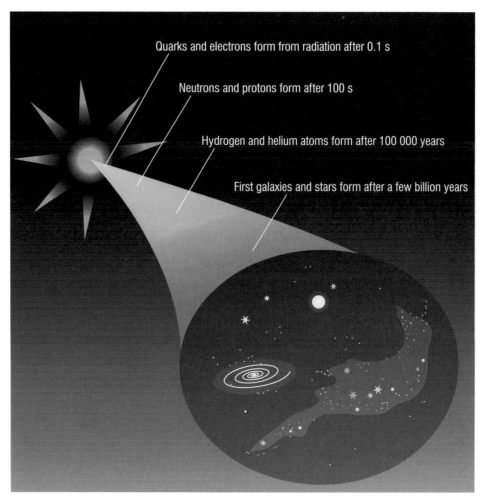

Quarks and electrons form from radiation after 0.1 s

Neutrons and protons form after 100 s

Hydrogen and helium atoms form after 100 000 years

First galaxies and stars form after a few billion years

Figure 1 Timeline for the universe

?? Did you know ...?

In the Cold War, US satellites detected bursts of gamma radiation from space. At first, the US military thought nuclear weapons were being tested in space by Russia. Then astronomers found the bursts were from violent events long ago in distant galaxies – maybe stars being sucked into black holes!

The stars we see in the night sky are all in the Milky Way galaxy, our home galaxy. The Sun is just one of billions of stars in the Milky Way galaxy. Using powerful telescopes, we can see many more stars in the Milky Way galaxy. We can also see individual stars in other galaxies.

We now know there are billions of galaxies in the universe. There is vast empty space between them. Light from the furthest galaxies that we can see has taken billions of years to reach us.

a Why do powerful telescopes give us a picture of the universe long ago?

Figure 2 Andromeda – the nearest big galaxy to the Milky Way

The Dark Age of the universe

As the universe expanded, it became transparent as radiation passed through the empty space between its atoms. The background microwave radiation that causes the spots on an untuned television was released at this stage. The Dark Age of the universe had begun!

For the next few billion years, the universe was a completely dark, patchy, expanding cloud of hydrogen and helium. Then the stars and galaxies formed and lit up the universe!

b How long, to the nearest billion years, has background microwave radiation been travelling for?

Figure 3 Arno Allan Penzias and Robert Woodrow Wilson standing on the radio antenna that unexpectedly discovered the universe's microwave background radiation

The force of gravity takes over

Uncharged atoms don't repel each other. But they can attract each other. During the Dark Age of the universe, the force of gravitational attraction was at work without any opposition from repulsive forces.

As the universe continued to expand, it became more patchy as the denser parts attracted nearby matter. Gravity pulled more matter into the denser parts and turned them into gigantic clumps.

Eventually, the force of gravity turned the clumps into galaxies and stars. A few billion years after the Big Bang, the Dark Age came to an end, as the stars lit up the universe.

c Why would the force of gravity between two helium nuclei be unable to pull the nuclei together?

Figure 4 The force of gravity takes over

Summary questions

1 Copy and complete **a** to **c** using the words below:

attracted   cooled   expanded   formed

  a As the universe .... , it ....
  b Uncharged atoms .... each other.
  c Galaxies and stars .... from uncharged atoms.

2 **a i** Why can't we take a photo of the Milky Way galaxy from outside?
   ii Why can't we take photos of a distant galaxy at different stages in its formation?
  b i Why do the stars in a galaxy not drift away from each other?
   ii Why are there vast spaces between the galaxies?

3 Put these events in the correct sequence with the earliest event first.
   1 Cosmic background radiation was released.
   2 Hydrogen nuclei were first fused to form helium nuclei.
   3 The Big Bang took place.
   4 Neutrons and protons formed.

Key points

- A galaxy is a collection of billions of stars held together by their own gravity.

- Before galaxies and stars formed, the universe was a dark patchy cloud of hydrogen and helium.

- The force of gravity pulled matter into galaxies and stars.

P2 7.5

The life history of a star

Learning objectives

- What is a protostar?
- What are the stages in the life of a star?
- What will eventually happen to the Sun?
- What is a supernova?

The birth of a star

Stars form out of clouds of dust and gas.

- The particles in the clouds are pulled together by their own gravitational attraction. The clouds merge together. They become more and more concentrated to form a **protostar**, the name for a star to be.
- As a protostar becomes denser, it gets hotter. If it becomes hot enough, the nuclei of hydrogen atoms and other light elements fuse together. Energy is released in this fusion so the core gets hotter and brighter and starts to shine. A star is born!
- Objects may form that are too small to become stars. Such objects may be attracted by a protostar to become **planets**.

a Where does the energy to heat a protostar come from?

Shining stars

Stars like the Sun radiate energy because of hydrogen fusion in the core. They are called **main sequence stars** because this is the main stage in the life of a star. It can maintain its energy output for millions of years until the star runs out of hydrogen nuclei to fuse together.

- Energy released in the core keeps the core hot so the process of fusion continues. Radiation flows out steadily from the core in all directions.
- The star is stable because the forces within it are balanced. The force of gravity that makes a star contract is balanced by the outward force of the radiation from its core. These forces stay in balance until most of the hydrogen nuclei in the core have been fused together.

b Why doesn't the Sun collapse under its own gravity?

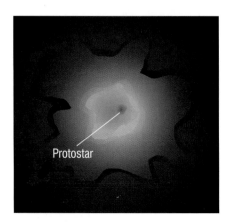

Figure 1 Star birth

The end of a star

When a star runs out of hydrogen nuclei to fuse together, it reaches the end of its main sequence stage and it swells out.

Stars about the same size as the Sun (or smaller) swell out, cool down and turn red.

- The star is now a **red giant**. At this stage, helium and other light elements in its core fuse to form heavier elements.
- When there are no more light elements in its core, fusion stops and no more radiation is released. Due to its own gravity, the star collapses in on itself. As it collapses, it heats up and turns from red to yellow to white. It becomes a **white dwarf**. This is a hot, dense white star much smaller in diameter than it was. Stars like the Sun then fade out, go cold and become **black dwarfs**.

Stars much bigger than the Sun end their lives much more dramatically.

- Such a star swells out to become a red **supergiant** which then collapses.
- In the collapse, the matter surrounding the star's core compresses the core more and more. Then the compression suddenly reverses in a cataclysmic explosion known as a **supernova**. Such an event can outshine an entire galaxy for several weeks.

?? Did you know ...?

- The Sun is about 5000 million years old and will probably continue to shine for another 5000 million years.
- The Sun will turn into a red giant bigger than the orbit of Mercury. By then, the human race will probably have long passed into history. But will intelligent life still exist?

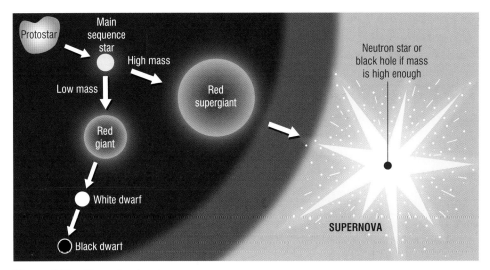

Figure 2 The life cycle of a star

c What force causes a red giant to collapse?

What remains after a supernova occurs?

The explosion compresses the core of the star into a **neutron star**. This is an extremely dense object composed only of neutrons. If the star is massive enough, it becomes a **black hole** instead of a neutron star. The gravitational field of a black hole is so strong that nothing can escape from it. Not even light, or any other form of electromagnetic radiation, can escape.

d What force causes matter to be dragged into a black hole?

Figure 3 M87 is a galaxy that spins so fast at its centre that it is thought to contain a black hole with a billion times more mass than the Sun

Summary questions

1 a The list below shows some of the stages in the life of a star like the Sun. Put the stages in the correct sequence.
   A main sequence
   B protostar
   C red giant
   D white dwarf
   b i Which stage in the above list is the Sun at now?
   ii What will happen to the Sun after it has gone through the above stages?

2 a Copy and complete **i** and **ii** using the words below. Each word can be used more than once.
   collapse   expand   explode
   i The Sun will eventually .... then ....
   ii A red supergiant will .... then ....
   b i What is the main condition needed for a supergiant to form a black hole?
   ii Why is it not possible for light to escape from a black hole?

3 a i What force makes a red supergiant collapse?
   ii What force prevents a main sequence star from collapsing?
   b Why does a white dwarf eventually become a black dwarf?

Key points

- A protostar is a gas and dust cloud in space that can go on to form a star.

Low mass star:
Protostar → main sequence star → red giant → white dwarf → black dwarf

High mass star:
Protostar → main sequence star → red supergiant → supernova → neutron star → black hole if sufficient mass

- The Sun will eventually become a black dwarf.

- A supernova is the explosion of a supergiant after it collapses.

P2 7.6 How the chemical elements formed

Learning objectives

- What chemical elements are formed inside stars?
- What chemical elements are formed in supernovas?
- Why does the Earth contain heavy elements?

The birthplace of the chemical elements

- **Light elements are formed as a result of fusion in stars.**

Stars like the Sun fuse hydrogen nuclei (i.e. protons) into helium and similar small nuclei, including carbon. When it becomes a red giant, it fuses helium and the other small nuclei into larger nuclei.

Nuclei larger than iron cannot be formed by this process because too much energy is needed.

- **Heavy elements are formed when a massive star collapses then explodes as a supernova.**

The enormous force of the collapse fuses small nuclei into nuclei larger than iron. The explosion scatters the star into space.

The debris from a supernova contains all the known elements from the lightest to the heaviest. Eventually, new stars form as gravity pulls the debris together.

Planets form from debris surrounding a new star. As a result, such planets will be composed of all the known elements too.

a Lead (Pb) is much heavier than iron (Fe). How did the lead we use form?

Did you know ...?

The Crab Nebula is the remnants of a supernova explosion that was observed in the 11th century. In 1987, a star in the southern hemisphere exploded and became the biggest supernova to be seen for four centuries. Astronomers realised that it was Sandaluk II, a star in the Andromeda galaxy millions of light years from Earth.

If a star near the Sun exploded, the Earth would probably be blasted out of its orbit. We would see the explosion before the shock wave hit us.

Figure 1 The Crab Nebula

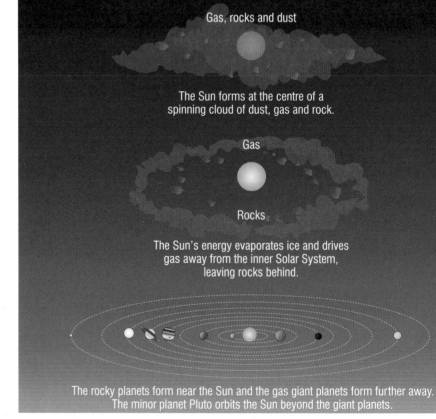

Gas, rocks and dust

The Sun forms at the centre of a spinning cloud of dust, gas and rock.

Gas

Rocks

The Sun's energy evaporates ice and drives gas away from the inner Solar System, leaving rocks behind.

The rocky planets form near the Sun and the gas giant planets form further away. The minor planet Pluto orbits the Sun beyond the giant planets.

Figure 2 Formation of the Solar System

Did you know ... ?

Molecules of carbon-based chemicals are present in space. Life on Earth probably developed from chemicals reacting in lightning storms.

So are we looking for any scientific evidence about life on other planets, either in our own Solar System or around other stars?

- **Space probes sent to Mars** have tested the atmosphere, rocks and soil on Mars looking for microbes or chemicals that might indicate life was once present on Mars. Water is necessary for life. Astronomers now have strong evidence of the presence of 'underground' water breaking through to the surface of Mars.

- **The search for extra-terrestrial intelligence,** known as **SETI**, has gone on for more than 40 years using radio telescopes. Signals from space would indicate the existence of living beings with technologies at least as advanced as our own. No signals have been detected – yet!

Figure 3 The NASA Exploration Rovers looked for signs of life on Mars

Planet Earth

The heaviest known natural element is uranium. It has a half-life of 4500 million years. The presence of uranium in the Earth is evidence that the Solar System must have formed from the remnants of a supernova.

Elements such as plutonium are heavier than uranium. Scientists can make these elements by bombarding heavy elements like uranium with high-speed neutrons. They would have been present in the debris which formed the Solar System. Elements heavier than uranium formed then have long since decayed.

b Plutonium-239 has a half-life of about 24 000 years. Why is it not found naturally like uranium?

c Why is carbon an important element?

Summary questions

1 Match each statement below with an element in the list.

helium hydrogen iron uranium

a Helium nuclei are formed when nuclei of this element are fused.

b This element is formed in a supernova explosion.

c Stars form nuclei of these two elements (and others not listed) by fusing smaller nuclei.

d The early universe mostly consisted of this element.

2 Copy and complete **a** to **c** using the words below. Each word can be used more than once.

galaxy planets stars supernova

a Fusion inside .... creates light elements. Fusion in a .... creates heavy elements.

b A .... scatters the elements throughout a ....

c .... and planets formed from the debris of a .... contain all the known elements.

3 Uranium-238 is a radioactive isotope found naturally in the Earth. It has a half-life of about 4500 million years. It was formed from lighter elements.

a **i** What is the name of the physical process in which this isotope is formed?

ii What is the name for the astronomical event in which the above process takes place?

b Why has all the uranium in the Earth not decayed by now?

Key points

- Elements as heavy as iron are formed inside stars as a result of nuclear fusion.

- Elements heavier than iron are formed in supernovas as well as light elements.

- The Sun and the rest of the Solar System were formed from the debris of a supernova.

Summary questions ⓚ

1 a Copy and complete **i** to **iii** using the words below:

decreases   increases   stays the same

When energy is released at a steady rate in a nuclear reactor,

- **i** the number of fission events each second in the core ....
- **ii** the amount of uranium-235 in the core ....
- **iii** the number of radioactive isotopes in the fuel rods ....

b Explain what would happen in a nuclear reactor if:
- **i** the coolant fluid leaked out of the core
- **ii** the control rods were pushed further into the reactor core.

2 a i What do we mean by nuclear fusion?
- **ii** Why do two nuclei repel each other when they get close?
- **iii** Why do they need to collide at high speed in order to fuse together?

b Give two reasons why nuclear fusion is difficult to achieve in a reactor.

3 a Copy and complete **i** to **iii** using the words below. Each word can be used more than once.

fission   fusion

- **i** In a .... reactor, two small nuclei join together and release energy.
- **ii** In a .... reactor, a large nucleus splits and releases energy.
- **iii** The fuel in a .... reactor contains uranium-235.

b State two advantages that nuclear fusion reactors would have in comparison with nuclear fission reactors.

4 a i What physical process causes energy to be released in the Sun?
- **ii** Which element is used in the physical process named in part **i** to release energy in the Sun?

b How will the Sun change in the next stage of its life cycle when it has used up all the element named in part **a ii**?

5 Copy and complete **a** to **d** using the words below. Each word can be used more than once.

galaxy   planet   stars

- **a** A .... isn't big enough to be a star.
- **b** The Sun is inside a ....
- **c** .... became hot after they formed from matter pulled together by the force of gravity.
- **d** The force of gravity keeps .... together inside a ....

6 a What force pulls dust and gas in space?

b Why do large planets like Jupiter not produce their own light?

c What is the name for the type of reaction that releases energy in the core of the Sun?

7 a The stages in the development of the Sun are listed below. Put the stages in the correct sequence.
- A   dust and gas
- B   present stage
- C   protostar
- D   red giant
- E   white dwarf

b i After the white dwarf stage, what will happen to the Sun?
- **ii** What will happen to a star that has much more mass than the Sun?

8 a i What is a supernova?
- **ii** How could we tell the difference between a supernova and a distant star like the Sun at present?

b i What is a black hole?
- **ii** What would happen to stars and planets near a black hole?

9 a i Which element as well as hydrogen is formed in the early universe?
- **ii** Which of the two elements is formed from the other one in a star?

b i Which two of the elements listed below is not formed in a star that gives out radiation at a steady rate?

carbon   iron   lead   uranium

- **ii** How do we know that the Sun formed from the debris of a supernova?

AQA Examination-style questions

1 a Copy and complete the following diagram to show how a chain reaction may occur inside a nuclear fuel rod containing many uranium-235 nuclei. (3)

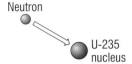

Neutron

U-235 nucleus

b Name the other fissionable substance that is used in some nuclear reactors. (1)

c The passages below reflect some of the conflicting opinions about nuclear power.

> Nuclear power is a low-emission source of energy and is the only readily available, large-scale alternative to fossil fuels for a continuous, reliable supply of electricity. The waste from nuclear power occupies a tiny volume and can be safely returned to the Earth for underground storage.

> A new generation of nuclear power stations will only reduce our emissions by four per cent by 2024: far too little, far too late, to stop global warming. They will create tens of thousands of tonnes of the most hazardous radioactive waste, which remains dangerous for up to a million years.

  i What are the 'emissions' that both sources refer to? (1)

  ii Why can nuclear waste remain dangerous for millions of years? (1)

  iii Give one advantage and one disadvantage of the storage of nuclear waste underground. (2)

  iv Explain why it would not be possible to replace fossil fuels with wind power alone. (2)

d For over 50 years scientists have been experimenting with fusion reactors with the aim of eventually generating electricity. The latest research project, called ITER, is scheduled to start operating in France in 2018 and is a collaboration between many countries.

  i State two of the potential benefits of fusion power. (2)

  ii Why are some people opposed to the research into fusion power? (2)

2 a Copy and complete the following sentences using the list of words and phrases below. Each one can be used once, more than once or not at all.

split   fusion   join   a larger one   fission two smaller nuclei

The Sun's energy is produced by nuclear .... This is where atomic nuclei .... to form .... (3)

b Which element was the first to form in the universe? (1)

c The red super giant star Betelgeuse is likely to explode as a supernova and then form a neutron star. The red supergiant VV Cephei is likely to explode as a supernova and become a black hole. What causes the fate of these two stars to be different? (2)

d Which type of star produces all the elements up to iron? (1)

e The diagram shows the forces acting within a star. The grey arrows show the outward force created by radiation. Star A is stable but in star B the outward force has become less.

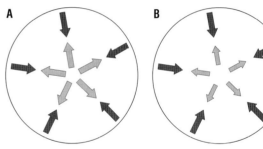

A   B

  i What type of force is counteracting the outward force from radiation? (1)

  ii What is about to happen to star B? (1)

  iii Suggest why the force from radiation may suddenly decrease. (2)

3 *In this question you will be assessed on using good English, organising information clearly and using specialist terms where appropriate.*

Explain how the solar system formed and why there were elements heavier than iron present when it formed. (6)

1 A toy cannon uses a spring to fire a metal ball bearing.

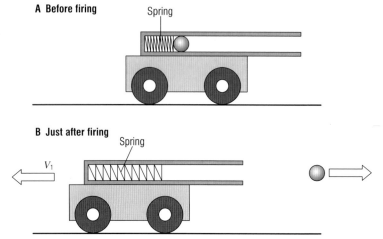

A Before firing

Spring

B Just after firing

Spring

v_1

a Calculate the deceleration of the cannon after it is fired and recoils to the left. The initial velocity of the cannon was −0.3 m/s, then it slows down and stops in 0.6 seconds. (2)

To find out how fast the ball bearing travels when it is fired, the student measures the recoil velocity (v_1) of the cannon using a light gate.

b What is meant by the conservation of momentum? (2)

c Calculate the velocity of the ball bearing if v_1 is −0.3 m/s.
   Mass of cannon = 0.15 kg
   Mass of ball bearing = 0.0045 kg
   Write down the equation you use. Show clearly how you work out the answer and give the unit. [H] (3)

d Calculate the kinetic energy of the cannon just after it is fired. (2)

e i Calculate the spring constant if the force required to compress the spring a distance of 2 cm was 23 N. (3)

   ii Describe the energy transfers that take place between diagram **A** and diagram **B**. [H] (2)

2

12V

S   18 Ω

18 Ω   L2

A   B   18 Ω C

L1   L3

a Explain why the resistance between B and C is less than the resistance between A and B when switch S is closed. (2)

b The potential difference between A and B is 8 V when switch S is closed. What is the potential difference between B and C? (1)

c i Calculate the current through bulb L1. (1)

   ii Calculate the current through bulb L3. (2)

d Switch S is opened.
   i Explain what effect this will have on the brightness of bulb L1. (3)
   ii Calculate the resistance between A and C. (1)
   iii Show that the current through L1 is now 0.33 A. (2)
   iv Calculate the total power delivered to both bulbs. (2)

e With the switch open the battery will deliver 500 C of charge before the bulbs start to dim. How long can the circuit be left on before this happens? (2)

3 Plutonium-239 has a half-life of 24 200 years and decays into uranium-235 with a half-life of 703 million years. These substances are both *fissionable*.

a i Explain what is meant by *fissionable*. (2)

ii What is meant by 'a half-life of 24 200 years'? (2)

iii A sample of plutonium-239 of mass 0.8 kg is being stored. How many years will pass before the sample contains 0.7 kg of uranium-235? Show clearly how you work out your answer. (3)

iv If the sample were kept at a higher temperature and pressure, what effect would this have on your answer to part **a iii**? (1)

b Explain how a small amount of uranium-235 is found in the Earth's crust in rocks such as granite, when hardly any plutonium is found occurring naturally and nearly all of it is formed in nuclear reactors. (3)

c Name one other natural source of background radiation that we are constantly exposed to, apart from rocks. (1)

d Plutonium (Pu) has 94 protons. Copy and complete the following decay equation to show how it decays into uranium-235. **[H]** (6)

$$\boxed{}^{\boxed{239}} Pu \rightarrow \boxed{}^{\boxed{}} U + \boxed{}^{\boxed{}}\boxed{}$$

e List the stages below in the correct order to describe the life cycle of a star that is about the same size as the Sun. One of the stages is not part of the life cycle of this type of star. (5)

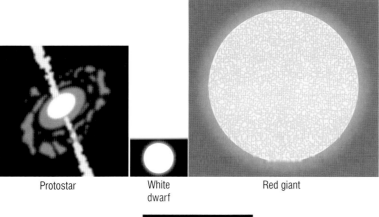

Protostar   White dwarf   Red giant

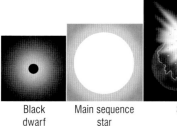

Black dwarf   Main sequence star   Supernova

4 *In this question you will be assessed on using good English, organising information clearly and using specialist terms where appropriate.*

In 1911 Ernest Rutherford published a scientific paper in which he suggested the existence of a very small region at the centre of every atom where most of the charge and mass is concentrated. Rutherford was interpreting the results of an experiment carried out by his research workers Geiger and Marsden in 1909.

Outline the main results of this experiment and explain why these results led Rutherford to suggest the existence of the atomic nucleus. (6)

AQA *Examiner's tip*

Tricky calculations involving half-life start to become quite straight-forward when you have practised a few.

When completing any decay equation, the atomic numbers on the right must add to give the atomic number on the left. The same rule applies to the mass numbers.

AQA *Examiner's tip*

A question that requires an extended written answer will have 6 marks available and the quality of your written English will influence your mark. Once you have written your answer, read the question again, before reading your answer through to check that all parts of the question have been answered.

Don't just know the results of the Rutherford and Marsden scattering experiments, know why these results were so important.

P3 1.1   X-rays ⓚ

Learning objectives

- What do we use X-rays in hospitals for?
- Why are X-rays dangerous?
- What can we say about the absorption of X-rays when they pass through the body?
- What is a CT scan?

Have you ever broken an arm or a leg? If you have, you will have gone to your local hospital for an X-ray photograph. X-rays are electromagnetic waves at the short-wavelength end of the electromagnetic spectrum. They are produced in an X-ray tube when fast-moving electrons hit a target. Their wavelengths are about the same as the diameter of an atom.

To make a **radiograph** or X-ray photograph, X-rays from an X-ray tube are directed at the patient. A lightproof cassette containing a photographic film or a **flat-panel detector** is placed on the other side of the patient.

- When the X-ray tube is switched on, X-rays from the tube pass through the part of the patient's body under investigation.
- X-rays pass through soft tissue but they are absorbed by bones, teeth and metal objects that are not too thin. The parts of the film or the detector that the X-rays reach become darker than the other parts. So the bones appear lighter than the surrounding tissue which appears dark. The radiograph shows a 'negative image' of the bones. A hole or a cavity in a tooth shows up as a dark area in the bright image of the tooth.
- An organ that consists of soft tissue can be filled with a substance called a **contrast medium** which absorbs X-rays easily. This enables the internal surfaces in the organ to be seen on the radiograph. For example, to obtain a radiograph of the stomach, the patient is given a barium meal before the X-ray machine is used. The barium compound is a good absorber of X-rays.
- Lead 'absorber' plates between the tube and the patient stop X-rays reaching other parts of the body. Lead is used because it is a good absorber of X-rays. The X-rays reaching the patient pass through a gap between the plates.
- A flat-panel detector is a small screen that contains a **CCD (charge-coupled device)**. The sensors in the CCD are covered by a layer of a substance that converts X-rays to light. The light rays then create electronic signals in the sensors that are sent to a computer which displays a digital X-ray image.

> **a** Why is a crack in a bone visible on a radiograph (X-ray image)?

a

b
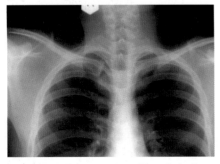

Figure 1 a Taking a chest X-ray **b** A chest X-ray

Safety matters

X-radiation, as well as gamma radiation, is dangerous because it ionises substances it passes through. High doses kill living cells. Low doses can cause cell mutation and cancerous growth. There is no evidence of a safe limit below which living cells would not be damaged.

Workers who use equipment or substances that produce X-radiation (or alpha, beta or gamma radiation) must wear a film badge. If the badge is overexposed to such radiation, its wearer is stopped from working with the equipment.

> **b** Why does a film badge have a plastic case, and not a metal case?

X-ray therapy

Doctors use X-ray therapy to destroy cancerous tumours in the body. Thick plates between the X-ray tube and the body stop X-rays from reaching healthy body tissues. A gap between the plates allows X-rays through to reach the tumour. X-rays for therapy are shorter in wavelength than X-rays used for imaging.

> **c** Why is it important to stop X-rays reaching healthy body tissues?

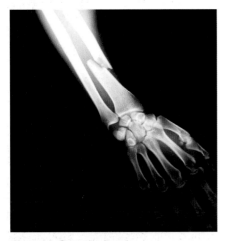

Figure 2 Spot the break

The CT scanner

A computerised tomography scanner (**CT scanner**) produces a digital image of any cross-section through the body. It can also be used to construct a three-dimensional (3-D) image of an organ.

Figure 3 shows an end-view of a CT scanner. The patient lies stationary on a bed that is in a ring of detectors.

- The X-ray tube automatically moves round the inside of the ring in small steps.
- At each position, X-rays from the tube pass through the patient and reach the detector ring.
- Electronic signals from the detector are recorded by a computer until the tube has moved round the ring.
- The computer displays a digital image of the scanned area.

Each detector receives X-rays that have travelled through different types of tissue. The detector signal depends on:

- the different types of tissue along the X-ray path
- how far the X-rays pass through each type of tissue.

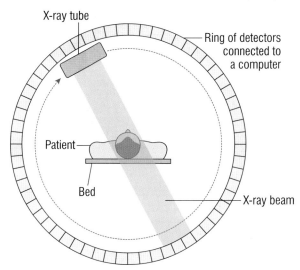

Figure 3 The CT scanner – it can distinguish between different types of soft tissue, as well as bone

Table 1 Comparison of a CT scanner with an ordinary X-ray machine

	CT scanner	Ordinary X-ray machine
Image distinguishes between bone and soft tissue	Yes	Yes
Image distinguishes between different types of soft tissue	Yes	No
Three-dimensional image	Yes	No
Radiation dose	CT scanner gives a much higher dose than an ordinary X-ray machine	
Cost	CT equipment cost is much greater than an ordinary X-ray machine	

∞ links

For more information about scanners used in hospitals, see P3 3.6 A physics case study.

Figure 4 A film badge tells you how much ionising radiation the wearer has received. Who might wear these?

Summary questions

1 Copy and complete **a** to **c** using the words below:

   absorb   damage   penetrate

   a X-rays .... thin metal sheets.
   b Thick lead plates will .... X-rays.
   c X-rays .... living tissue.

2 When an X-ray photograph is taken, why is it necessary:
   a to place the patient between the X-ray tube and the film cassette?
   b to have the film in a lightproof cassette?
   c to shield those parts of the patient not under investigation from X-rays? Explain what would happen to healthy cells.

3 State one advantage and one disadvantage of a CT scanner in comparison with an ordinary X-ray machine.

Key points

- X-rays are used in hospitals:
  1 to make images and CT scans
  2 to destroy tumours at or near the body surface.

- X-rays can damage living tissue when they pass through it.

- X-rays are absorbed more by bones and teeth than by soft tissues.

- CT scans distinguish between different types of soft tissue as well as between bone (or teeth) and soft tissue.

P3 1.2

Ultrasound ⓚ

Learning objectives

- What are ultrasound waves?
- What are ultrasound waves used for in medicine?
- Why can ultrasound waves be used to scan the human body?
- Why is an ultrasound scan safer than taking an X-ray photograph?

∞ links

For more information on the use of an oscilloscope, look back at P2 5.1 Alternating current.

The human ear can detect sound waves in the frequency range from about 20 Hz to about 20 000 Hz. Sound waves above the highest frequency that humans can detect are called **ultrasound waves**.

Practical

Testing ultrasound

High frequency signal generator · Loudspeaker · Oscilloscope · Microphone

Figure 1 Testing ultrasounds

Use a loudspeaker connected to a signal generator to produce ultrasound waves. Connect a microphone to an oscilloscope to detect the waves and display them. You can use the apparatus to:

- measure the frequency of the ultrasound waves
- test different materials to see if they absorb ultrasound waves
- show that ultrasound waves can be partly reflected.

a When a layer of material is placed between the loudspeaker and the microphone, the waves on the screen become smaller in amplitude. What conclusions can you draw from this?

Ultrasound scanners

Ultrasound waves are used for prenatal scans of a baby in the womb. They are also used to 'see' organs in the body such as a kidney or damaged ligaments and muscles. An ultrasound scanner consists of a **transducer** placed on the body surface, a control system and a display screen. The transducer produces and detects pulses of ultrasound waves.

Each pulse from the transducer:

- is partially reflected from the different tissue boundaries in its path
- returns to the transducer as a sequence of reflected pulses from the boundaries, arriving back at different times.

The transducer is moved across the surface of part of the body. The pulses are then detected by the transducer. They are used to build up an image on a screen of the internal tissue boundaries in the body.

The advantages of using ultrasound waves instead of X-rays for medical scanning are that ultrasound waves (unlike X-rays) are:

- non-ionising and therefore harmless when used for scanning
- reflected at boundaries between different types of tissue so they can be used to scan organs and other soft tissues in the body.

a
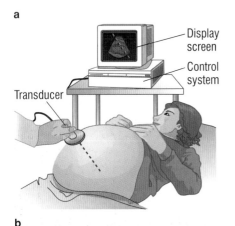

b
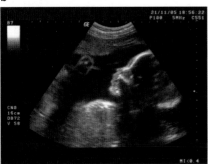

Figure 2 a An ultrasound scanner system **b** An ultrasound image of a baby in the womb

Distance measurements

Sight can sometimes be restored to a blind person by replacing the eye lens with an artificial lens. Before this is done, the eye surgeon needs to know how long the eyeball is. This is to make sure the new lens gives clear vision. Figure 3 shows how ultrasound is used to measure the length of the eyeball. This type of scan is called an **A-scan**.

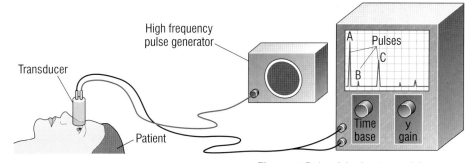

A transducer at the front of the eye sends ultrasound pulses into the eye. The reflected pulses are detected by the transducer and displayed on an oscilloscope screen or on a computer monitor, as shown in Figure 3.

Figure 3 Pulse A is due to partial reflection at the front surface of the eye. Pulse B is due to partial reflection at the surfaces of the eye lens. Pulse C is due to reflection at the back of the eye. Some further pulses are present due to partial reflection beyond the back of the eye.

  b In Figure 3, how can we tell that the eye lens is nearer to the front of the eye than it is to the back of the eye?

We can use the oscilloscope to measure the 'transit time' of each pulse. This is the time taken by the pulse to travel from the transmitter at the surface to and from the boundary that reflected it. To calculate distance travelled:

$$\frac{\text{the distance travelled}}{\text{by the pulse}} = \frac{\text{speed of ultrasound}}{\text{waves in body tissue}} \times \frac{\text{its transit}}{\text{time}}$$

Since the pulse travels from the surface to the boundary then back to the surface, the depth of the boundary below the surface is therefore half the distance travelled by each pulse to and from the boundary. So:

$$\frac{\text{the depth of the boundary}}{\text{below the surface}} = \frac{1}{2} \times \frac{\text{speed of the}}{\text{ultrasound waves}} \times \frac{\text{transit}}{\text{time}}$$

Maths skills

The distance equation in symbols:
$$s = v \times t$$
Where:
s is the distance in metres, m
v is the speed in metres per second, m/s
t is the time taken in seconds, s.

Ultrasound therapy

Kidney stones can be very painful. Powerful ultrasound waves can be used to break a kidney stone into tiny bits. The fragments are small enough to leave the kidney naturally. The transmitter is used in an A scan system so that the waves are aimed exactly at the kidney stone.

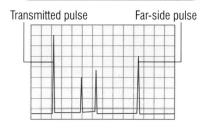

Transmitted pulse   Far-side pulse

Figure 4 The screen of an oscilloscope connected to an ultrasound detector on the surface of a patient's body

Summary questions

1 **a** Why are ultrasound waves partly reflected by body organs?
  b Why is an ultrasound scanner better than an X-ray scanner for scanning a body organ?
2 Look at the screen in Figure 4. It shows the reflected pulses that are detected for each transmitted pulse.
  a How many internal boundaries are present according to this display?
  b The oscilloscope beam takes 32 millionths of a second to travel across each grid square on the screen.
    i How long does each pulse take to travel from the body surface to the nearest internal boundary?
    ii The speed of ultrasound in the body is 1500 m/s. What is the distance from the body surface to the nearest tissue boundary?
3 **a** In an A-scan of a 'model' eye, the distance from the front to the back of the eye was known to be 48 mm. If Figure 3 represented the oscilloscope display for the model eye, what would be the distance from the eye lens to the front of the model eye?
  b Estimate the accuracy of the distance you calculated in **a**.

Key points

- Ultrasound waves are sound waves of frequency above 20 000 Hz.
- Ultrasound waves are used in medicine for ultrasonic scanning and the destruction of kidney stones.
- Ultrasound waves are partly reflected at a boundary between two different types of body tissue.
- An ultrasound scan is non-ionising so it is safer than an X-ray.

P3 1.3

Refractive index ⓚ

Learning objectives

- What do we mean by the refractive index of a transparent substance?

- How can we calculate the refractive index from the angles of incidence and refraction of a light ray?

Figure 1 A laser beam entering water

Maths skills

Using a calculator

To find the value of the sine of a given angle in degrees or the angle in degrees for a given sine value, make sure your calculator is in degree mode. Key the angle in degrees into your calculator then press the button marked 'sin' (or on some calculators press 'sin' first). The calculator will then display the sine of the angle.

To find the angle for a given sine value, key the sine value into your calculator and press the button marked 'inv sin' (or 'sin⁻¹ 'on some calculators).

When a light ray travels from air into a transparent substance, its direction may change. We say **refraction** of the light ray takes place at the boundary between the air and the transparent substance. Figure 1 shows the refraction of a light ray travelling from air into water. If the incident light ray had been directed at right angles to the surface (i.e. along the normal), no change of direction would have occurred.

Practical

Investigating how the angle of refraction varies with the angle of incidence ⓚ

We can use a semicircular transparent glass block as shown in Figure 2.

- Measure the angle of refraction, r, for different angles of incidence, i.

- Record all your measurements in a table.

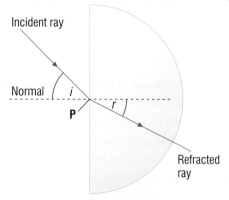

Figure 2 Using a semicircular block

For a light ray travelling into glass from air as in Figure 2, your results should show that:

1 the angle of refraction is always less than the angle of incidence

2 **the greater the angle of incidence, i, the greater the angle of refraction, r.**

Snell's law

Some typical results are shown below in Table 1.

Table 1

i (°)	r (°)	$\sin i$	$\sin r$	$\dfrac{\sin i}{\sin r}$
10.0	6.5	0.174	0.113	1.54
20.0	13.0	0.342	0.225	1.52
30.0	19.0			

The measurements can be used to show that $\dfrac{\sin i}{\sin r}$ always has the same value, regardless of the angle of incidence. This relationship was first discovered in 1618 and is known as **Snell's law** after its discoverer. Calculate the mean value of $\dfrac{\sin i}{\sin r}$ for your own measurements. This is the **refractive index** of the block you tested.

The law of refraction

For a light ray travelling from air into a transparent substance, the ratio of $\frac{\sin i}{\sin r}$ is always the same for the same substance.

This ratio is called the refractive index of the substance. In other words:

the refractive index of the substance, $n = \dfrac{\sin i}{\sin r}$

where i is the angle of incidence and r is the angle of refraction.

Rearranging the above equation to make $\sin i$ the subject gives $\sin i = n \sin r$

Rearranging the above equation to make $\sin r$ the subject gives $\sin r = \dfrac{\sin i}{n}$

a In Table 1, the angle of refraction is 19.0° for an angle of incidence of 30°. Use this information to calculate the refractive index of the block.

b Give one reason why the refractive index values from **a** differ from those given in Table 1.

When a light ray travels from a transparent substance into air at a non-zero angle of incidence:

- the light ray is refracted **away** from the normal, see point P in Figure 3.
- the larger the angle of incidence is, the larger the angle of refraction.

If the light ray in Figure 3 were reversed, the direction arrows would be reversed but the path would be the same. We can adapt the law of refraction to cover both situations by writing it as:

the sine of the angle in air = $n \times$ the sine of the angle in glass

---

Maths skills

Worked example

A light ray travels from glass into air across a straight boundary, as shown at P in Figure 3. The angle of incidence of the light ray in the glass is 32.0°. The refractive index of the glass is 1.55. Calculate the angle of refraction of the light ray in the air.

Solution

Let r be the angle of refraction in air.

The sine of the angle in air = n × the sine of the angle in glass

hence $\sin r = 1.55 \times \sin 32.0°$

$= 0.821$

Therefore $r = 55.1°$

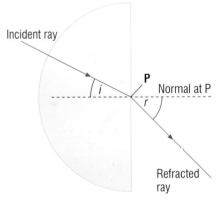

Figure 3 From glass to air

Summary questions

1 Copy and complete **a** to **c** using the words below. Each word can be used more than once.

less more

a When a light ray travels from air to glass, the angle of refraction is always .... than the angle of incidence.

b When a light ray travels from water into air, the angle of incidence in the water is .... than the angle of refraction in air.

c When a light ray travelling from air into glass is refracted, reducing the angle of incidence makes the angle of refraction ....

2 In Table 1, when $i = 40°$, $r = 26°$.

a Show that the value of refractive index given by the above values of angles i and r is 1.47.

b Give one possible reason why the value of refractive index in **a** differs so much from the values shown in the table.

3 The refractive index of water is 1.33.

a A light ray enters a flat water surface at an angle of incidence of 35.0°. Calculate the angle of refraction of the light ray.

b A light ray travels from water into air. The angle of incidence of the light ray in the water is 45.0°. Calculate the angle of refraction of the light ray in the air.

Key points

- Refractive index, **n**, is a measure of how much a substance can refract a light ray.

- $n = \dfrac{\sin i}{\sin r}$

P3 1.4

The endoscope ⓚ

Learning objectives

● What do we mean by the critical angle of a substance?

● How is the critical angle related to the refractive index of the substance?   [H]

● What do we mean by total internal reflection?

● What do doctors use an endoscope for?

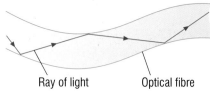

Ray of light   Optical fibre

Figure 1 Light rays in an optical fibre

Optical fibres are very thin glass fibres that are designed to transmit light or infrared radiation. We use them in medicine to see inside the body without cutting the body open. In telecommunications they are used to send signals securely. The light rays can't escape from the fibre. Each light ray entering a fibre at one end leaves the fibre at the other end even if the fibre bends round. This is because a light ray in the fibre is **totally internally reflected** each time it reaches the fibre's boundary (Figure 1).

Investigating total internal reflection

In the previous topic, we saw that a light ray travelling from glass into air at a non-zero angle of incidence is refracted away from the normal. A partially reflected ray is also seen, as shown in Figure 2. The angle of reflection of this ray in the glass is the same as the angle of incidence.

● If the angle of incidence in the glass is gradually increased, the angle of refraction increases until the refracted ray emerges along the boundary, as shown in Figure 3. The angle of incidence at this position is referred to as the **critical angle,** labelled c in Figure 3.

a What is the angle between the normal and the refracted ray when the angle of incidence is equal to the critical angle?

● If the angle of incidence is increased beyond the critical angle, the light ray is **totally internally reflected** at P, as shown in Figure 4. When total internal reflection occurs, the angle of reflection r, at P, is equal to the angle of incidence, i.

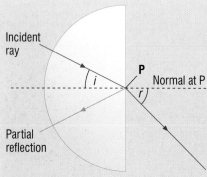

Figure 2 Partial reflection and refraction

If the light ray in Figure 2 was reversed, the direction arrows would be reversed but the path would be the same. We can adapt the law of refraction to cover this situation by writing it as:

the sine of the angle in air = n × the sine of the angle in glass

We can apply this equation to the critical ray in Figure 3. The angle in air is 90° and the angle in glass is c,

$$\sin 90° = n \times \sin c$$

Because sin 90° = 1, the equation above becomes 1 = n × sin c. Rearranging this equation gives

$$n = \frac{1}{\sin c} \quad \text{or} \quad \sin c = \frac{1}{n}$$

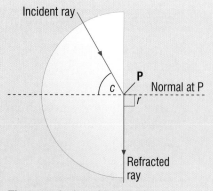

Figure 3 At the critical angle

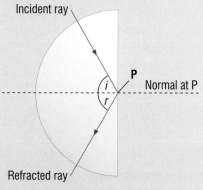

Figure 4 Total internal reflection

Higher

🖩 Maths skills

Worked example

Calculate the critical angle for glass of refractive index 1.59.

Solution

$\sin c = \dfrac{1}{n} = \dfrac{1}{1.59} = 0.629$

Therefore c = 39.0°

b The critical angle for a certain type of glass is 43.0°. Calculate the refractive index of this glass.

The endoscope

The **endoscope** is used by a surgeon to see inside a body cavity, such as the stomach, without cutting the body open. The endoscope is inserted into the stomach via the patient's throat. The endoscope contains two bundles of optical fibres alongside each other. One of the bundles is used to shine light into the cavity and the other to see the internal surfaces of the cavity. A tiny lens over the second bundle is used to form an image on the ends of the fibres in the bundle. The image can then be seen directly or by using a digital camera at the other end of the fibre bundle.

For example, the endoscope can be used to observe a stomach ulcer or a bone fragment in the knee joint. The surgeon can then use keyhole surgery to remove them.

Laser light may be used as a source of energy in an endoscope to carry out some surgical procedures. It can cut away or burn away and destroy diseased tissue. It can also seal off (cauterise) leaking blood vessels. This is possible with laser light because the energy can be focused on to a very small area of a surface. In addition, the colour of laser light can be matched to the type of tissue. By choosing an appropriate laser source, it ensures the most effective absorption. Eye surgery on the retina can be carried out by applying the laser light through the pupil of the eye for a very short time.

Safety note: Never look into or along a laser beam, even after reflection. It will damage the retina and may cause permanent blindness. Special safety goggles should always be worn in the presence of a laser beam.

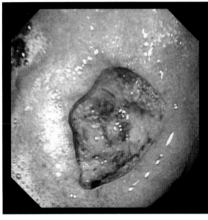

Figure 5 A stomach ulcer viewed through an endoscope

Summary questions

1 Copy and complete **a** to **d** using the words below:

refraction   reflection   partial reflection   total internal reflection

  a When a light ray travels from air into glass and changes direction, it undergoes .... ,

  b When a light ray inside glass reaches the surface with air and stays in the glass, it undergoes ....

  c When a light ray passes from glass to air, some of the light undergoes ....

  d When a light ray inside glass reaches the surface with air, it undergoes .... and .... if the angle of incidence is less than the critical angle.

2 **a** Figure 6 shows a light ray in an optical fibre. The angle of incidence of the light ray at **P** is greater than the critical angle of the optical fibre. Copy the diagram and complete the path of the light ray inside the optical fibre.

  b State two advantages of using an endoscope instead of X-rays to observe fragments of bone in a knee joint?

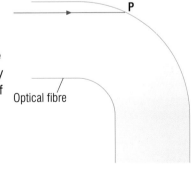

Figure 6

3 **a** The refractive index of water is 1.33. Calculate the critical angle of water.

  b The critical angle for a certain type of glass is 42.0°. Calculate the refractive index of the glass. **[H]**

Key points

- The critical angle is the angle of incidence of a light ray in a transparent substance that produces refraction along the boundary.

- Refractive index = $\dfrac{1}{\sin c}$

  where c is the critical angle **[H]**

- Total internal reflection occurs when the angle of incidence of a light ray in a transparent substance is greater than the critical angle.

- An endoscope is used to see inside the body directly.

P3 1.5 Lenses

Learning objectives

- What is a converging (convex) lens?
- What is a diverging (concave) lens?
- What is a real image and what is a virtual image?
- What do we mean by magnification?

Figure 2 A digital camera

AQA Examiner's tip

Remember that a diverging lens makes light rays from a point object spread out (diverge) more.

Lenses are used in optical devices such as the camera. Although a digital camera is very different from the first cameras made over 160 years ago, they both contain a lens that is used to form an image.

Types of lenses

A lens works by changing the direction of light passing through it. Figure 1 shows the effect of a lens on the light rays from a ray box.

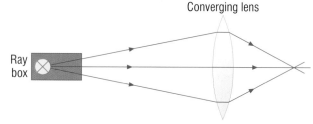

Figure 1 Investigating lenses

The curved shape of the lens surfaces refracts the rays so they meet at a point.

Different lens shapes can be tested using this arrangement.

- A **converging (convex) lens** makes parallel rays converge to a focus. The point where *parallel* rays are focused to is the **principal focus** (or focal point) of the lens. See Figure 3. We use a converging lens as a **magnifying glass** and in a camera to form a clear image of a distant object.

- A **diverging (concave) lens** makes parallel rays diverge (spread out). The point where the rays appear to come from is the principal focus of the lens. See Figure 4. We use diverging lenses to correct short sight.

- In both cases, the distance from the centre of the lens to the principal focus is the **focal length** of the lens. Notice that the principal focus is usually shown in ray diagrams on each side of the lens.

a Which is more powerful, a lens with a focal length of 5 cm or one with a focal length of 50 cm?

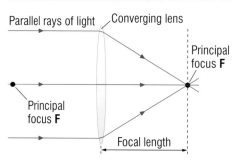

Figure 3 The focal length of a converging (or convex) lens

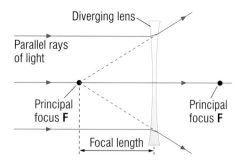

Figure 4 The focal length of a diverging (or concave) lens

The converging lens

Practical

Investigating the converging lens

- Use the arrangement in Figure 5 to investigate the image formed by a converging lens.

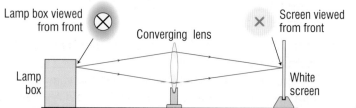

Figure 5 Investigating images

1 *With the object at different distances beyond the principal focus of the lens*, adjust the position of the screen until you see a clear image of the object on it. This is called a **real image** because it is formed on the screen where the light rays meet.

- When the object is a long distance away, the image is formed at the principal focus of the lens. This is because the rays from any point on the object are effectively parallel to each other when they reach the lens.

- If the object is moved nearer the lens towards its principal focus, the screen must be moved further from the lens to see a clear image. The nearer the object is to the lens, the larger the image is.

b Is the image inverted or upright in Figure 6a?

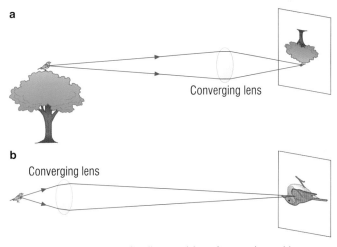

Figure 6 a The image of a distant object, **b** an enlarged image

2 *With the object nearer to the lens than the principal focus*, a magnified image is formed. The image is called a **virtual image** because it is formed where the rays appear to come from. But you can only see the image when you look into the lens from the side opposite to the object. The lens acts as a magnifying glass in this situation.

c Why is a diamond usually inspected with a magnifying glass?

Magnification

The **magnification** produced by a lens = $\dfrac{\textbf{image height}}{\textbf{object height}}$

If the image is larger than the object, as in Figure 6b and Figure 7, the magnification is greater than 1.

If the image is smaller than the object, as in Figure 6a, the magnification is less than 1.

Figure 7 A magnifying glass

Summary questions

1 Copy and complete **a** and **b** using the words below:

converging   diverging   real   virtual

a A .... lens can be used to focus light from an object on to a screen. The image of the object is a .... image.

b A .... lens can be used to make light rays from a point object spread out more. The image of the object is a .... image.

2 a A postage stamp is inspected using a converging lens as a magnifying glass. Describe the image.

b A converging lens is used to form a magnified image of a slide on to a screen.

   i Describe the image formed by the lens.

   ii The screen is moved away from the lens. What adjustment must be made to the position of the slide to focus its image on the screen again?

3 a Describe the image of the bird in Figure 6b and estimate the magnification of the lens.

b Describe how the image changes if the lens is moved further away from the bird and the card is moved to obtain a new clear image.

Key points

- A converging (convex) lens focuses parallel rays to a point called the principal focus.

- A diverging (concave) lens makes parallel rays spread out as if they came from a point called the principal focus.

- A real image is formed by a converging lens if the object is further away than the principal focus.

- A virtual image is formed by a diverging lens and, if the object is nearer to the lens than the principal focus, by a converging lens.

- Magnification = $\dfrac{\textbf{image height}}{\textbf{object height}}$

P3 1.6

Using lenses

Learning objectives

- How can we find the position and nature of an image formed by a lens?

- What type of image is formed by a converging lens when the object is between the lens and its principal focus?

- What type of lens is used in a camera and in a magnifying glass?

- What type of image is formed in a camera and in a magnifying glass, and by a diverging (concave) lens?

The position and nature of the image formed by a lens depends on:

- the focal length, *f,* of the lens
- the distance from the object to the lens.

If we know the focal length and the object distance, we can find the position and nature of the image by drawing a ray diagram.

Formation of a real image by a converging lens

To form a real image using a converging (convex) lens, the object must be beyond the principal focus, F, of the lens. See Figure 1. The image is formed on the other side of the lens to the object.

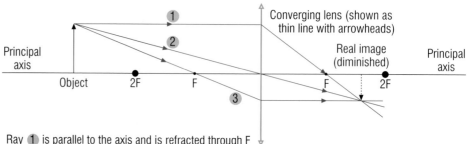

Ray ① is parallel to the axis and is refracted through F
Ray ② passes straight through the centre of the lens
Ray ③ passes through F and is refracted parallel to the axis

Figure 1 Formation of a real image by a converging lens

The diagram shows that we can use three key 'construction' rays from a single point of the object to locate the image.

- The **principal axis** of the lens is the straight line that passes along the normal at the centre of each lens surface. Notice we draw the lens as a straight line with 'outward' arrows to show it is a converging lens.

- The image is real, inverted and smaller than the object.

Notice that:

1 *ray 1* is refracted through F, the principal focus of the lens, because it is parallel to the principal axis of the lens before it passes through the lens

2 *ray 2* passes through the centre of the lens (its pole) without change of direction; this is because the lens surfaces at the principal axis are parallel to each other

3 *ray 3* passes through F, the principal focus of the lens, before the lens, so it is refracted by the lens parallel to the principal axis.

The image is smaller than the object because the object distance is greater than twice the focal length (*f*) of the lens. This is how a **camera** is used.

    a i Draw a ray diagram to show that a real, inverted and magnified image is produced if the object is between F and 2F.

    ii What optical device projects a magnified image on to a screen?

AQA *Examiner's tip*

Make sure your ray diagrams are neat and that you put arrows on the rays. You need to be able to draw a scale diagram to find the focal length of a lens for a particular magnification. See Summary question 3.

The camera

In a camera, a converging lens is used to produce a real image of an object on a film (or on an array of 'pixels' in the case of a digital camera). The position of the lens is adjusted to focus the image on the film.

- For a distant object, the distance from the lens to the film must be equal to the focal length of the lens.
- The nearer an object is to the lens, the greater the distance from the lens to the film.

b If an object moves closer to the camera, does the lens of a camera need to be moved towards or away from the object?

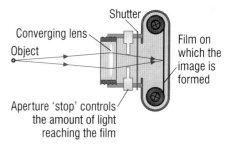

Figure 2 The camera

Formation of a virtual image by a converging lens

The object must be between the lens and its principal focus, as shown in Figure 3. The image is formed on the same side of the lens as the object.

The image is virtual, upright and larger than the object.

The image can only be seen by looking at it through the lens. This is how a magnifying glass works.

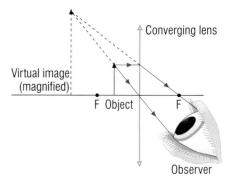

Figure 3 Formation of a virtual image by a converging lens

Formation of a virtual image by a diverging lens

The image formed by a diverging (concave) lens is always virtual, upright and smaller than the object. Figure 4 shows why. A diverging lens is shown as a line with 'inward' arrows.

c Why is a diverging lens no use as a magnifying glass?

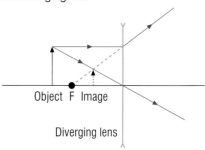

Figure 4 Image formation by a diverging lens

Summary questions

1 a Copy and complete the ray diagram in Figure 5 to show how a converging lens forms an image of an object which is smaller than the object, as in a camera.

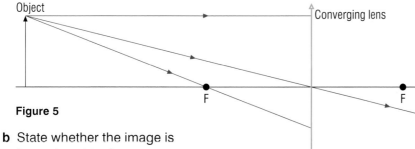

Figure 5

b State whether the image is
 i real or virtual
 ii magnified or diminished
 iii upright or inverted.

2 a Draw a ray diagram to show how a converging lens is used as a magnifying glass.
 b State whether the image is **i** real or virtual **ii** magnified or diminished **iii** upright or inverted.

3 A converging lens produces a magnification of × 2 when it is used to form a real image that is at a distance of 8.0 cm from the object.
 a Draw a scale ray diagram to show the formation of this image.
 b Use your diagram to find the focal length of the lens.

Key points

- A ray diagram can be drawn to find the position and nature of an image formed by a lens.

- When an object is placed between a converging lens and F, the image formed is virtual, upright, magnified and on the same side of the lens as the object.

- A camera contains a converging lens that is used to form a real image of an object.

- A magnifying glass is a converging lens that is used to form a virtual image of an object.

P3 1.7 The eye

Learning objectives

- How does the eye work?
- What is the range of vision of a normal human eye?
- What do we mean by the power of a lens?

Inside the eye

Figure 1 shows the inside of a human eye. Light enters the eye through a tough transparent layer called the **cornea.** This protects the eye and helps to focus light onto the **retina**. This is a layer of light-sensitive cells at the back of the inside of the eye.

The amount of light entering the eye is controlled by the **iris** which adjusts the size of the **pupil** – the circular opening at the centre of the iris. The **eye lens** focuses light to give a sharp image on the retina. Although the image on the retina is inverted, the brain interprets it so you can see it the right way up.

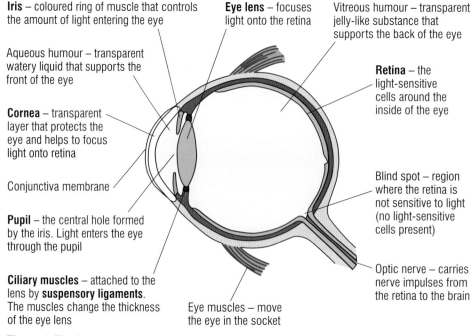

Iris – coloured ring of muscle that controls the amount of light entering the eye

Eye lens – focuses light onto the retina

Vitreous humour – transparent jelly-like substance that supports the back of the eye

Aqueous humour – transparent watery liquid that supports the front of the eye

Retina – the light-sensitive cells around the inside of the eye

Cornea – transparent layer that protects the eye and helps to focus light onto retina

Conjunctiva membrane

Blind spot – region where the retina is not sensitive to light (no light-sensitive cells present)

Pupil – the central hole formed by the iris. Light enters the eye through the pupil

Optic nerve – carries nerve impulses from the retina to the brain

Ciliary muscles – attached to the lens by **suspensory ligaments**. The muscles change the thickness of the eye lens

Eye muscles – move the eye in the socket

Figure 1 The human eye

> **a** You need to know what the parts listed below are for. Copy the list and see if you can remember what each part of the eye is for.
>
> Ciliary muscles   Eye lens   Pupil   Suspensory ligaments
> Cornea   Iris   Retina

How does the eye focus on objects at different distances? If you look up from this book and gaze out of a window, your eye lens automatically becomes thinner to keep what you see in focus. The **ciliary muscles** alter the thickness of the eye lens. They are attached to the edge of the lens by means of the **suspensory ligaments**. The fibres of the ciliary muscles are parallel to the circular edge of the eye lens. When they contract, they shorten and squeeze the eye lens, making it thicker.

The normal human eye has a **range of vision** from 25 cm to infinity. This means it can see clearly any object that is 25 cm or more from the eye. In other words, the normal eye has a **near point** of 25 cm and a **far point** at infinity.

> **b** Why does the pupil of the eye appear much wider in darkness than in daylight?

To see a nearby object clearly, the eye lens has to be thicker than if the object is far away. Figures 2a and b show this.

a Near point of the normal eye

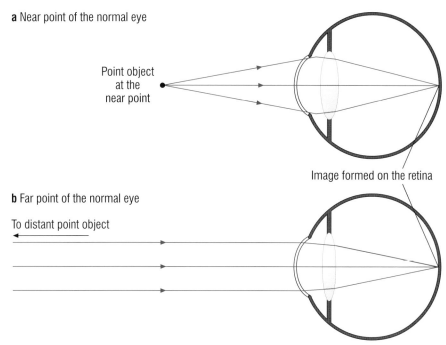

Point object at the near point

Image formed on the retina

b Far point of the normal eye

To distant point object

Figure 2 The normal eye

Lens power

The **power of a lens** is defined as $\dfrac{1}{\text{focal length in metres}}$

The unit of power is the **dioptre** (D). The type of lens is indicated by:

- a positive value for the power of a converging lens (e.g. + 5.0 D for a converging lens of focal length 0.20 m)
- a negative value for a diverging lens (e.g. − 4.0 D for a diverging lens of focal length 0.25 m)

c What is the power of a converging lens of focal length 0.05 m?

 Maths skills

We can write this word equation in symbols as follows:

lens power $P = \dfrac{1}{f}$

Where:

P = lens power in dioptres, D
f = focal length in metres, m.

Summary questions

1 Copy and complete **a** to **c** using the words below:

lens   iris   cornea   retina

   a The front of the eye is protected by the ....
   b The pupil of the eye is at the centre of the ....
   c The .... of the eye focuses light onto the ....

2 What is the power in dioptres of:
   a a converging lens of focal length 0.50 m?
   b a diverging lens of focal length 0.40 m?

3 A person with normal eyesight who is reading a book looks up to observe a distant object.
   a Describe what happens to the shape of each eye lens in this change.
   b What change takes place in the power of each eye lens?

Key points

- Light is focused on to the retina by the cornea and the eye lens, which is a variable focus lens.

- The normal human eye has a range of vision from 25 cm to infinity.

- $P = \dfrac{1}{f}$

P3 1.8

More about the eye

Learning objectives

- What is short sight and how do we correct it?
- What is long sight and how do we correct it?
- Why is the refractive index of glass important in making spectacle lenses? **[H]**

Figure 2 A contact lens

Sight defects

Short sight occurs when an eye cannot focus on distant objects. The **uncorrected** image is formed in front of the retina, as shown in Figure 1. This is because the eyeball is too long or the eye lens is too powerful. The eye muscles cannot make the eye lens thin enough to focus the image of a faraway object on the retina of the eye. The eye can focus nearby objects so the defect is referred to as 'short sight'.

Short sight is corrected by placing a diverging lens of a suitable focal length in front of the eye as shown in Figure 1. The diverging lens counteracts some of the 'excess' focusing power of the eye lens.

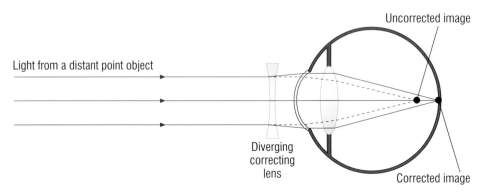

Figure 1 Short sight and its correction

Long sight occurs when an eye cannot focus on nearby objects. The uncorrected image is formed behind the retina, as shown in Figure 3. The eye lens cannot be made thick enough to focus an image on the retina. The eye can focus distant objects so the defect is referred to as 'long sight'.

Long sight is corrected by placing a converging lens of a suitable focal length in front of the eye, as shown in Figure 3. The correcting lens makes the rays from the object diverge less. The eye lens can then focus the rays onto the retina. The correcting lens adds to the focusing power of the eye lens.

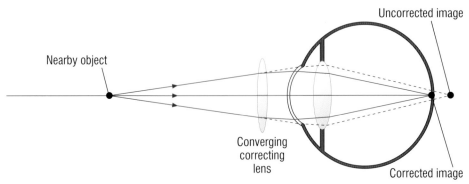

Figure 3 Long sight and its correction

a A student is unable to see clearly the number plate on a car 15 m away with his left eye, which has a normal near point. Is the student's left eye short-sighted or long-sighted?

Comparison of the eye and the camera

How do the eye and the camera compare as optical instruments? They are similar in that they both contain a converging lens which forms a real image. Look at Table 1 to see how they compare in other ways.

Table 1 Comparison of the optics of the eye and a camera.

	The eye	**The camera**
Type of lens	Variable focus converging lens	Fixed focus converging lens
Focusing adjustment	Ciliary muscle alters the lens thickness	Adjustment of lens position
Image	Real, inverted, magnification less than 1	
Image detection	Light sensitive cells on the retina	Photographic film (or CCD sensors in a digital camera)
Brightness control	Iris controls the width of the eye pupil	Adjustment of aperture 'stop'

Lens makers at work

The eye lens is a remarkable optical device, as it has a variable focal length that depends on its thickness. Lens makers working for opticians need to make contact lenses and spectacle lenses exactly the right shape to obtain the exact focal length for each lens.

The focal length of a lens depends on the refractive index of the lens material and the curvature of the two lens surfaces.

The larger the refractive index or the greater the curvature of the lens surfaces, the greater the power of the lens (and the shorter its focal length).

Higher

For a lens of a given focal length, the greater the refractive index of the lens material, the flatter and thinner the lens can be manufactured. This is because the lens surfaces would be less curved.

b After photographing a distant object, a photographer moves the camera lens away from the CCD in the camera to take a close-up photograph. What difference would it make to the image if she did not move the lens?

Figure 4 An eye test

Summary questions

1 Using his left eye, a student can only see the writing on a board at the front of the class if he sits near the board.
   a What sight defect is he suffering from in this eye?
   b What type of lens should be used to correct this defect?
2 An optician prescribes a lens of power +2.0 dioptres to correct a sight defect.
   a What type of lens is it and what is its focal length?
   b State what the sight defect is and give one possible cause of the defect.
3 A lens of a given focal length can be made using two materials that each have a different refractive index.
   How would the lens differ if it was made of the higher refractive index material instead of the lower refractive index material? **[H]**

Key points

- A short-sighted eye is an eye that can only see near objects clearly. We use a diverging lens to correct it.

- A long-sighted eye is an eye that can only see distant objects clearly. We use a converging lens to correct it.

- The higher the refractive index of the glass used to make a spectacle lens, the flatter and thinner the lens can be. **[H]**

Summary questions

1 Figure 1 shows an X-ray source which is used to direct X-rays at a broken leg. A photographic film in a light-proof wrapper is placed under the leg. When the film is developed, an image of the broken bone is observed.

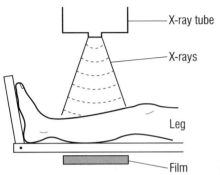

Figure 1

a i Explain why an image of the bone is seen on the film.

  ii Why is it possible to see the fracture on the image?

b When an X-ray photograph of the stomach is taken, the patient is given food containing barium before the photograph is taken.

  i Why is it necessary for the patient to be given this food before the photograph is taken?

  ii The exposure time for a stomach X-ray must be shorter than the exposure time for a limb X-ray. Why?

c An ultrasonic scanner is used to observe an unborn baby. Why is ultrasound instead of X-rays used to observe an unborn baby?

2 Ultrasonic waves used for medical scanners have a frequency of 2000 kHz.

a Use the equation 'speed = frequency × wavelength' to calculate the wavelength of these ultrasonic waves in human tissue. (The speed of ultrasound in human tissue is 1500 m/s.)

b Ultrasonic waves of this frequency in human tissue are not absorbed much. Why is it important in a medical scanner that they are not absorbed?

3 a Figure 2 shows a light ray entering a glass block of refractive index 1.50 at an angle of incidence of 40° at point P.

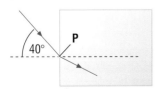

Figure 2

Show by calculation that the angle of refraction at P is 25°.

b i Show by calculation that the critical angle of the glass is 42°. **[H]**

  ii Copy the diagram and continue the path of the ray in the glass until it reaches a point Q at the bottom of the block. Explain why the angle of incidence at Q is 65°.

  iii Explain why the light ray does not enter the air at Q.

4 An endoscope is used to see inside the body.

Figure 3

a Copy and complete Figure 3 to show the path of the light ray along the optical fibre.

b Explain why an endoscope needs to have two bundles of optical fibres.

c Give one reason why it is better to observe an endoscope image using a CCD camera and a TV monitor rather than observing the image directly.

5 Figure 4 shows an incomplete ray diagram of image formation by a lens.

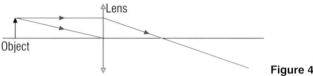

Figure 4

a i What type of lens is shown in this diagram?

  ii Copy the diagram and mark the focal point of the lens on the diagram.

b i Complete the ray diagram and label the image.

  ii Describe the image and state an application of the lens used in this way.

6 a State two optical differences between the eye lens and a film camera lens.

b State two advantages of a digital camera compared with a film camera.

7 An object of height 50 mm is placed perpendicular to the principal axis of a converging lens at a distance of 100 mm from the pole of the lens. The focal length of the lens is 150 mm.

a Draw a scale ray diagram to show where the image of the object is formed.

b State whether the image is

  i real or virtual

  ii upright or inverted.

c Determine the magnification produced by the lens.

AQA Examination-style questions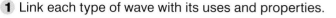

1 Link each type of wave with its uses and properties. (8)

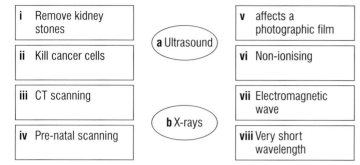

i	Remove kidney stones
ii	Kill cancer cells
iii	CT scanning
iv	Pre-natal scanning

a Ultrasound

b X-rays

v	affects a photographic film
vi	Non-ionising
vii	Electromagnetic wave
viii	Very short wavelength

2 The diagram shows the oscilloscope trace for an ultrasound A-scan for part of a person's thigh. The trace is used to measure the thickness of the person's fat and muscle.

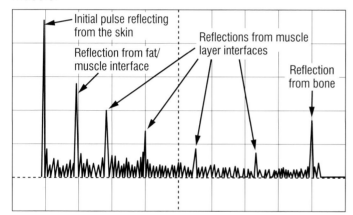

Each square on the screen represents a time of 10 millionths of a second (0.000010 s).

a How much time does it take for the ultrasound pulse to travel to the patient's bone and back? The speed of ultrasound through tissue is 1540 m/s. (1)

b Calculate the thickness of the layer of fat. The speed of ultrasound in the fat = 1540 m/s. Give your answer in millimetres. (2)

c Why do the reflections from the different layers of muscle get weaker with depth? (1)

3 The diagram shows a converging lens.

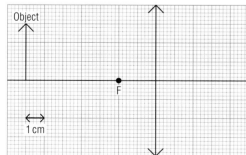

a Copy and complete the ray diagram to show the position of the image. (3)

b Describe the image that is formed. (3)

c Calculate the magnification produced by this lens. (2)

d Where would a lens be used in this way? Give a reason for your answer. (2)

The following diagram shows a diverging lens. Copy the diagram on to graph paper.

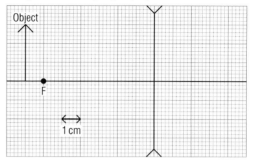

e Calculate the power of the lens. Show your working clearly and give the unit. (3)

f Complete the ray diagram to show the position of the image. (3)

g Describe the image that is formed. (3)

4 The diagram shows a ray of light travelling through a semicircular glass block at various angles of incidence, A, B and C.

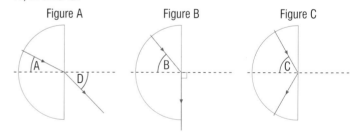

Figure A   Figure B   Figure C

a Which of the angles **A**, **B**, **C** or **D** represents the critical angle? (1)

b What is happening in **Figure C**? (1)

c If angle **A** is 30° and angle **D** is 54°, calculate the refractive index of the glass.

Write down the equation you use. Show clearly how you work out your answer. (3)

d Calculate the critical angle.

Show clearly how you work out your answer. [H] (2)

P3 2.1 Moments ⓚ

To undo a very tight wheel-nut on a bicycle, you need a spanner. The force you apply to the spanner has a turning effect on the nut. You couldn't undo a tight nut with your fingers but you can with the spanner. The spanner exerts a much larger turning effect on the nut than the force you apply to the spanner.

If you had a choice between a long-handled spanner and a short-handled one, which would you choose? The longer the spanner handle, the less force you need to exert on it to loosen the nut.

In this example, the turning effect of the force, called the **moment** of the force, can be increased by:

- increasing the size of the force
- using a spanner with a longer handle.

a What happens if a nut won't undo and you apply too much force to it?

Figure 1 A turning effect

Levers

A crowbar is a lever that can be used to raise one edge of a heavy object. Look at Figure 2.

The weight of the object is called the **load**. The force the person applies to the crowbar is called the **effort**. Using the crowbar, the effort needed to lift the same object is only a small fraction of its weight. The point about which the crowbar turns is called the **pivot** or the **fulcrum**.

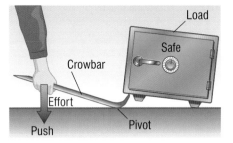

Figure 2 Using a crowbar

b Would you choose a long crowbar or a short crowbar to shift a heavy weight?

Practical

Investigating the turning effect of a force

The diagram in Figure 3 shows one way to investigate the turning effect of a force. The weight W is moved along the metre ruler.

- How do you think the reading on the newtonmeter compares with the weight?

You should find that the newtonmeter reading (i.e. the force needed to support the ruler) increases if the weight is increased.

- How does this reading change as the weight is moved away from the pivot?

You should find that the newtonmeter reading increases as the weight is moved away from the pivot.

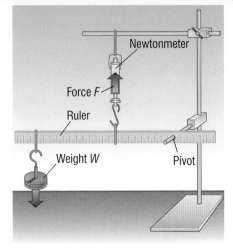

Figure 3 Investigating turning forces

The line along which a force acts is called its **line of action**.

We work out the moment of a force using this equation:

moment = **force** × **perpendicular distance from the**
(newton-metres, N m)   (newtons, N)   **line of action of the force to the**
pivot (metres, m)

c How does the moment of the weight W in Figure 3 change as it is moved away from the pivot?

Look at Figure 4. The claw hammer is being used to remove a nail from a wooden beam.

● The applied force F on the claw hammer tries to turn it clockwise about the pivot.

● The moment of force F about the pivot is $F \times d$, where d is the perpendicular distance from the pivot to the line of action of the force.

● The effect of the moment is to cause a much larger force to be exerted on the nail.

Maths skills

Worked example

A force of 50 N is exerted on a claw hammer of length 0.30 m, as shown in Figure 4. Calculate the moment of the force.

Solution

Force = 50 N × 0.30 m = 15 N m

d Calculate the moment if the force on the claw hammer had been 70 N.

Summary questions

1 Copy and complete **a** to **c** using the words below:

larger   smaller   unchanged

A force acts on an object and makes it turn about a fixed point.

a If the force is increased without changing its line of action, the moment of the force is ....

b If the force is doubled and the perpendicular distance from its line of action to the pivot is halved, the moment is ....

c If the force is reduced and the perpendicular distance from its line of action to the pivot is reduced, the moment is ....

2 In Figure 1, a force is applied to a spanner to undo a nut. State whether the moment of the force is:

a clockwise or anticlockwise,

b increased or decreased by:

  i increasing the force   **ii** exerting the force nearer the nut.

3 Explain each of the following statements:

a It is easier to remove a nail with a claw hammer if the hammer has a long handle.

b A door with rusty hinges is more difficult to open than a door of the same size with lubricated hinges.

4 A spanner of length 0.25 m is used to turn a nut as in Figure 1. Calculate the force that needs to be applied to the end of the spanner if the moment it exerts is to be 18 N m.   **[H]**

Maths skills

The word equation can be written using symbols, as follows:

moment $M = F \times d$

Where: M = moment in newton-metres, N m

F = force in newtons, N

d = perpendicular distance from the line of action of the force to the pivot, in metres, m.

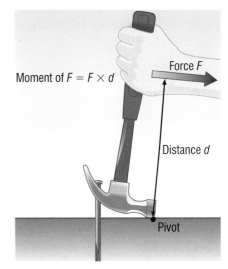

Moment of $F = F \times d$

Force F

Distance d

Pivot

Figure 4 Using a claw hammer

 Did you know … ?

A patient fitted with a replacement hip joint has to be very careful at first. A slight movement can cause a turning effect that pulls the hip joint apart.

Key points

● The moment of a force is a measure of the turning effect of the force on an object.

● The moment of a force F about a pivot = $F \times d$, where d is the perpendicular distance from the line of action of the force to the pivot.

● To increase the moment of a force F, increase F or increase d.

P3 2.2 Centre of mass ⓚ

Learning objectives

- What is the centre of mass of an object?
- What can we say about the centre of mass of an object that is suspended from a fixed point?
- What can we say about the centre of mass of a symmetrical object?

The design of racing cars has changed a lot since the first models. But one thing that has not changed is the need to keep the car near the ground. The weight of the car must be as low as possible. Otherwise the car would overturn when cornering at high speeds.

a

b

Figure 1 Racing cars: **a** 1920s racing car design **b** modern racing car design

We can think of the weight of an object as if it acts at a single point. This point is called the **centre of mass** (or the centre of gravity) of the object.

The centre of mass of an object is that point at which its mass may be thought to be concentrated.

> **a** Balance a ruler on the tip of your finger. The point of balance is at the centre of mass of the ruler. How far is the centre of mass from the middle of the ruler?

Practical

Suspended equilibrium

If you suspend an object and then release it, it will sooner or later come to rest with its centre of mass directly below the point of suspension, as shown in Figure 2a. The object is then in **equilibrium**. Its weight does not exert a turning effect on the object because its centre of mass is directly below the point of suspension.

If the object is turned from this position and then released, it will swing back to its equilibrium position. This is because its weight has a turning effect that returns the object to equilibrium, as shown in Figure 2b. We say the object is *freely suspended* if it returns to its equilibrium position.

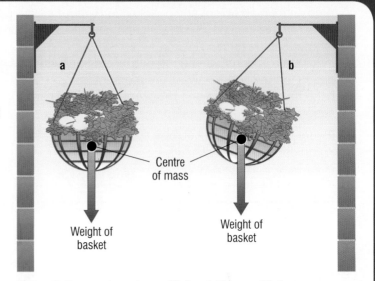

Figure 2 Suspension **a** In equilibrium **b** Non-equilibrium

The centre of mass of a symmetrical object

For a symmetrical object, its centre of mass is along the axis of symmetry. You can see this in Figure 3.

If the object has more than one axis of symmetry, its centre of mass is where the axes of symmetry meet.

- A rectangle has two axes of symmetry, as shown Figure 3a. The centre of mass is where the axes meet.
- The equilateral triangle in Figure 3b has three axes of symmetry, each bisecting one of the angles of the triangle. The three axes meet at the same point. This is where the centre of mass of the triangle is.

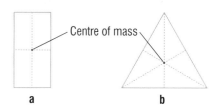

Figure 3 Symmetrical objects

Practical

A centre of mass test

Figure 4 shows how to find the centre of mass of an irregular-shaped flat card. The card is at rest, freely suspended from a rod.

Its centre of mass is directly below the rod. A 'plumbline' can be used to draw a vertical line on the card from the rod downwards.

The procedure is repeated with the card suspended from a second point to give another similar line. The centre of mass of the card is where the two lines meet.

Try drawing a third line to see if all three cross at the same point.

- What can you say about the accuracy of your experiment?

Test your results to see if you can balance the card at this point on the end of a pencil.

- Now find the centre of mass of a semicircular card of radius 100 mm.

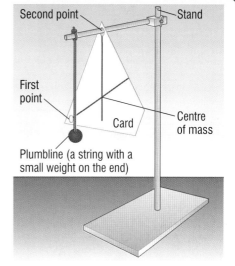

Figure 4 Finding the centre of mass of a card

b Every diameter of a flat circular card is an axis of symmetry. Where is its centre of mass?

Summary questions

1 Sketch each of the objects shown and mark its centre of mass.

a

b

c

2 Explain why a child on a swing comes to rest directly below the top of the swing.

3 Describe how you would find the centre of mass of a flat semicircular card.

AQA Examiner's tip

Make sure you can describe all the steps in the practical to find the centre of mass.

Key points

- The centre of mass of an object is that point where its mass may be thought to be concentrated.

- When a suspended object is in equilibrium, its centre of mass is directly beneath the point of suspension.

- The centre of mass of a symmetrical object is along the axis of symmetry.

P3 2.3 — Moments in balance

Learning objectives

- What can we say about the moments of the forces acting on an object that isn't turning?

- How can we use our knowledge of forces and moments to explain why objects at rest don't turn?

- How can we calculate the size of a force (or its perpendicular distance from a pivot) acting on an object that is balanced? **[H]**

A seesaw is an example in which clockwise and anticlockwise moments might balance each other out. The girl in Figure 1 sits near the pivot to balance her younger brother at the far end of the seesaw. Her brother is not as heavy as his big sister. She sits nearer the pivot than he does. That means her anticlockwise moment about the pivot balances his clockwise moment.

A model seesaw

Look at the model seesaw in Figure 2. The ruler is balanced horizontally by adjusting the position of the two weights. When it is balanced:

- the anticlockwise moment due to W_1 about the pivot = $W_1 d_1$, and
- the clockwise moment due to W_2 about the pivot = $W_2 d_2$

The anticlockwise moment due to W_1 = the clockwise moment due to W_2 therefore,

$$W_1 d_1 = W_2 d_2$$

a Use the equation to explain why the girl in Figure 1 needs to sit nearer the pivot than her younger brother.

Figure 1 The seesaw

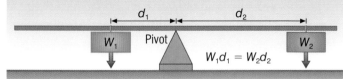

Figure 2 The principle of moments

The seesaw is an example of the **principle of moments**. This states that, for an object in equilibrium:

the sum of all the clockwise = the sum of all the anticlockwise moments about any point   moments about that point

b Calculate W_1, if W_2 = 5.0 N, d_1 = 0.30 m and d_2 = 0.15 m.

 Examiner's tip

Make sure the units in your calculations are consistent.

Practical

Measuring an unknown weight

You can use the arrangement in Figure 1 to find an unknown weight, W_1, if we know the other weight, W_2, and we measure the distances d_1 and d_2. Then you can calculate the unknown weight using the equation $W_1 d_1 = W_2 d_2$

Maths skills

Worked example

Calculate W_1 in Figure 2, if W_2 = 4.0 N, d_1 = 0.25 m and d_2 = 0.20 m.

Solution

Rearranging $W_1 d_1 = W_2 d_2$ gives

$$W_1 = \frac{W_2 d_2}{d_1} = \frac{4.0\,\text{N} \times 0.20}{0.25\,\text{m}} = 3.2\,\text{N}$$

[H]

Practical

Measuring the weight of a beam

Figure 3 shows how we can measure the weight of a beam by balancing it off-centre using a known weight. The weight of the beam acts at its centre of mass, which is at distance d_0 from the pivot.

● The moment of the beam about the pivot = $W_0 d_0$ clockwise, where W_0 is the weight of the beam.

● The moment of W_1 about the pivot = $W_1 d_1$ anticlockwise, where d_1 is the perpendicular distance from the pivot to the line of action of W_1.

Applying the principle of moments gives $W_1 d_1 = W_0 d_0$.

So we can calculate W_0 if we know W_1 and distances d_1 and d_0.

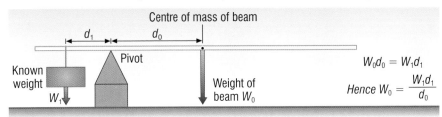

$W_0 d_0 = W_1 d_1$

Hence $W_0 = \dfrac{W_1 d_1}{d_0}$

Figure 3 Finding the weight of a beam

c Calculate the weight of the beam above if $W_1 = 2.0\,N$, $d_1 = 0.15\,m$ and $d_0 = 0.25\,m$. **[H]**

Summary questions

1 Copy and complete **a** to **c** using the words below:

equal to   greater than   less than

Jack and Ahmed are sitting on a seesaw which is pivoted at its centre.
a They are at the same distance from the pivot on opposite sides. The side Jack is on is lower than the other side. This is because Jack's weight is .... than Ahmed's weight.
b Jack moves towards the pivot until the seesaw is horizontal. This is because Jack's moment about the pivot is now .... Ahmed's moment about the pivot.

2 Dawn sits on a seesaw 2.50 m from the pivot. Jasmin balances the seesaw by sitting 2.00 m on the other side of the pivot.
a Who is lighter, Dawn or Jasmin?
b Jasmin weighs 425 N. What is Dawn's weight?
c Dawn gets off the seesaw so John can sit on it to balance Jasmin. His weight is 450 N. How far from the pivot should he sit? **[H]**

3 For the balanced beam in the figure, work out its weight, W.

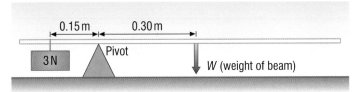

[H]

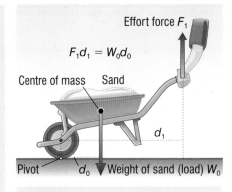

Effort force F_1

$F_1 d_1 = W_0 d_0$

Centre of mass   Sand

Pivot d_0 Weight of sand (load) W_0

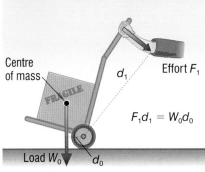

Centre of mass

d_1 Effort F_1

$F_1 d_1 = W_0 d_0$

Load W_0 d_0

Figure 4 Using moments

If you have to move a heavy load, think beforehand about how to make the job easier. Figure 4 shows a wheelbarrow and a trolley being used to move a load. The load (weight W_0) is lifted and moved using a much smaller effort (force F_1).

d In Figure 4 explain why the effort is smaller than the load.

Key points

● For an object in equilibrium, the sum of the anticlockwise moments about any point = the sum of the clockwise moments about that point.

● If an object at rest doesn't turn, the above statement must apply.

● To calculate the force needed to stop an object turning we use the equation above. We need to know all the forces that don't act through the pivot and their perpendicular distances from the line of action to the pivot. **[H]**

P3 2.4 Stability ⓚ

Learning objectives

- What factors affect the stability of an object?
- What will make a body topple over when it is tilted?
- What can we say about the moments on a body when it topples over? **[H]**

Figure 1 At the bowling alley

Practical

Tilting and toppling tests

How far can you tilt something before it topples over? Figure 2 shows how you can test your ideas using a tall box or a brick on its end.

- If you tilt the brick slightly, as in Figure 2a, and release it, the turning effect of its weight returns it to its upright position.
- If you tilt the brick more, you can just about balance it on one edge, as in Figure 2b. Its centre of mass is then directly above the edge on which it balances. Its weight has no turning effect in this position.
- If you tilt the brick even more, as in Figure 2c, it will topple over if it is released. This is because the line of action of its weight is 'outside' its base. So its weight has a turning effect that makes it topple over.

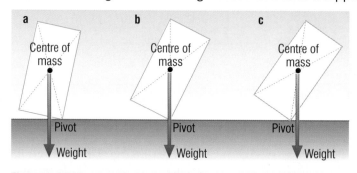

Figure 2 Tilting and toppling **a** Tilted **b** At balance **c** Toppled over

Investigate the factors that affect the stability of an object.

a Why does the weight of the brick have no turning effect in Figure 2b?

Stability and safety

Look around you and see how many objects could topple over. Bottles, table lamps and floor-standing bookcases are just a few objects that can easily topple over. Lots of objects are designed for stability so they can't topple over easily.

b Why do tall cabinets sometimes topple over when a top drawer is opened?

1 Tractor safety

Look at the tractor on a hillside in Figure 3. It doesn't topple over because the line of action of its weight acts within its wheelbase.

If it is tilted more, it will topple over when the line of action of its weight acts outside its wheelbase. This is because the moment about the lower wheel of the weight is clockwise.

The moment of the support force from the ground on the upper wheel is clockwise.

So the resultant moment about the lower wheel makes it topple over.

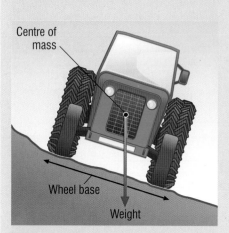

Figure 3 Forces on a tilting tractor

c Why is the engine of a tractor as low as possible?

Higher

2 Bus tests

Look at the double-decker bus in Figure 4. It is being tested on a platform to see how much it can tilt without toppling. Such tests are important to make sure buses are safe to travel on, especially when they go round bends and on hilly roads. The explanation of why it topples over when the platform is tilted too much is the same as for the tractor.

> **d** Would a double-decker bus be more stable or less stable if everyone on it sat upstairs?

Figure 4 A toppling test

3 High chairs

A high chair for a young infant needs to have a wide base. When the child is sitting in it, the centre of mass is above the seat. If the base were narrow and the child was strapped in, the chair would topple over when the child leant sideways too much. That's why a highchair should have a wide base (see Figure 5).

The chair topples over if the child's weight acts outside the chair base on one side. The high chair will turn about the position where the chair legs on that side are in contact with the floor.

Toppling happens if the moment of the child's weight about this position is greater than the moment of the chair's weight.

> **e** Why are stabiliser wheels fitted to bicycles designed for young children?

How to stop an object toppling over

To prevent toppling, the centre of mass of the object should be as low as possible and:

- either the base should be wide enough to prevent toppling when the object is tilted or knocked sideways
- or the base should be bolted or clamped down.

Figure 5 A high chair needs a wide base

Suppose an object is not clamped or bolted down. If the line of action of its weight lies outside its base, the object will topple over. This is because there is a **resultant moment** on the object. In other words, the object topples over because the sum of the clockwise moments about any point is not equal to the sum of the anticlockwise moments about that point.

Summary questions

1 **a** Make a list of objects that are designed to be difficult to knock over.
   b Think of an object that needs to be redesigned because it is knocked over too easily. Sketch the object and explain how it could be redesigned to make it more stable.

2 A well-designed baby chair has a wide base and a low seat.
   a If the base of a baby chair was too narrow, why would the chair be unsafe?
   b Why is a baby chair with a low seat safer than one with a high seat?

3 Explain why a tall plastic bottle is less stable when it is empty than when it is half-full of water using the idea of moments. **[H]**

Key points

- The stability of an object is increased by making its base as wide as possible and its centre of mass as low as possible.

- An object will tend to topple over if the line of action of its weight is outside its base. **[H]**

- An object topples over if the resultant moment about its point of turning is not zero. **[H]**

P3 2.5

Hydraulics ⓚ

Learning objectives

- What do we mean by pressure?
- What can we say about the pressure in a fluid?
- How does a hydraulic system work?
- What does the force exerted by a hydraulic system depend on?

Figure 1 Caterpillar tracks

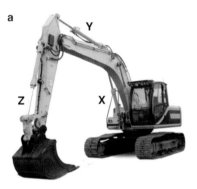

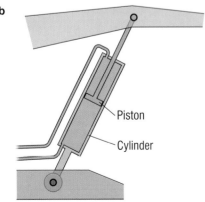

Figure 3 a A mechanical digger
b A hydraulic system

Caterpillar tracks fitted to vehicles are essential on sandy, muddy or snow-covered ground. The reason is that the contact area of the tracks on the ground is much greater than it would be if the vehicle had wheels instead. The tracks therefore reduce the pressure of the vehicle on the ground. That's because its weight is spread over a much greater contact area.

About pressure

Pressure is defined as force per unit area. The unit of **pressure** is the **pascal (Pa)**, which is equal to one newton per square metre (N/m²).

For a force F acting evenly on a surface of area A at right angles to the surface, the pressure p on the surface is given by the equation:

$$\text{pressure} = \frac{\text{force}}{\text{area}}$$

a Camels have much wider feet than horses. Why are they better at walking in the desert than horses?

Pressure in a liquid

The pressure in a liquid acts equally in all directions. Figure 2 shows how we can demonstrate this. There are several holes around the bottle at the same depth below the water level in the bottle. The jets from these holes hit the bench at the same distance from the bottle. So they are at the same pressure.

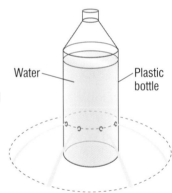

Figure 2 Pressure in a liquid at rest

Hydraulic machines

Mechanical diggers are used to remove large quantities of earth. An example is when soil has to be removed from above an underground pipe to reach the pipe. The 'grab' of the digger is operated by a **hydraulic pressure** system. The hydraulic system of a machine is its 'muscle power'.

Look at the hydraulic system shown in Figure 3b. Oil is pumped into the upper or lower part of the cylinder to make the piston move in or out of the cylinder.

A liquid is virtually incompressible. This means that its volume does not change when it is under pressure. This is why a force exerted on one part of a liquid is transmitted to other parts. In other words, the pressure in a hydraulic system is transmitted through the oil.

b In Figure 3b, which direction, up or down, does the piston move when oil is forced into the lower end of the cylinder?

A hydraulic car jack can be used to lift a car. When the handle is pressed down, oil is forced out of a narrow cylinder and into a wider cylinder. The pressure of the oil forces the piston in the wider cylinder outwards. As a result, the piston forces the pivoted lever to raise the car.

The force of a hydraulic system is much greater than the force applied to it.

In Figure 4, the force F_1 applied to the system is called the **effort.** The force F_2 exerted by the system is called the **load**. As explained below, the load is moved by a much smaller effort.

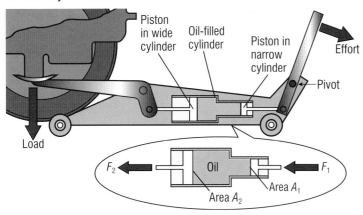

Figure 4 A hydraulic car jack

- The force F_1 acts on the piston in the narrow cylinder. This creates a pressure $P = \dfrac{F_1}{A_1}$ in the oil, where A_1 is the piston area.
- This pressure in the oil acts on the wide cylinder.
- The force F_2 on the wider piston $= PA_2$ where A_2 is the piston area.

Therefore $F_2 = \dfrac{F_1}{A_1} \times A_2$

The force F_2 is therefore much greater than F_1 because area A_2 is much greater than area A_1. In other words, the hydraulic system is a **force multiplier**.

Summary questions

1 Explain each of the following:
   a When you do a handstand, the pressure on your hands is greater than the pressure on your feet when you stand upright.
   b A sharp knife cuts more easily than a blunt knife.

2 a Write down as many machines as you can think of that are operated hydraulically.
   b Figure 3a shows the arm of a mechanical digger. It is controlled by three hydraulic pistons called 'rams', labelled X, Y and Z.
      i Explain why the arm is raised when compressed air is released into ram X so it extends.
      ii State and explain what happens to the 'bucket' on the end of the arm when rams Y and Z are both extended.

3 The hydraulic lift shown in Figure 5 is used to raise a vehicle so its underside can be inspected.
The lift has 4 pistons, each of area 0.01 m², to lift the platform. The pressure in the system must not be greater than 5.0×10^5 kPa. The platform weight is 2000 N. Calculate the maximum load that can be lifted on the platform.

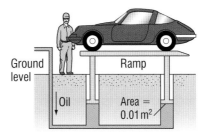

Figure 5 [H]

Key points

- Pressure is force divided by the area which the force acts on. The unit of pressure is the pascal (Pa) which is equal to 1 N/m²
- The pressure in a fluid acts equally in all directions.
- A hydraulic system uses the pressure in a fluid to exert a force.
- The force exerted by a hydraulic system depends on the force exerted on the system, the area of the cylinder which this force acts on and the area of the cylinder that exerts the force.

P3 2.6

Circular motion ⓚ

Learning objectives

- How can an object moving on a circular path at constant speed be accelerating?

- What do we mean by centripetal acceleration?

- What factors affect the centripetal force on an object in circular motion?

Figure 1 A hammer thrower

Practical

Testing circular motion

An object whirled round on the end of a string moves in a circle, as shown in Figure 2. The pull force on the object from the string changes the object's direction of motion. What factors affect the centripetal force?

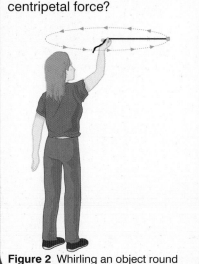

Figure 2 Whirling an object round

Fairground rides whirl you round in circles and make your head spin. But you don't need to go to a fairground to see objects moving in circles.

- A vehicle on a roundabout or moving round a corner travels on a circular path.
- A satellite moving across the sky moves on a circular orbit round the Earth.
- An athlete throwing the hammer spins round in a circle before releasing it.

For an object moving in a circle at constant speed, at any instant:

- the object's velocity is directed along a tangent to the circle
- its velocity changes direction as it moves round
- the change of velocity is towards the centre of the circle.

The object therefore accelerates continuously towards the centre of the circle. The acceleration changes the direction of motion of the object, not its speed.

Because the acceleration always acts towards the centre of the circle, we call it a **centripetal acceleration**.

So the resultant force on the object acts towards the centre of the circle.

> **a** In Figure 2, which direction would the object move if the string suddenly snapped at the position shown?

Centripetal force

Any object moving in a circle must be acted on by a resultant force that acts towards the centre of the circle. We call the resultant force a **centripetal force** because it *always* acts towards the centre of the circle.

- The centripetal force on a vehicle moving round a roundabout is due to friction between the tyres and the road.
- A person in a capsule on the London Eye moves round at a constant speed. The centripetal force on the person acts towards the centre of the 'wheel'. This force is the resultant force of the person's weight and the support force from the floor.
- A fairground 'gravity wheel' starts off spinning horizontally. When it is spinning fast enough, the wheel is turned until it is spinning vertically. The riders are strapped to the inside of the wheel. **When a rider is at or near the top of the wheel,** the rider experiences a downward push from the wheel to keep him or her moving round the circle. The centripetal force is due to the weight of the rider and the downward push from the wheel on the rider.

a

b

Figure 3 a The London Eye **b** A fairground gravity wheel

b Why does each rider on the 'gravity wheel' need to be strapped in?

Centripetal force factors

How is centripetal force affected by the speed of the object and the radius of the circle?

You could find out using a radio-controlled model car.

● If it goes too fast, the car will skid off in a straight line. The centripetal force needed increases if the speed is increased.

● If the circle is too small, the car will skid off. So the centripetal force needed increases if the radius of the circle is decreased.

c Why is the speed for no skidding much less on an icy roundabout?

How is the centripetal force affected by the mass of the object?

If you whirl a rubber bung round on the end of a thread, you can feel the force on it. If you tie another rubber bung on, you will find the force (for the same speed and radius) has increased. This shows that the greater the mass of the object, the greater the centripetal force is.

Figure 4 Centripetal force factors

??? Did you know ...?

A spin drier contains a drum that rotates very fast. When the drum spins, the cylindrical sides of the drum keep the wet clothing inside the drum.

● The force of the drum on the clothing provides the necessary centripetal force to keep the wet clothing moving in a circle.

● Water from the wet clothing leaves the spinning drum through tiny holes in the sides of the drum.

Summary questions

1 Figure 5 shows an object moving clockwise in a circle at constant speed.
   Copy and complete **a** and **b** below using directions A, B, C or D, as shown in Figure 5.

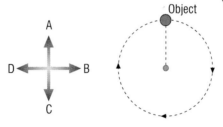

Figure 5

   a When the object is at the position shown, its velocity is in direction .... and the force on it is in direction ....

   b When the object has moved round by 90° from the position shown in the diagram, its velocity is in direction .... and the force on it is in direction ....

2 In each of the following situations, a single force acts as the centripetal force. Match each situation with the force, **a** to **d**, that causes the circular motion.

   electrostatic force   friction   gravity   pull (tension)

   a A car travelling round a bend.
   b A stone being whirled round on the end of a string.
   c A planet moving round the Sun.
   d An electron orbiting the nucleus of an atom.

3 **a** A student testing a model car measured its speed and found it could go round a bend at that speed without skidding off, provided its speed was no more than 2.2 m/s. What would happen to the model car if the test were repeated at the same speed on a bend that was more curved? Give a reason for your answer.

   b Explain why a high-speed railway track is sloped or banked where there is a curve.

Key points

● The velocity of an object moving in a circle at constant speed is continually changing as the object's direction is continually changing.

● Centripetal acceleration is the acceleration towards the centre of the circle of an object that is moving round the circle.

● The centripetal force on an object depends on its mass, its speed and the radius of the circle.

P3 2.7 The pendulum ⓚ

Learning objectives

- What affects the time period of a pendulum?

- How do we measure the time period of a pendulum accurately?

- How does the motion of a playground swing compare with the motion of a pendulum?

Figure 1 On a playground swing

When you sit on a playground swing moving backwards and forwards, you move fastest when the swing's seat is nearest the ground. This position is also the **equilibrium position** of the swing. This is its position when it eventually stops moving altogether. The motion of a swing is an example of **oscillating motion**. This is the motion of any object that moves to and fro along the same line.

The motion of a pendulum

Figure 2 shows a snapshot of a pendulum in motion. This type of pendulum is called a **simple pendulum**. It moves like a playground swing along the line between the bob's highest positions at A and B on each side. There is very little air resistance on it or friction at the point of suspension so it takes a long time to stop swinging.

The **amplitude of an oscillating object** is the distance it moves from its equilibrium position O to its highest position A or B on either side.

Its **time period** is the time taken for one complete cycle of its motion. In other words, it is:

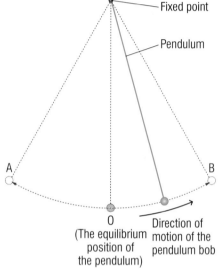

Figure 2 The pendulum

- the time it takes to swing across and back from its highest position on one side, for example from A to B and back to A, or

- the time between successive passes in the same direction through its equilibrium position O.

Practical

Investigating the time period of a pendulum

Task 1 To measure the time period, use a stopwatch to time 20 complete cycles of the pendulum's motion. Repeat this twice more and find the average of the three timings. Divide this timing by 20 to obtain the time period.

Each cycle takes the pendulum from when it passes through 'equilibrium' to the next time it passes through 'equilibrium' moving in the same direction. You could use a 'marker' to see exactly when it passes through equilibrium.

Task 2 Find out if the time period depends on the amplitude of the oscillations by measuring the time period for different amplitudes.

- Use a millimetre ruler to measure the amplitude.
- Record your measurements and plot a suitable graph.

Task 3 Find out if the time period depends on the length of the pendulum by measuring the time period for different lengths.

- The length of a simple pendulum is the distance from the middle of the bob to the point of suspension on the string. Use a metre ruler marked in millimetres to measure the length of the pendulum.
- Record your measurements and plot a suitable graph.

a You could measure the time period by timing the pendulum from its highest position to when it next returns. Try this and explain why the 'equilibrium' method used in Task 1 is better than this method.

b What conclusion can you draw from the measurements obtained in Task 2?

Your results from the last investigation should show that the time period of a pendulum depends on its length. Further investigations show that the time period of a pendulum depends only on its length and increases as its length increases.

The frequency of the oscillations is the number of complete cycles of oscillation per second.

The unit of frequency is the hertz (Hz) where 1 hertz is 1 cycle per second.

Note that the greater the frequency of the oscillations, the shorter the time period is. This is because:

the time period (in seconds, s) = $\dfrac{1}{\text{frequency of the oscillations (in hertz, Hz)}}$

Practical

Testing a model swing

1 Make a model swing with a plasticine person fixed to the seat. Measure the time period of small oscillations.

2 Repeat the test with a 'person' with a higher centre of mass on the seat. Use the same lump of plasticine so the mass of the person is unchanged.

3 Record your results.

● What conclusions can you draw from your results?

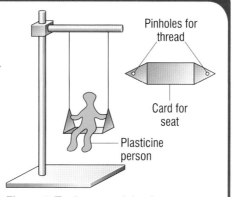

Pinholes for thread

Card for seat

Plasticine person

Figure 3 Testing a model swing

Summary questions

1 Copy and complete **a** to **c** using the words below:

decrease   increase   stay the same

a The time period of a simple pendulum will .... if its length is made longer.

b If a playground swing in motion is not pushed repeatedly, the amplitude of its motion will gradually ....

c If the bob of a simple pendulum is replaced by a bob of smaller mass and the pendulum length is unchanged, its time period will ....

2 Three timings were taken for 20 oscillations of a simple pendulum, as follows: 37.95 s, 37.73 s, 38.12 s

a Calculate the mean value of these timings.

b Calculate frequency and the time period of the oscillations.

3 State one similarity and one difference between the motion of a playground swing carrying a child and the motion of a simple pendulum of the same length as the swing.

Maths skills

We can write the word equation for the time period of a pendulum using symbols, as follows:

$$T = \frac{1}{f}$$

Where:
T = time period in seconds, s
f = frequency in hertz, Hz.

Note: Rearranging $T = \frac{1}{f}$ to make f the subject gives $f = \frac{1}{T}$

[H]

Worked example

A pendulum undergoes 20 complete cycles of oscillation in 4.0 seconds.

Calculate **a** the frequency of the oscillations, **b** the time period.

Solution

a For 20 complete cycles in 4 seconds, there must be 5 cycles each second (20 ÷ 4 = 5). The frequency is therefore 5 Hz.

b $T = \frac{1}{f}$ so the time period is $\frac{1}{5}$ = 0.2 seconds.

Key points

● The time period of a simple pendulum depends only on its length.

● To measure the time period of a pendulum, we can measure the average time for 20 oscillations and divide the timing by 20.

● Friction at the top of a playground swing and air resistance will stop it oscillating if it is not pushed repeatedly.

Summary questions

1 The bottle opener in Figure 1 is being used to force the cap off a bottle.

a Copy the diagram and add to it to show:
  i where the force is applied to the bottle opener by the person opening the bottle. Show the direction of this force.
  ii where the force of the bottle opener acts on the bottle top. Show the direction of this force.
  iii where the pivot (fulcrum) is.

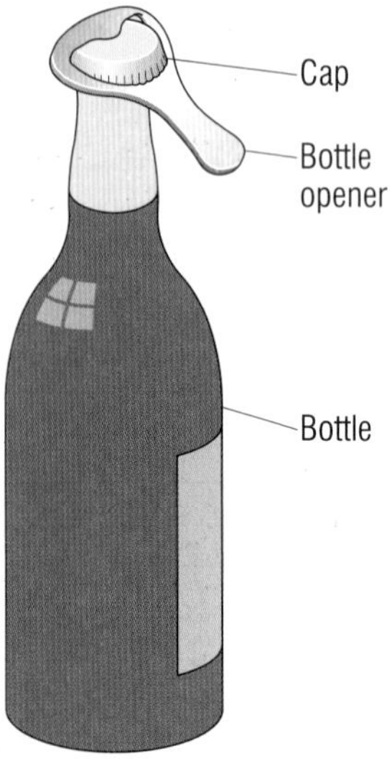

b Explain why the force of the bottle opener on the cap is much larger than the force applied to the bottle opener by the person opening the bottle.

c If a shorter bottle opener were used, would the force needed to open the bottle be smaller, larger or the same? Give a reason for your answer.

Cap
Bottle opener
Bottle

Figure 1

2 Figure 2 shows a toy suspended from a ceiling.

a How would the stability of the toy be affected if the Sun were removed from it?

b The star on the toy has a weight of 0.04 N and is a distance of 0.30 m from the point P where the thread is attached to the toy. Calculate the moment of the star about P. **[H]**

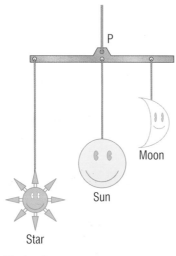

P
Moon
Sun
Star

Figure 2

3 The crescent moon attached to the toy in question 2 is at a distance of 0.20 m from P. Calculate the weight of the crescent. **[H]**

4 a A satellite is moving at constant speed in a circular orbit above the Earth.

Copy and complete **i** and **ii** using the words below. Each word can be used more than once.

acceleration force velocity

  i The centripetal .... on the satellite is equal to the .... of gravity on it.

  ii The .... of the satellite is in a direction at right angles to the direction of the .... of the satellite and to the .... on it.

b Explain why a high-sided lorry on a roundabout might topple over if its speed is too great.

5 A hammer thrower whirls a 'hammer' around anticlockwise and releases it when he is facing due south.

a i Which direction does the hammer go in when it is released?
  ii Which direction is the acceleration of the hammer just before it is released?

b Discuss whether or not the hammer thrower would be able to throw a hammer with a longer handle further. Assume the mass of the hammer is the same.

6 When the foot brake of a vehicle is applied, a force is applied to the piston in the master cylinder of the brake system, as shown in Figure 3.

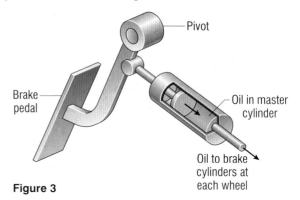

Pivot
Brake pedal
Oil in master cylinder
Oil to brake cylinders at each wheel

Figure 3

a The cylinder has an area of cross-section of 0.0006 m². Calculate the pressure in the brake system when a force of 20 N is applied to the piston in the cylinder.

b The master cylinder is connected by pipes to a brake cylinder at each wheel. Each brake cylinder is much wider than the master cylinder. Explain why the force of the brake cylinder on each wheel is much greater than 20 N.

7 a What is meant by the time period of a pendulum?

b The data below shows the time taken for 10 cycles of a pendulum that consists of a metal bob on a thread.

30.5 s, 29.8 s, 30.6 s

  i Use this information to calculate the time period of the pendulum.

  ii If the length of the pendulum is shortened, state how its time period and its frequency would change.

AQA Examination-style questions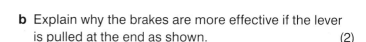

1 A boy is making a mobile of birds in flight. The diagram shows a thin sheet of card that he has cut in the shape of a bird. There are holes in the card at **A** and **B**. He decides to find the *centre of mass* of the shape.

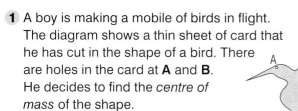

a What is meant by the *centre of mass* of an object? (1)

b *In this question you will be assessed on using good English, organising information clearly and using specialist terms where appropriate.*

Describe how you would find the centre of mass of the mobile shape. (6)

c The wrestler is shown in a 'ready' stance. Which two features of this stance make the wrestler very stable? **[H]** (2)

d A waiter trips while carrying a tray of mugs and the tray starts to tilt. The white dots indicate the position of the centre of mass of each mug.

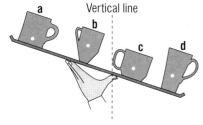

i Which mug will have started to topple over already? (1)

ii Which mug will topple over next if the tray is tilted even more? (1)

iii Which mug would be the last to topple? **[H]** (1)

2 The diagram shows a brake lever on a bicycle.

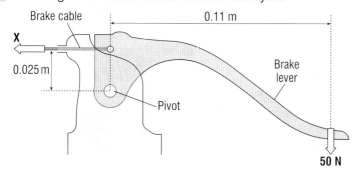

The lever is pulled with a force of 50 N and then stops turning.

a Calculate the moment of the force about the pivot.

Write down the equation you use. Show clearly how you work out your answer and give the unit. (3)

b Explain why the brakes are more effective if the lever is pulled at the end as shown. (2)

c There is a clockwise and an anticlockwise moment about the pivot. What can be said about these moments if the lever is not moving? (1)

d Calculate **X**, the force the cable exerts on the lever.

Show clearly how you work out your answer and give the unit. **[H]** (3)

3 The 'pirate ship' is a very common amusement park ride. The ride is simply a giant pendulum.

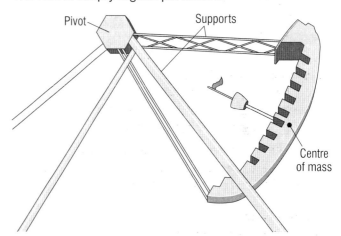

The designers of the ride wanted there to be three seconds between the highest points on each side of the ride.

a What would the time period of this ride be? (1)

b Calculate the frequency of the ride.

Write down the equation you use. Show clearly how you work out your answer and give the unit. (2)

c When the ride was sold to another amusement park, the dimensions of the ride were reduced. As a result of this, the distance between the pivot and the centre of mass of the ship was reduced. How would this affect the time period? (1)

The ship swings through an angle of 65° to the vertical on each side. While it does so, it is moving in a circular path.

d Describe the direction of the centripetal force acting on the ship when it is moving. (1)

e What provides the centripetal force in this case? (1)

f The new operators of the ride use the drive motors to increase the maximum speed of the ship to its original value as it travels past its lowest point. Explain why they should have the ride carefully checked before they do so. (3)

P3 3.1 Electromagnets

Learning objectives

- What can we say about the force between two magnets?
- What do we mean by magnetic field lines?
- How do we make electromagnets?
- What do we use electromagnets for?

About magnets

A magnetic compass is a tiny magnetic needle pivoted at its centre. Because of the Earth's magnetic field, one end of the compass always points north and the other south. The end that points north is called the **north pole** and the other end the **south pole.** Using two magnets, it is easy to show that:

like poles repel; unlike poles attract

a How could you compare the strength of two magnets?

b What happens if you hold the south pole of a bar magnet near the south pole of a different bar magnet?

Practical

Investigating magnetic fields

Hold a magnet near any iron or steel object and you should find that the magnet attracts the object. Iron and steel are examples of magnetic materials. A magnet is a piece of steel that has been magnetised. See for yourself the effect of a magnet on different objects including paper clips and iron filings on paper.

Figure 1 shows the effect of a magnet on iron filings sprinkled on a piece of paper.

- The paper clips (and the iron filings) stick to the ends of the magnet. We refer to the ends as **magnetic poles**.

- The iron filings near the magnet form a pattern of lines. The lines loop from one pole to the other. We say that there is a magnetic field around the magnet. We refer to the lines as **lines of force** or **magnetic field lines**.

A plotting compass placed in the magnetic field will always point along a field line. The direction the compass points in tells us the direction of the field line. Use a plotting compass to see the direction of the lines of force in the magnetic field.

- What do you find?

You should see that the lines always loop round from the north pole of the magnet to its south pole.

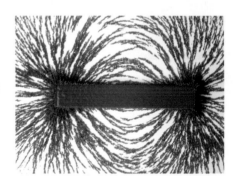

Figure 1 The magnetic field near a bar magnet

Electromagnets

An electromagnetic is made of insulated wire wrapped round an iron bar (the core). When a current is passed along the wire, a magnetic field is created around the wire. As a result, the magnetic field of the wire magnetises the iron bar strongly. When the current is switched off, the iron bar loses most of its magnetism. Iron easily loses its magnetism when the current is switched off. On the other hand, steel is unsuitable as the core of an electromagnet. That's because steel keeps its magnetism when the current is switched off.

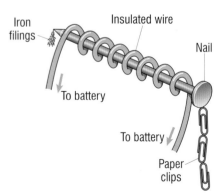

Figure 3 A simple electromagnet

Figure 2 Using an electromagnet

We use electromagnets in many devices. Four such devices are described below.

1 The scrapyard crane

Scrap vehicles are lifted in a scrap yard using powerful electromagnets attached to cranes. The steel frame of a vehicle sticks to the electromagnet when current passes through the coil. When the current is switched off, the steel frame drops off the electromagnet.

2 The circuit breaker

This is a switch in series with an electromagnet. The switch is normally held in place by an iron catch. When too much current passes through the electromagnet, the switch is pulled open by the electromagnet, and the switch opens. It stays open until it is reset manually.

c Why is a spring needed in a circuit breaker?

3 The electric bell

When the bell is connected to the battery, the iron armature is pulled on to the electromagnet. This opens the make-and-break switch and the electromagnet is switched off. As a result, the armature springs back and the make-and-break switch closes again so the whole cycle repeats itself.

4 The relay

A relay is used to switch an electrical machine, such as a motor, on or off. Figure 6 shows what a relay consists of. When current passes through the electromagnet, the armature is pulled on to the electromagnet. As a result the armature turns about the pivot and closes the switch gap. In this way, a small current (through the electromagnet) is used to switch on a much larger current.

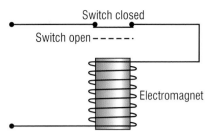

Figure 4 A circuit breaker

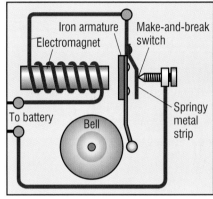

Figure 5 An electric bell

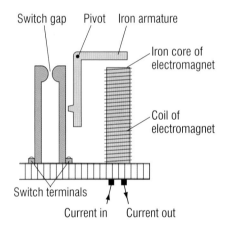

Figure 6 The construction of a relay

Summary questions

1 a i Which material, iron or steel, is used to make a permanent magnet?
  ii Which material, iron or steel, is used as the core of an electromagnet?
  b State whether there is a force of attraction or repulsion when:
   i a magnet is held near an unmagnetised iron bar
   ii the north pole of a bar magnet is placed near the north pole of a second bar magnet
   iii the north pole of a bar magnet is placed near the south pole of a second bar magnet.

2 List the statements A–E below in the correct order to explain how the circuit breaker in Figure 4 works. Statement E is third in the correct order.
A The current is cut off.
B The iron core of the electromagnet is magnetised strongly.
C Too much current passes through the coil.
D The circuit breaker switch is opened.
E The switch is attracted onto the core of the electromagnet.

3 The construction of a buzzer is like that of the electric bell except it does not have a striker or a bell.
  a Explain why the armature of the buzzer vibrates when the buzzer is connected to a battery.
  b In a bell and a buzzer, the armature vibrates continuously. The buzzer armature does not have a striker attached to it though. Explain why the buzzer vibrates at a higher frequency than the electric bell.

Key points

- The force between two magnets: like poles repel; unlike poles attract.

- A magnetic field line is the line along which a plotting compass points.

- An electromagnet consists of a coil of insulated wire wrapped round an iron core.

- Electromagnets are used in scrapyard cranes, circuit breakers, electric bells and relays.

P3 3.2

The motor effect ⓚ

Learning objectives

- How can we change the size of the force on a current-carrying wire in a magnetic field?

- How can we reverse the direction of the force on a current-carrying wire in a magnetic field?

- How do we use the motor effect to make objects move?

We use electric motors lots of times every day. Using a hairdryer, an electric shaver, a refrigerator pump and a computer hard drive are just a few examples. All these electrical appliances contain an electric motor. The electric motor works because a force can act on a wire in a magnetic field when we pass a current through the wire. This is called the **motor effect**.

Practical

Investigating the motor effect

Figure 1 shows how you can investigate the motor effect. You should find that a force acts on the wire unless the wire is parallel to the magnetic field lines.

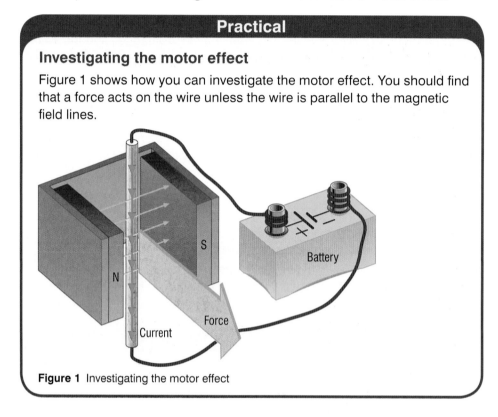

Figure 1 Investigating the motor effect

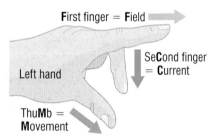

Figure 2 Fleming's left-hand rule. Hold the fingers at right angles to each other. You can use this rule to work out the direction of the force (i.e. movement) on the wire.

Force factors

Your investigations should show that:

- The size of the force can be increased by:
  - increasing the current
  - using a stronger magnet.
- The size of the force depends on the angle between the wire and the magnetic field lines;
  the force is:
  - greatest when the wire is perpendicular to the magnetic field
  - zero when the wire is parallel to the magnetic field lines.
- The direction of the force is always at right angles to the wire and the field lines. Also, the direction of the force is reversed if the direction of the current or the magnetic field is reversed. Figure 2 shows Fleming's left-hand rule which tells us how these directions are related.

a What happens if the current and the magnetic field are both reversed?

The electric motor

An electric motor is designed to use the motor effect. We can control the speed of an electric motor by changing the current. Also, we can reverse the direction the motor turns in by reversing the current.

The simple motor shown in Figure 3 consists of a rectangular coil of insulated wire (the armature coil) that is forced to rotate. The coil is connected via two metal or graphite 'brushes' to the battery. The brushes press onto a metal **'split-ring' commutator** fixed to the coil.

When a current is passed through the coil, the coil spins because:

● a force acts on each side of the coil due to the motor effect

● the force on one side is in the opposite direction to the force on the other side.

The split-ring commutator reverses the current round the coil every half turn of the coil. Because the sides swap over each half-turn, the coil is pushed in the same direction every half-turn.

b Why are the brushes made of metal or graphite?

The loudspeaker

A loudspeaker is designed to make a diaphragm attached to a coil vibrate when alternating current passes through the coil.

● When a current passes through the coil, a force due to the motor effect makes the coil move.

● Each time the current changes its direction, the force reverses its direction. So the coil is repeatedly forced backwards and forwards. This motion makes the diaphragm vibrate so sound waves are created.

c Why does a loudspeaker not produce sound when direct current is passed through it?

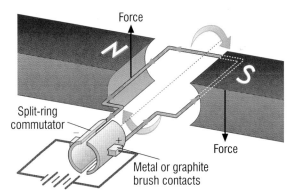

Figure 3 The electric motor

Practical

Make and test a simple electric motor like the one in Figure 3.

??? Did you know ... ?

Graphite is a form of carbon that conducts electricity and is very slippery. It causes very little friction when in contact with the rotating commutator.

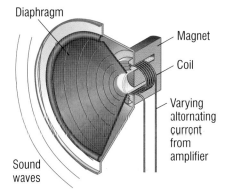

Figure 4 A loudspeaker

Summary questions

1 Copy and complete **a** and **b** using the words below. Each word can be used more than once.

coil   current   force   magnet

a When a .... passes through the .... of an electric motor, a .... acts on each side of the ....

b The .... along each side is in opposite directions so the .... is in opposite directions and the .... turns.

2 **a** Explain why a simple electric motor connected to a battery reverses if the battery connections are reversed.

b Discuss whether or not an electric motor would run faster if the coil was wound on:

i a plastic block

ii an iron block instead of a wooden block.

3 A force is exerted on a straight wire when a current is passed through it and it is at right angles to the lines of a magnetic field. Describe how the force changes if the wire is turned through 90° until it is parallel to the field lines.

Key points

● In the motor effect, the force:

– is increased if the current or the strength of the magnetic field is increased,

– is at right angles to the direction of the magnetic field and to the wire,

– is reversed if the direction of the current or the magnetic field is reversed.

● An electric motor has a coil which turns when a current is passed through it.

P3 3.3 Electromagnetic induction

Learning objectives

- What do we mean by electromagnetic induction?

- How can we use a magnet to induce a potential difference across the ends of a conductor?

- How can we induce a potential difference if we use an electromagnet instead of a magnet?

Figure 1 A standby generator

A hospital has its own electricity generator always 'on standby' in case of a power-cut. Patients' lives would be put at risk if the mains electricity supply failed and there was no standby generator.

A generator contains coils of wire that spin in a magnetic field. A potential difference (pd), or voltage, is created or **induced** across the ends of the wire when it cuts across the magnetic field lines. We call this process **electromagnetic induction**. If the wire is part of a complete circuit, the induced pd makes an electric current pass round the circuit.

Practical

Investigating a simple generator

Connect some insulated wire to an ammeter as shown in Figure 2. Move the wire between the poles of a U-shaped magnet and observe the ammeter. You should discover the ammeter pointer deflects as a current is generated when the wire cuts across the magnetic field.

- What difference is made if:
  a the wire is held stationary between the poles of the magnet?
  b the direction of motion of the wire is reversed?
  c the wire is moved parallel to the lines of the magnetic field?

Make the wire into a coil and you should find the current is bigger.

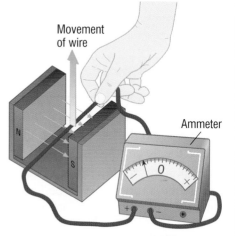

Figure 2 Electromagnetic induction

a In the previous experiment, what can you say about the current when:
  i the wire was stationary
  ii the direction of motion of the wire was reversed?

A generator test

Look at Figure 3. It shows a coil of insulated wire connected to a centre-reading ammeter. When one end of a bar magnet is pushed into the coil, the ammeter pointer deflects.

This is because:

- the movement of the bar magnet causes an induced pd in the coil

- the induced pd causes a current, because the coil is part of a complete circuit.

b What do you think happens if:
  i the magnet is left at rest in the coil
  ii there were more turns of wire around the cardboard tube?

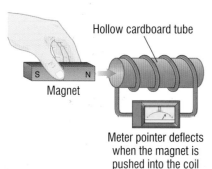

Meter pointer deflects when the magnet is pushed into the coil

Testing electromagnetic

Practical

A magnetic puzzle

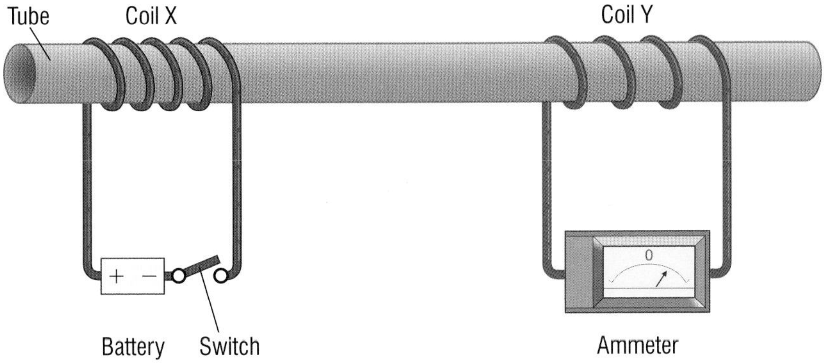

Tube Coil X Coil Y

+ −

Battery Switch Ammeter

Figure 4 A magnetic puzzle

Use two separate lengths of insulated wire to make two coils on a cardboard tube as shown in Figure 4. Connect one of the coils (X) to a battery in series with a switch and connect the other coil (Y) to an ammeter.

1 Close the switch and you should discover that the ammeter pointer deflects briefly. Switching the current on creates a magnetic field that passes through coil Y as well as coil X. The effect on Y is the same as pushing a magnet into it. So a pd is induced in coil Y.

2 Keep the switch closed and observe the ammeter. You should see that the ammeter pointer does not deflect. This is because the current in X is now constant so the magnetic field does not change. The effect on Y is the same as when a magnet is held stationary in it. The magnetic field of the electromagnet has to be changing to induce a pd.

3 Repeat the first test with an iron bar in the tube. You should find that the deflection is much bigger.

c In the previous experiment, explain why the deflection was much bigger with the iron bar in the tube.

Summary questions

1 Copy and complete sentences **a** and **b** using the words below. Each word may be used more than once.

magnetic field conductor current potential difference

  a A .... is induced in a .... when it cuts across the .... lines.

  b If a .... connected to an ammeter moves parallel to the lines of a .... no .... passes through the ammeter.

2 A coil of wire is connected to a centre-reading ammeter. A bar magnet is inserted into the coil, making the ammeter pointer flick briefly. What would you observe if :

  a the magnet was then held at rest in the coil

  b the coil had more turns of wire wrapped round the tube?

3 **a** In Figure 4, explain why the ammeter pointer deflects when the switch is closed.

  b Explain why the pointer does not deflect when there is a constant current in coil X.

Key points

- Electromagnetic induction is the process of creating a potential difference using a magnetic field.

- When a conductor cuts the lines of a magnetic field, a potential difference is induced across the ends of the conductor.

- When an electromagnet is used, it needs to be switched on or off to induce a pd.

P3 3.4 Transformers ⓚ

Learning objectives

- Why do transformers only work with ac?
- What is the core of a transformer made from?
- How does a switch mode transformer differ from an ordinary transformer?

⊂⊃ **links**

For more information on the National Grid, look back at P1 4.5 The National Grid.

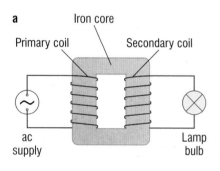
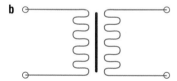

a

Iron core

Primary coil   Secondary coil

ac supply   Lamp bulb

b

Figure 1 Transformer action **a** in a circuit **b** circuit symbol

A typical power station generator produces an alternating potential difference (pd) of about 25 000 volts. Mains electricity to homes is at 230 volts.

When you plug an appliance into the mains, the electricity to run it comes from a power station. The electricity arrives via a network of cables called the **National Grid**. The alternating pd of the cables (the grid voltage) is typically 132 000 volts. A **transformer** is used to change the size of the alternating pd.

How a transformer works

A transformer has two coils of insulated wire, both wound round the same iron core, as shown in Figure 1. The primary coil is connected to an alternating current supply. When alternating current passes through the primary coil, an alternating pd is induced in the secondary coil.

This happens because:

- alternating current passing through the primary coil produces an alternating magnetic field
- the lines of the alternating magnetic field pass through the secondary coil
- the magnetic field is changing.

This creates an alternating potential difference between the terminals of the secondary coil. We say an alternating potential difference is 'induced' in the secondary coil.

If a bulb is connected across the secondary coil, the induced pd causes an alternating current in the secondary circuit. So the bulb lights up. Electrical energy is therefore transferred from the primary to the secondary coil. This happens even though they are **not** electrically connected in the same circuit.

- A **step-up transformer** makes the pd across the secondary coil greater than the pd across the primary coil. Its secondary coil has more turns than its primary coil.
- A **step-down transformer** makes the pd across the secondary coil less than the pd across the primary coil. Its secondary coil has fewer turns than its primary coil.

For example, we use a step-down transformer in a low-voltage supply to step the mains pd down from 230 V.

Practical

Make a model transformer

Wrap a coil of insulated wire round the iron core of a model transformer as the primary coil. Connect the coil to a 1 V ac supply. Then connect a second length of insulated wire to a 1.5 V torch bulb. When you wrap enough turns of the second wire round the iron core, the bulb should light up.

- Test if cores made from different materials affect the transformer.

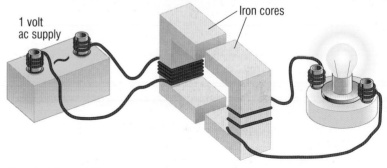

1 volt ac supply

Iron cores

Figure 2 A model transformer

a Why would the lamp not light up as brightly if the iron core was replaced with a wooden core?

b What happens if you wrap more turns on the secondary coil?

c What happens if you use a cell or a battery instead of the ac supply?

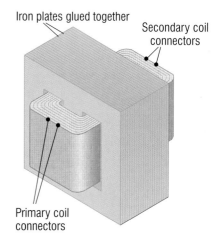

Figure 3 A practical transformer

Transformers in action

Transformers only work with alternating current. With a direct current, there is no changing magnetic field so the secondary pd is zero.

In the type of transformer described above, the core of the transformer 'guides' the field lines in a loop through the coils. But the field must be changing to induce a pd in the secondary coil.

Figure 3 shows a practical transformer.

The primary and secondary coils are both wound round the same part of the core. The core is layered (laminated) to cut out induced currents in the iron layers. If it were not laminated, the efficiency of the transformer would be greatly reduced.

Switch mode transformers

A **switch mode transformer** works in a different way to the traditional transformer described above. It operates at frequencies between 50 000 Hz (50 kHz) and 200 000 Hz (200 kHz). Its main features listed below make it very suitable for use in mobile phone chargers.

- It is lighter and smaller than a traditional transformer which works at 50 Hz.
- It uses very little power when there is no device connected across its output terminals.

A mobile phone charger has three main circuits. Figure 4 shows you what each circuit does.

The switch mode transformer has a ferrite core. This is much lighter than an iron core and, unlike an iron core, can work at high frequency. The circuits convert the mains pd (at 230 V and 50 Hz in Europe) to a much lower direct pd.

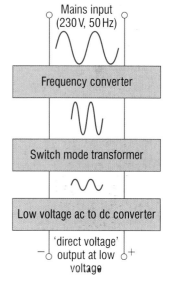

Figure 4 Block diagram of a mobile phone charger

Summary questions

1 Copy and complete the paragraph using the words below. Each word can be used more than once.

current   magnetic field   pd   primary   secondary

In a transformer, an alternating .... is passed through the .... coil. This coil creates an alternating .... that passes through the .... coil. As a result, an alternating .... is induced in the .... coil.

2 a Why does a transformer not work with direct current?

b Why is it important that the coil wires of a transformer are insulated?

c Why is the core of a transformer made of iron?

3 a A laptop computer can operate with a 14 V battery or with a mains transformer.

i What is the benefit of having a dual power supply?

ii Does the transformer step up or step down the pd applied to it?

b Why is a switch mode transformer lighter than an ordinary transformer?

Key points

- A transformer only works on ac because a changing magnetic field is necessary to induce ac in the secondary coil.

- A transformer has an iron core unless it is a switch mode transformer which has a ferrite core.

- A switch mode transformer is lighter and smaller than an ordinary transformer. It operates at high frequency.

P3 3.5 Transformers in action ⓚ

Learning objectives

- Why are transformers used in the National Grid?

- How does the ratio of the primary pd to the secondary pd depend on the number of turns on each coil?

- What is the difference between a step-up and a step-down transformer?

- What can we say about a transformer which is 100% efficient?

Figure 1 A power transformer under inspection

When we use mains appliances, the electricity is supplied to us through the National Grid from distant power stations. Figure 2 shows how the grid system is used to supply industry as well as homes.

The higher the grid pd, the greater the efficiency of transferring electrical power through the grid.

This is why transformers are used to step up the pd from a power station to the grid pd and to step the grid pd down to the mains voltage. The grid pd is at least 132 000 V. So what difference would it make if the grid pd were much lower? Much more current would be needed to deliver the same amount of power. The grid cables would therefore heat up more and waste more energy.

The transformer equation

The secondary pd of a transformer depends on the primary pd and the number of turns on each coil.

We can use the following equation to calculate any one of these factors if we know the other ones.

$$\frac{\text{pd across primary, } V_P}{\text{pd across secondary, } V_S} = \frac{\text{number of turns on primary, } n_p}{\text{number of turns on secondary, } n_S}$$

- **For a step-up transformer**, the number of secondary turns, n_S, is greater than the number of primary turns, n_p. Therefore V_S is greater than V_P.

- **For a step-down transformer**, the number of secondary turns, n_S is less than the number of primary turns, n_p. Therefore V_S is less than V_P.

> **Maths skills**
>
> #### Worked example
>
> A transformer is used to step a pd of 230 V down to 10 V. The secondary coil has 60 turns. Calculate the number of turns of the primary coil.
>
> #### Solution
>
> $V_P = 230\,V$, $V_S = 10\,V$, $n_S = 60$ turns
>
> Using $\dfrac{V_P}{V_S} = \dfrac{n_P}{n_S}$ gives $\dfrac{230}{10} = \dfrac{n_P}{60}$ Therefore $n_P = \dfrac{230 \times 60}{10} = 1380$ turns

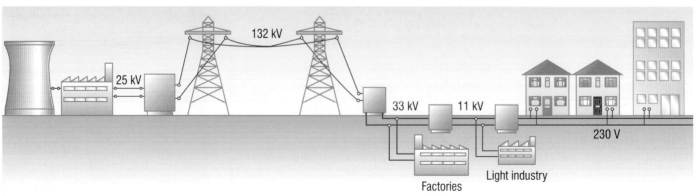

Figure 2 The grid system

a A transformer with 1200 turns in the primary coil is used to step a pd of 120 V down to 6 V. Calculate the number of turns on the secondary coil.

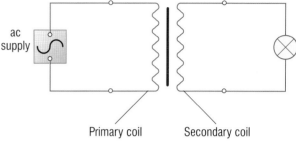
Transformer efficiency

Transformers are almost 100% efficient. When a device is connected to the secondary coil (see Figure 3), almost all the electrical power supplied to the transformer is delivered to the device. If we know how much electrical power a device requires to work normally, this tells us how much electrical power must be supplied to the transformer.

- Power supplied to the transformer
  = primary current, I_P × primary pd, V_P
- Power delivered by the transformer
  = secondary current, I_S × secondary pd, V_S

ac supply

Primary coil   Secondary coil

Figure 3 Transformer efficiency

Therefore, for 100% efficiency:

power supplied to the transformer = power delivered by the transformer

primary pd × primary current = secondary pd × secondary current

$$V_P \times I_P \qquad = \qquad V_S \times I_S$$

b A 230 V, 60 W bulb lights normally when it is connected to the secondary coil of a transformer and a 10 V ac supply is connected to the primary coil. Assume the transformer is 100% efficient.
Calculate:
i the primary current
ii the bulb current.

Summary questions

1 Copy and complete **a** and **b** using the words below:

down   primary   secondary   up

a In a step-up transformer, the pd across the .... coil is greater than the pd across the .... coil.

b The pd from a power station is stepped .... so the same amount of power can be delivered through the cables as a result of stepping the current ....

2 A transformer with a secondary coil of 100 turns is to be used to step a pd down from 240 V to 12 V.

a Calculate the number of turns on the primary coil of this transformer.

b A 12 V 36 W bulb is connected to the secondary coil. Calculate the current in the following. Assume the transformer is 100% efficient.

i the bulb
ii the primary coil.

3 Two separate cables A and B deliver the same amount of electrical power to two factories. A is at a higher pd than B.

a What can we say about the current in cable A compared with that in B?

b Why is less power wasted in A than B?

Key points

- Transformers are used to step potential differences up or down.

- The transformer equation is:
  $$\frac{\text{primary pd, } V_P}{\text{secondary pd, } V_S} = \frac{n_P}{n_S}$$
  Where:
  n_P = number of primary turns
  n_S = number of secondary turns

- For a step-down transformer, n_S is less than n_P
  For a step-up transformer, n_S is greater than n_P

- For a 100% efficient transformer,
  $V_P \times I_P = V_S \times I_S$
  Where:
  I_P = primary current
  I_S = secondary current

P3 3.6

A physics case study

Learning objectives

- How do we use physics in hospitals?

- What physics measurements do we make in a hospital?

- Why is a CT scan not harmless?

⚙️ **links**

For more information on the endoscope, look back to P3 1.4 The endoscope.

⁇ Did you know …?

An electronic blood pressure gauge contains a sensor that produces a pd when pressure is applied to it. When the gauge is started, the cuff round the upper arm inflates with air and cuts off the blood flow to the lower arm. The cuff pressure is then released and the sensor detects when the flow starts again (the systolic pressure) and at a lower pressure when the flow becomes normal again (the diastolic pressure).

We use many devices in hospitals to find out why patients are unwell and to help them recover. In this case study, we look at the physics of some of these devices.

A patient has a swollen neck and the doctors are doing tests to find out what has caused it. The swelling is affecting his breathing so they need to monitor his condition closely as well.

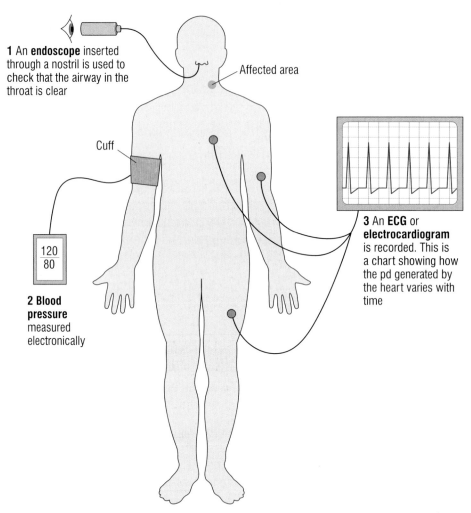

1 An **endoscope** inserted through a nostril is used to check that the airway in the throat is clear

Affected area

Cuff

120 / 80

2 Blood pressure measured electronically

3 An **ECG** or **electrocardiogram** is recorded. This is a chart showing how the pd generated by the heart varies with time

Figure 1 Initial tests

a When the cuff is inflated and the blood supply to the lower arm is cut off, what can we say about blood pressure and the pressure of the cuff?

The doctor in charge of the patient tells the patient his airway is clear and his blood pressure is normal. The ECG test shows there are no problems with his heart and his blood circulation.

A sample of fluid is taken from the affected area using a fine needle. The sample is tested in the pathology lab. It indicates that there may be a cyst or fluid in the affected area. So the patient needs to have some scans of the affected area.

⚙️ **links**

For more information on the CT scanner, look back at P3 1.1 X-rays.

Different types of scans include ultrasonic scans, X-ray CT scans, and magnetic resonance (MR) scans. The patient needs:

1 **an X-ray CT scan** to see if there are other affected areas in the neck or the chest. The operators or radiographers must wear film badges to monitor and record their exposure to ionising radiation.

2 **an MR scan** to see exactly where the swelling is located; an MR scan uses radio waves. These waves, unlike X-rays, are non-ionising.

The CT scan shows that no other areas are affected. With the exact position of the cyst known from the MR scan, the cyst is then removed surgically.

b Why does a radiographer using a CT scanner need to wear a film badge?

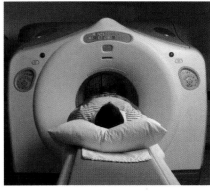

Figure 2 An X-ray CT scanner

 Did you know … ?

The magnetic resonance (MR) scanner is used in hospitals to scan the hydrogen content in the body. It can distinguish different types of tissue. Hydrogen atoms are present in the body, mostly in water molecules but also in many other molecules in the body.

The nucleus of a hydrogen atom (and certain other atoms) is like a tiny bar magnet. In a strong magnetic field, such nuclei can be made to emit radio waves. The process is called **magnetic resonance.** The radio waves can be detected and used to pinpoint the location of the hydrogen atoms.

The strong magnetic field in an MR scanner is produced by a superconducting coil of wire. The patient lies on a bed which moves through the coil. A *superconductor* has zero electrical resistance. A very large current can therefore pass along the wire and generate a very strong magnetic field. To be in a superconducting state, the wire needs to be very cold.

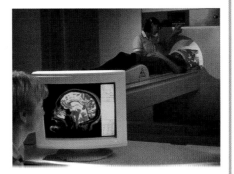

Figure 3 An MR scanner

⚬⚬ links

For more information on magnetic fields produced by electric currents, look back at P3 3.1 Electromagnets.

c What would happen to an ordinary non-superconducting wire if a very large current was passed along it?

Summary questions

1 Match each device below to the correct function.

A CT scanner   An endoscope   An ECG machine

a .... is used to record voltage.

b .... is used to observe cavities in the body.

c .... is used to give a picture of organs in the body.

2 **a** What is a film badge used for?
  b Why is it important for a radiographer but not a patient to wear a film badge?

3 Which of the following forms of radiation are ionising radiation?

gamma rays   radio waves   infrared radiation   X-rays

Key points

- We use physics in hospital whenever:
  - blood pressure (or temperature) is measured
  - an ECG recording is made
  - an endoscope is used
  - a scanner is used.

- We measure blood pressure, ECG potential differences and exposure to ionising radiation in hospitals.

- A CT scanner uses X-rays, which are ionising radiation and can therefore damage living tissue.

Summary questions

1 Copy and complete **a** and **b** using the words below. Each word can be used more than once.

field force lines current

a A vertical wire is placed in a horizontal magnetic field. When a .... is passed through the wire, a .... acts on the wire.

b A force acts on a wire in a magnetic field when a .... passes along the wire and the wire is not parallel to the .... of the ....

2 a Figure 1 shows the construction of a relay. Explain why the switch closes when a current passes through the coil of the electromagnet.

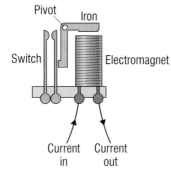

Figure 1

b Figure 2 shows the relay coil in a circuit that is used to switch on the starter motor of a car. Explain why the motor starts when the ignition switch is closed.

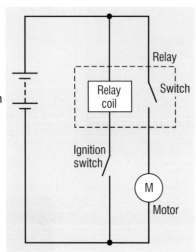

Figure 2

3 Figure 3 shows a rectangular coil of wire in a magnetic field viewed from above. When a direct current passes clockwise round the coil, an upward force acts on side X of the coil.

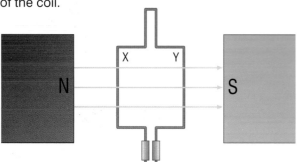

Figure 3

a What is the direction of the force on side Y of the coil?

b What can you say about the force on each side of the coil parallel to the magnetic field lines?

c What is the effect of the forces on the coil?

4 a Copy and complete **i** and **ii** using the words below.

step-down step-up

   i A transformer that changes an alternating pd from 12 V to 120 V is a .... transformer.

   ii A .... transformer has more turns on the primary coil than on the secondary coil.

b Explain why a transformer does not work on direct current.

5 a Cables at a potential difference of 100 000 V are used to transfer 1 million watts of electrical power in part of a grid system.

   i Show that the current in the cables is 10 A.

   ii If the potential difference had been 10 000 V, how much current would be needed to transfer the same amount of power?

b Explain why power is transmitted through the National Grid at a high pd rather than a low pd.

6 A transformer has 50 turns in its primary coil and 500 turns in its secondary coil. It is to be used to light a 120 V, 60 W bulb connected to the secondary coil. Assume the transformer is 100% efficient.

a Calculate the primary pd.

b Calculate the current in the bulb.

c Calculate the current in the primary coil.

7 A transformer has 3000 turns on its primary coil. An alternating pd of 240 V is to be connected to the primary coil and a 12 V bulb is to be connected to the secondary coil.

a Calculate the number of turns the secondary coil should have.

b What would be the current through the primary coil if the current through the lamp is to be 3.0 A. Assume the transformer is 100% efficient.

AQA Examination-style questions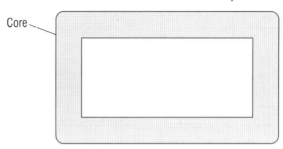

1 A magnet is held suspended near a wire that has a magnetic field around it.

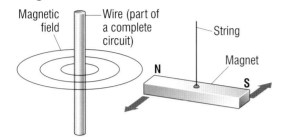

Magnetic field — Wire (part of a complete circuit) — String — Magnet — N — S

a What causes the magnetic field to be produced around a wire? (1)

b When released, the magnet rotates slowly in the direction shown. What is the name for this effect? (1)

c Give **two** ways in which the force on the magnet could be reversed. (2)

d Give **two** ways of increasing the force on the magnet. (2)

The wire and the magnet are now arranged as shown below. A current passes through the wire and the wire experiences a force. Only part of the magnetic field from the magnet is shown.

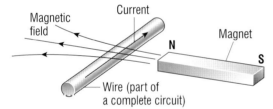

Current — Magnetic field — N — Magnet — S — Wire (part of a complete circuit)

e Use Fleming's left-hand rule to find the direction of the force on the wire. Explain fully how you used the rule. (3)

2 The diagram shows a wire being moved upwards through the magnetic field between two poles of a magnet. A voltmeter is connected across the wire.

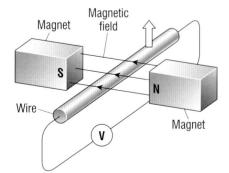

Magnet — Magnetic field — S — N — Wire — V — Magnet

a Explain why the voltmeter gives a reading when the wire is being moved upwards but not when it is stationary. (2)

b How does the potential difference change if the wire is moved
  i downwards (1)
  ii sideways? (1)

3 a Copy and complete the diagram to show the basic structure of a transformer. Label each part. (3)

Core

b Explain the difference between a step-up transformer and a step-down transformer. (2)

c What must be happening in the iron core of the transformer when it is producing a potential difference on the output? (1)

d The following equation can be used when designing transformers.

$$\frac{V_P}{V_S} = \frac{n_P}{n_S}$$

  i Give the meaning of each of the symbols in the equation. (4)
  ii Calculate V_P if $V_S = 55\,V$, $n_P = 320$ and $n_S = 40$
    Write down the equation you use. Show clearly how you work out your answer. (3)
  iii Calculate n_S if $V_P = 11\,000\,V$, $V_S = 230\,V$ and $n_P = 55\,000$
    Write down the equation you use. Show clearly how you work out your answer. (3)
  iv Calculate the output current from this transformer when the input current is 2.5 A.
    Show clearly how you work out your answer and give the unit. (3)

e Give a typical frequency at which a switch mode transformer works. (1)

f Give **two** advantages of a switch mode transformer compared with an iron-core transformer. (2)

1 The diagram shows two images of a normal eye.

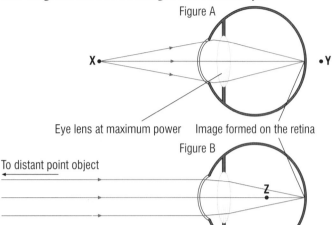

Figure A

X•

•Y

Eye lens at maximum power   Image formed on the retina

Figure B

To distant point object

Z

AQA **Examiner's tip**

To remember the difference between short- and long-sightedness practise writing these out:
Short sight – only **see short** distance – **focus short** of retina – must **diverge**
Long sight – only **see long** distance – **focus** too **long** – must **converge**

a Give the name of the point labelled **X**. (1)
b In a normal eye, approximately how far is this point from the eye? (1)
c Name the muscle that changes the shape of the lens. (1)
d How does this muscle change the shape of the lens between **Figure A** and **Figure B**? (1)
e If the rays of light focused at point **Y** in diagram **A**, what sight defect would this show? (1)
f If the rays of light focused at point **Z** in diagram **B**, what sight defect would this show? (1)

2 A converging lens is placed between an object and a screen which are 16 cm apart. The position of the lens is adjusted until the magnification produced by the lens is ×3. Copy and complete the scale ray diagram of this arrangement below to find the focal length of the lens.

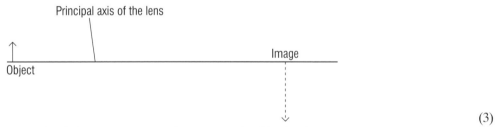

Principal axis of the lens

Image

Object

AQA **Examiner's tip**

Remember a ray that goes through the centre of a lens travels straight; all rays from the top of the object will arrive at the top of the image.

(3)

3 A teacher balances a metre rule on a wooden block using two masses.

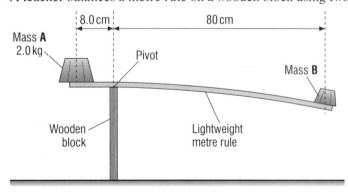

8.0 cm

80 cm

Mass **A**
2.0 kg

Pivot

Mass **B**

Wooden block

Lightweight metre rule

AQA **Examiner's tip**

You may be able to do the calculations in your head but you should still write down plenty of working or you could lose marks.

a i Calculate the weight of mass **A**. Write down the equation you use. Show clearly how you work out your answer and give the unit. (2)
   ii Calculate the anticlockwise moment about the pivot caused by the weight of mass **A**. Write down the equation you use. Show clearly how you work out your answer and give the unit. (3)

b In terms of moments, explain why the rule is not turning. (2)

c Find the mass of mass **B**. Write down the equation you use. Show clearly how you work out your answer and give the unit. **[H]** (3)

d The rule is replaced with a much heavier rule of the same size. Explain how mass **B** must be moved so that the rule is still balanced. **[H]** (3)

4 Mechanical diggers use hydraulic cylinders. The input force on the main cylinder is produced by the engine. Several further cylinders connected to the main cylinder move the arm of the digger.

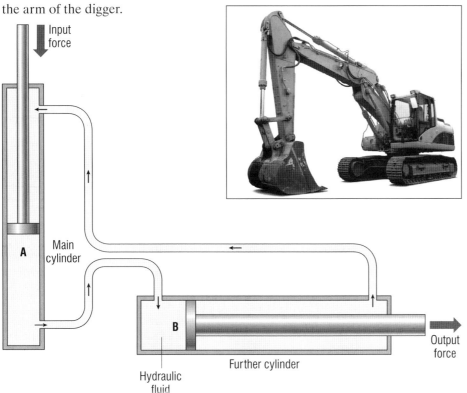

a Why are liquids more effective than gases for use in applications such as this? (1)

b Calculate the pressure in the main cylinder at **A** when the force on the piston is 10 000 N.
Area of main cylinder = 0.0004 m²
Write down the equation you use. Show clearly how you work out your answer and give the unit. (3)

c What will the pressure be in the piston at **B**? (1)

d Calculate the output force from the piston at **B**.
Write down the equation you use. Show clearly how you work out your answer and give the unit.
Area of further cylinder = 0.0025 m². (3)

5 *In this question you will be assessed on using good English, organising information clearly and using specialist terms where appropriate.*
Explain how the motor effect could be used to move a length of copper wire. Include details of how the wire should be positioned to maximise the force. (6)

Glossary

A

Acceleration Change of velocity per second (in metres per second per second, m/s²).

Acid rain Rain that is acidic due to dissolved gases, such as sulfur dioxide, produced by the burning of fossil fuels.

Activity Number of atoms of a radioactive substance that decay each second.

Alpha radiation Alpha particles, each composed of two protons and two neutrons, emitted by unstable nuclei.

Alternating current Electric current in a circuit that repeatedly reverses its direction.

Amplitude The height of a wave crest or a wave trough of a transverse wave from the rest position. Of oscillating motion, is the maximum distance moved by an oscillating object from its equilibrium position.

Angle of incidence Angle between the incident ray and the normal.

Angle of reflection Angle between the reflected ray and the normal.

Anomalous results Results that do not match the pattern seen in the other data collected or are well outside the range of other repeat readings. They should be retested and if necessary discarded.

A-scan An ultrasound scan used to measure the distance between two boundaries that partially reflect ultrasound.

Atomic nucleus Tiny positively charged object composed of protons and neutrons at the centre of every atom.

Atomic number The number of protons (which equals the number of electrons) in an atom. It is sometimes called the proton number.

B

Band The part of the radio and microwave spectrum used for communications.

Bar chart A chart with rectangular bars with lengths proportional to the values that they represent. The bars should be of equal width and are usually plotted horizontally or vertically. Also called a bar graph.

Base load Constant amount of electricity generated by power stations.

Beta radiation Beta particles that are high-energy electrons created in and emitted from unstable nuclei.

Big Bang theory The theory that the universe was created in a massive explosion (the Big Bang) and that the universe has been expanding ever since.

Black dwarf A star that has faded out and gone cold.

Black hole An object in space that has so much mass that nothing, not even light, can escape from its gravitational field.

Blue-shift Decrease in the wavelength of electromagnetic waves emitted by a star or galaxy due to its motion towards us. The faster the speed of the star or galaxy, the greater the blue-shift is.

Braking distance The distance travelled by a vehicle during the time its brakes act.

C

Cable Two or three insulated wires surrounded by an outer layer of rubber or flexible plastic.

Camera An instrument for photographing an object by using a converging lens to form a real image of the object on a film (or on electronic pixels) in a lightproof box.

Carbon capture and storage Capture and storage of carbon dioxide produced in fossil fuel power stations. Old gas or oil fields are suitable places to store the carbon dioxide.

CCD (charge-coupled device) Used to record and display an image.

Centre of mass The point where an object's mass may be thought to be concentrated.

Centripetal acceleration The acceleration of an object moving in a circle at constant speed. Centripetal acceleration always acts towards the centre of the circle.

Centripetal force The resultant force towards the centre of a circle acting on an object moving in a circular path.

Chain reaction Reactions in which one reaction causes further reactions, which in turn cause further reactions, etc. A nuclear chain reaction occurs when fission neutrons cause further fission, so more fission neutrons are released. These go on to produce further fission.

Chemical energy Energy of an object due to chemical reactions in it.

Circuit breaker An electromagnetic switch that opens and cuts the current off if too much current passes through it.

Compression Squeezing together.

Condense Turn from vapour into liquid.

Conservation of energy Energy cannot be created or destroyed.

Conservation of momentum In a closed system, the total momentum before an event is equal to the total momentum after the event. Momentum is conserved in any collision or explosion provided no external forces act on the objects that collide or explode.

Control group If an experiment is to determine the effect of changing a single variable, a control is often set up in which the independent variable is not changed, thus enabling a comparison to be made. If the investigation is of the survey type a control group is usually established to serve the same purpose.

Control rod Metal rod (made of boron or cadmium) used to absorb excess fission neutrons in a nuclear reactor so that only one fission neutron per fission on average goes on to produce further fission.

Converging lens A lens that makes light rays parallel to the principal axis converge to (i.e. meet at) a point; also referred to as a convex lens.

Coolant Fluid in a sealed circuit pumped through the core of a nuclear reactor to remove energy to a heat exchanger.

Cosmic microwave background radiation Electromagnetic radiation that has been travelling through space ever since it was created shortly after the Big Bang.

Coulomb (C) The unit of electrical charge, equal to the charge passing a point in a (direct current) circuit in 1 second when the current is 1 A.

CT scanner A medical scanner that uses X-rays to produce a digital image of any cross-section through the body or a three-dimensional image of an organ.

D

Data Information, either qualitative or quantitative, that have been collected.

Deceleration Change of velocity per second when an object slows down.

Density Mass per unit volume of a substance.

Diffraction The spreading of waves when they pass through a gap or around the edges of an obstacle which has a similar size as the wavelength of the waves.

Diffusion Spreading out of particles away from each other.

Dioptre The unit of lens power, D.

Direct current Electric current in a circuit that is in one direction only.

Directly proportional A graph will show this if the line of best fit is a straight line through the origin.

Diverging lens A lens that makes light rays parallel to the axis diverge (i.e. spread out) as if from a single point; also referred to as a concave lens.

Doppler effect The change of wavelength (and frequency) of the waves from a moving source due to the motion of the source towards or away from the observer.

Drag force A force opposing the motion of an object due to fluid (e.g. air) flowing past the object as it moves.

E

Earthed Connected to the ground by means of a conducting lead or wire.

Echo Reflection of sound that can be heard.

Efficiency Useful energy transferred by a device ÷ total energy supplied to the device.

Effort The force applied to a device used to raise a weight or shift an object.

Elastic A material is elastic if it is able to regain its shape after it has been squashed or stretched.

Elastic potential energy Energy stored in an elastic object when work is done to change its shape.

Electric current Flow of electric charge. The size of an electric current (in amperes, A) is the rate of flow of charge.

Electrical energy Energy transferred by the movement of electrical charge.

Electricity meter Meter in a home that measures the amount of electrical energy supplied.

Electromagnetic induction The process of inducing a potential difference in a wire by moving the wire so it cuts across the lines of force of a magnetic field.

Electromagnetic waves Electric and magnetic disturbances that transfer energy from one place to another. The spectrum of electromagnetic waves, in order of increasing wavelength, is as follows: gamma and X-rays, ultraviolet radiation, visible light, infrared radiation, microwaves, radio waves.

Equilibrium The state of an object when it is at rest.

Errors Sometimes called uncertainties.

Error – human Often present in the collection of data, and may be random or systematic. For example, the effect of human reaction time when recording short time intervals with a stopwatch.

Error – random Causes readings to be spread about the true value, due to results varying in an unpredictable way from one measurement to the next. Random errors are present when any measurement is made, and cannot be corrected. The effect of random errors can be reduced by making more measurements and calculating a new mean.

Error – systematic Causes readings to be spread about some value other than the true value, due to results differing from the true value by a consistent amount each time a measurement is made. Sources of systematic error can include the environment, methods of observation or instruments used. Systematic errors cannot be dealt with by simple repeats. If a systematic error is suspected, the data collection should be repeated using a different technique or a different set of equipment, and the results compared.

Error – zero Any indication that a measuring system gives a false reading when the true value of a measured quantity is zero, for example, the needle on an ammeter failing to return to zero when no current flows.

Evaporate Turn from liquid into vapour.

Evidence Data which have been shown to be valid.

F

Fair test A fair test is one in which only the independent variable has been allowed to affect the dependent variable.

Far point The furthest point from an eye at which an object can be seen in focus by the eye. The far point of a normal eye is at infinity.

Field lines See lines of force.

Fluid A liquid or a gas.

Focal length The distance from the centre of a lens to the point where light rays parallel to the principal axis are focused (or, in the case of a diverging lens, appear to diverge from).

Force A force can change the motion of an object (in newtons, N).

Free electron Electron that moves about freely inside a metal and is not held inside an atom.

Frequency The number of wave crests passing a fixed point every second.

Frequency of an alternating current The number of complete cycles an alternating current passes through each second. The unit of frequency is the hertz (Hz).

Frequency of oscillating motion Number of complete cycles of oscillation per second, equal to 1/the time period. The unit of frequency is the hertz (Hz).

Fuse A fuse contains a thin wire that melts and cuts the current off if too much current passes through it.

G

Gamma radiation Electromagnetic radiation emitted from unstable nuclei in radioactive substances.

Geothermal energy Energy from hot underground rocks.

Gradient (of a straight line graph) Change of the quantity plotted on the y-axis divided by the change of the quantity plotted on the x-axis.

Gravitational field strength, g The force of gravity on an object of mass 1 kg (in newtons per kilogram, N/kg).

Gravitational potential energy Energy of an object due to its position in a gravitational field. Near the Earth's surface, change of GPE (in joules, J) =

weight (in newtons, N) × vertical distance moved (in metres, m).

Ground heat Geothermal energy that heats buildings directly.

H

Half-life of a radioactive isotope Average time taken for the number of nuclei of the isotope (or mass of the isotope) in a sample to halve.

Hazard A hazard is something (for example, an object, a property of a substance or an activity) that can cause harm.

High mass star A star that has a much greater mass than the Sun.

Hooke's law The extension of a spring is directly proportional to the force applied, provided its limit of proportionality is not exceeded.

Hydraulic pressure The pressure in the liquid in a hydraulic arm.

Hypothesis A proposal intended to explain certain facts or observations.

I

Infrared radiation Electromagnetic waves between visible light and microwaves in the electromagnetic spectrum.

Interval The quantity between readings, for example, a set of 11 readings equally spaced over a distance of 1 m would give an interval of 10 cm.

Ionisation Any process in which atoms become charged.

J

Joule (J) The unit of energy.

Kilowatt (kW) 1000 watts.

K

Kilowatt-hour (kW h) Electrical energy supplied to a 1 kW electrical device in 1 hour.

Kinetic energy Energy of a moving object due to its motion; kinetic energy (in joules, J) = mass (in kilograms, kg) × (speed)^2 (in m^2/s^2).

L

Limit of proportionality The limit for Hooke's law applied to the extension of a stretched spring.

Line graph Used when both variables are continuous. The line should normally be a line of best fit, and may be straight or a smooth curve. (Exceptionally, in some (mainly biological) investigations, the line may be a 'point-to-point' line.)

Line of action The line along which a force acts.

Line of force Line in a magnetic field along which a magnetic compass points; also called a magnetic field line.

Load The weight of an object raised by a device used to lift the object, or the force applied by a device when it is used to shift an object.

Long sight An eye that cannot focus on nearby objects but can focus on distant objects.

Longitudinal waves Waves in which the vibrations are parallel to the direction of energy transfer.

Low mass star A star that has a much smaller mass than the Sun.

M

Magnetic field line Line in a magnetic field along which a magnetic compass points; also called a line of force.

Magnetic poles Ends of a bar magnet or a magnetic compass.

Magnification The image height ÷ the object height.

Magnifying glass A converging lens used to magnify a small object which must be placed between the lens and its focal point.

Main sequence star The main stage is the life of a star during which it radiates energy because of fusion of hydrogen nuclei in its core.

Mass The quantity of matter in an object; a measure of the difficulty of changing the motion of an object (in kilograms, kg).

Mean The arithmetical average of a series of numbers.

Mechanical wave Vibration that travels through a substance.

Moderator A solid or liquid used in a nuclear reactor to slow fission neutrons down so they can cause further fission.

Moment The turning effect of a force defined by the equation: Moment of a force (in newton-metres) = force (in newtons) × perpendicular distance from the pivot to the line of action of the force (in metres).

Momentum This equals mass (in kg) × velocity (in m/s). The unit of momentum is the kilogram metre per second (kg m/s).

Monitor Observations made over a period of time.

Motor effect When a current is passed along a wire in a magnetic field and the wire is not parallel to the lines of the magnetic field, a force is exerted on the wire by the magnetic field.

N

National Grid The network of cables and transformers used to transfer electricity from power stations to consumers (i.e. homes, shops, offices, factories, etc.).

Near point The nearest point to an eye at which an object can be seen in focus by the eye. The near point of a normal eye is 25 cm from the eye.

Neutral wire The wire of a mains circuit that is earthed at the local sub-station so its potential is close to zero.

Neutron star The highly compressed core of a massive star that remains after a supernova explosion.

Newton (N) The unit of force.

Normal Straight line through a surface or boundary perpendicular to the surface or boundary.

North pole North-pointing end of a freely suspended bar magnet or of a magnetic compass.

Nuclear fission The process in which certain nuclei (uranium-235 and plutonium-239) split into two fragments, releasing energy and two or three neutrons as a result.

Nuclear fusion The process in which small nuclei are forced together so they fuse with each other to form a larger nucleus.

O

Ohm's law The current through a resistor at constant temperature is directly proportional to the potential difference across the resistor.

Ohmic conductor A conductor that has a constant resistance and therefore obeys Ohm's law.

Opinion A belief not backed up by facts or evidence.

Optical fibre Thin glass fibre used to send light signals along.

Oscillate Move to and fro about a certain position along a line.

Oscillating motion Motion of any object that moves to and fro along the same line.

Oscilloscope A device used to display the shape of an electrical wave.

P

Parallel Components connected in a circuit so that the potential difference is the same across each one.

Pascal (Pa) The unit of pressure, equal to 1 newton per square metre.

Perpendicular At right angles.

Pitch The pitch of a sound increases if the frequency of the sound waves increases.

Pivot The point about which an object turns when acted on by a force that makes it turn.

Plane mirror A flat mirror.

Planet A large object that moves in an orbit round a star. A planet reflects light from the star and does not produce its own light.

Plug A plug has an insulated case and is used to connect the cable from an appliance to a socket.

Potential difference A measure of the work done or energy transferred to the lamp by each coulomb of charge that passes through it. The unit of potential difference is the volt (V).

Power The energy transformed or transferred per second. The unit of power is the watt (W).

Power of a lens The focal length of the lens in metres. The unit of lens power is the dioptre, d.

Precise A precise measurement is one in which there is very little spread about the mean value. Precision depends only on the extent of random errors – it gives no indication of how close results are to the true value.

Prediction A forecast or statement about the way something will happen in the future. In science it is not just a simple guess, because it is based on some prior knowledge or on a hypothesis.

Pressure Force per unit area for a force acting on a surface at right angles to the surface. The unit of pressure is the pascal (Pa).

Principal axis A straight line that passes along the normal at the centre of each lens surface.

Principal focus The point where light rays parallel to the principal axis of a lens are focused (or, in the case of a diverging lens, appear to diverge from).

Principle of moments For an object in equilibrium, the sum of all the clockwise moments about any point = the sum of all the anticlockwise moments about that point.

Protostar The concentration of dust clouds and gas in space that forms a star.

R

Radiation dose Amount of ionising radiation a person receives.

Radiograph An X-ray picture.

Random Cannot be predicted and has no recognisable cause.

Range The maximum and minimum values of the independent or dependent variables; important in ensuring that any pattern is detected.

Range of vision Distance from the near point of an eye to its far point.

Rarefaction Stretched apart.

Real image An image formed where light rays meet.

Real image An image formed by a lens that can be projected on a screen.

Red giant A star that has expanded and cooled, resulting in it becoming red and much larger and cooler than it was before it expanded.

Red-shift Increase in the wavelength of electromagnetic waves emitted by a star or galaxy due to its motion away from us. The faster the speed of the star or galaxy, the greater the red-shift is.

Refraction The change of direction of a light ray when it passes across a boundary between two transparent substances (including air).

Refractive index Refractive index, n, of a transparent substance is a measure of how much the substance can refract a light ray.

Relationship The link between the variables that were investigated. These relationships may be: causal, i.e. changing x is the reason why y changes; by association, i.e. both x and y change at the same time, but the changes may both be caused by a third variable changing; by chance occurrence.

Renewable energy Energy from sources that never run out including wind energy, wave energy, tidal energy, hydroelectricity, solar energy and geothermal energy.

Repeatable A measurement is repeatable if the original experimenter repeats the investigation using same method and equipment and obtains the same results.

Reproducible A measurement is reproducible if the investigation is repeated by another person, or by using different equipment or techniques, and the same results are obtained.

Residual Current Circuit Breaker (RCCB) An RCCB cuts off the current in the live wire when it is different from the current in the neutral wire.

Resistance Resistance (in ohms). = potential difference (in volts, V) ÷ current (in amperes, A).

Resolution This is the smallest change in the quantity being measured (input) of a measuring instrument that gives a perceptible change in the reading.

Resonate When sound vibrations build up in a musical instrument and cause the sound from the instrument to become much louder.

Resultant force The combined effect of the forces acting on an object.

Resultant moment The difference between the sum of the clockwise moments and the anticlockwise moments about the same point if they are not equal.

Risk The likelihood that a hazard will actually cause harm. We can reduce risk by identifying the hazard and doing something to protect against that hazard.

S

Sankey diagram An energy transfer diagram.

Series Components connected in a circuit so that the same current passes through them are in series with each other.

Short sight An eye that cannot focus on distant objects but can focus on near objects.

Simple pendulum A pendulum consisting of a small spherical bob suspended by a thin string from a fixed point.

Snell's law Refractive index n = sin i/sin r, where i is the angle of incidence and r is the angle of refraction of a light ray passing from air into the substance.

Socket A mains socket is used to connect the mains plug of a mains appliance to the mains circuit.

Solar cell Electrical cell that produces a voltage when in sunlight; solar cells are usually connected together in solar cell panels.

Solar heating panel Sealed panel designed to use sunlight to heat water running through it.

Solar power tower Tower surrounded by mirrors that reflect sunlight onto a water tank at the top of the tower.

South pole South-pointing end of a freely suspended bar magnet or of a magnetic compass.

Specific heat capacity Energy needed by 1 kg of the substance to raise its temperature by 1 °C.

Speed Distance moved ÷ time taken.

Split-ring commutator Metal contacts on the coil of a direct current motor that connects the rotating coil continuously to its electrical power supply.

Static electricity Charge 'held' by an insulator or an insulated conductor.

Step-down transformer Electrical device that is used to step down the size of an alternating voltage.

Step-up transformer Electrical device that is used to step up the size of an alternating voltage.

Stopping distance Thinking distance + braking distance.

Supergiant A massive star that becomes much larger than a giant star when fusion of helium nuclei commences.

Supernova The explosion of a massive star after fusion in its core ceases and the matter surrounding its core collapses on to the core and rebounds.

Switch mode transformer A transformer that works at much higher frequencies than a traditional transformer. It has a ferrite core instead of an iron core.

T

Terminal velocity The velocity reached by an object when the drag force on it is equal and opposite to the force making it move.

Thinking distance The distance travelled by the vehicle in the time it takes the driver to react.

Three-pin plug A three-pin plug has a live pin, a neutral pin and an earth pin. The earth pin is used to earth the metal case of an appliance so the case cannot become live.

Time base control An oscilloscope control used to space the waveform out horizontally.

Time period Time taken for one complete cycle of oscillating motion.

Transducer A device that produces and detects ultrasound waves.

Transformer Electrical device used to change an (alternating) voltage. See also Step-up transformer and Step-down transformer.

Transverse wave Wave in which the vibration is perpendicular to the direction of energy transfer.

Trial run Preliminary work that is often done to establish a suitable range or interval for the main investigation.

Turbine A machine that uses steam or hot gas to turn a shaft.

U

Ultrasonic wave Sound wave at frequency greater than 20 000 Hz, which is the upper frequency limit of the human ear.

Useful energy Energy transferred to where it is wanted in the form it is wanted.

V

Valid Suitability of the investigative procedure to answer the question being asked.

Variable Physical, chemical or biological quantity or characteristic.

Variable – categoric Categoric variables have values that are labels. For example, names of plants or types of material.

Variable – continuous Can have values (called a quantity) that can be given by measurement (for example, light intensity, flow rate, etc.).

Variable – control A variable which may, in addition to the independent variable, affect the outcome of the investigation and therefore has to be kept constant or at least monitored.

Variable – dependent The variable for which the value is measured for each and every change in the independent variable.

Variable – independent The variable for which values are changed or selected by the investigator.

Velocity Speed in a given direction (in metres/second, m/s).

Vibrate Oscillate rapidly (or move to and fro rapidly about a certain position).

Virtual image An image, seen in a lens or a mirror, from which light rays appear to come after being refracted by the lens or reflected by the mirror.

Volt (V) The unit of potential difference, equal to energy transfer per unit charge in joules per coulomb.

W

Wasted energy Energy that is not usefully transferred.

Watt (W) The unit of power.

Wavelength The distance from one wave crest to the next wave crest (along the waves).

Weight The force of gravity on an object (in newtons, N).

White dwarf A star that has collapsed from the red giant stage to become much hotter and denser than it was.

White light Light that consists of all the colours of the visible spectrum.

Work Energy transferred by a force, given by:
Work done (in joules, J) = force (in newtons, N) × distance moved in the direction of the force (in metres, m).

Y

Y-gain control An oscilloscope control used to adjust the height of the waveform.

Index